Regional Sustainable Development of Yangtze River Delta, China II

Regional Sustainable Development of Yangtze River Delta, China II

Guest Editors

Wei Sun
Zhaoyuan Yu
Kun Yu
Weiyang Zhang
Jiawei Wu

Basel • Beijing • Wuhan • Barcelona • Belgrade • Novi Sad • Cluj • Manchester

Guest Editors

Wei Sun
Nanjing Institute of
Geography and Limnology
Chinese Academy of Sciences
Nanjing
China

Zhaoyuan Yu
School of Geography
Nanjing Normal University
Nanjing
China

Kun Yu
Institute of Germplasm
Resources and Biotechnology
Jiangsu Academy of
Agricultural Sciences
Nanjing
China

Weiyang Zhang
School of Geographic
Sciences
East China Normal
University
Shanghai
China

Jiawei Wu
Nanjing Institute of
Geography and Limnology
Chinese Academy of Sciences
Nanjing
China

Editorial Office
MDPI AG
Grosspeteranlage 5
4052 Basel, Switzerland

This is a reprint of the Special Issue, published open access by the journal *Land* (ISSN 2073-445X), freely accessible at: https://www.mdpi.com/journal/land/special_issues/RTVA4S10LB.

For citation purposes, cite each article independently as indicated on the article page online and as indicated below:

Lastname, A.A.; Lastname, B.B. Article Title. *Journal Name* **Year**, *Volume Number*, Page Range.

ISBN 978-3-7258-2451-9 (Hbk)
ISBN 978-3-7258-2452-6 (PDF)
https://doi.org/10.3390/books978-3-7258-2452-6

© 2024 by the authors. Articles in this book are Open Access and distributed under the Creative Commons Attribution (CC BY) license. The book as a whole is distributed by MDPI under the terms and conditions of the Creative Commons Attribution-NonCommercial-NoDerivs (CC BY-NC-ND) license (https://creativecommons.org/licenses/by-nc-nd/4.0/).

Contents

About the Editors vii

Chen Chen
Changes in the Spatial Distribution of the Employed Population in the Yangtze River Delta Region since the 21st Century: An Analysis and Discussion Based on Census Data
Reprinted from: *Land* 2023, 12, 1249, https://doi.org/10.3390/land12061249 1

Xianzhong Cao, Bo Chen, Yi Guo and Zhenzhen Yi
The Impact of Intra-City and Inter-City Innovation Networks on City Economic Growth: A Case Study of the Yangtze River Delta in China
Reprinted from: *Land* 2023, 12, 1463, https://doi.org/10.3390/land12071463 25

Yihang Zhao, Jing Xiong and De Hu
Reputation, Network, and Performance: Exploring the Diffusion Mechanism of Local Governments' Behavior during Inter-Governmental Environmental Cooperation
Reprinted from: *Land* 2023, 12, 1466, https://doi.org/10.3390/land12071466 42

Chiming Guan, Liuying Chen and Danyang Li
Does the Opening of High-Speed Railway Improve High-Quality Economic Development in the Yangtze River Delta, China?
Reprinted from: *Land* 2023, 12, 1629, https://doi.org/10.3390/land12081629 59

Jiayu Kang, Xuejun Duan and Ruxian Yun
The Impact of Urbanization on Food Security: A Case Study of Jiangsu Province
Reprinted from: *Land* 2023, 12, 1681, https://doi.org/10.3390/land12091681 87

Wenqin Yan and Dongsheng Yan
The Regional Effect of Land Transfer on Green Total Factor Productivity in the Yangtze River Delta: A Spatial Econometric Investigation
Reprinted from: *Land* 2023, 12, 1794, https://doi.org/10.3390/land12091794 103

Xufeng Cao, Jiqin Han and Xueying Li
Analysis of the Impact of Land Use Change on Grain Production in Jiangsu Province, China
Reprinted from: *Land* 2024, 13, 20, https://doi.org/10.3390/land13010020 121

Qing Wang and Chunmei Mao
Evolutionary Game Analysis of Ecological Governance Strategies in the Yangtze River Delta Region, China
Reprinted from: *Land* 2024, 13, 212, https://doi.org/10.3390/land13020212 137

Min Wang, Yu Lan, Huayu Li, Xiaodong Jing, Sitong Lu and Kexin Deng
Spatial–Temporal Differentiation and Trend Prediction of Coupling Coordination Degree of Port Environmental Efficiency and Urban Economy: A Case Study of the Yangtze River Delta
Reprinted from: *Land* 2024, 13, 374, https://doi.org/10.3390/land13030374 159

Ming Wei, Wen Chen and Yi Wang
Assessment of the Implementation Effects of Main Functional Area Planning in the Yangtze River Economic Belt
Reprinted from: *Land* 2024, 13, 940, https://doi.org/10.3390/land13070940 178

Hui Guo and Wei Sun
Carbon Balance Zoning and Spatially Synergistic Carbon Reduction Pathways—A Case Study in the Yangtze River Delta in China
Reprinted from: *Land* 2024, 13, 943, https://doi.org/10.3390/land13070943 196

About the Editors

Wei Sun

Wei Sun is a professor and doctoral supervisor at the Nanjing Institute of Geography and Limnology, Chinese Academy of Sciences (NIGLAS), and the deputy director of the Yangtze River Delta Integration and Development Research Center of NIGLAS, Chinese Academy of Sciences (CAS). His research areas include regional development and planning, regional spatial growth management, and spatial functional zoning theory and methods. He has presided over two projects of the National Natural Science Foundation of China (NSFC), one sub-project of the CAS—a self-deployed project—and one key project of the aforementioned institute; additionally, he has participated in more than 30 national and provincial research projects, such as the key projects of the NSFC, the Science and Technology Service Network Program (STS) of the Chinese Academy of Sciences, the Pilot Project of Beautiful China, and the Major Demonstration Project of Social Development of Jiangsu Province. He has published three monographs and more than 90 academic papers in English and Chinese, and he has also been awarded First Prize for the 16th Philosophy and Social Science Outstanding Achievements of Jiangsu Province, the Science and Technology for Development Award of the Chinese Academy of Sciences, and the Young Geoscientist Award of Jiangsu Province.

Zhaoyuan Yu

Zhaoyuan Yu is a professor and doctoral supervisor at the School of Geography, Nanjing Normal University. He serves as the executive director of the Jiangsu Center for Collaborative Innovation in Geographical Information Resource Development and Application and holds the position of deputy secretary-general of the Information Geography Committee of the Geographical Society of China. His research areas include GIS modeling, the sustainable development of coastal zones with digital twins, geometric algebra, and quantum geo-computing research. He has presided over seven national research projects, participated in more than 20 major/key projects of the Ministry of Science and Technology and the Foundation Committee as a key member, has published more than 40 papers as the first/corresponding author, participated in the editing of three national standards, and authorized 12 Chinese invention patents, one Australian patent, and two Japanese patents; he has also disclosed eight international patents of the PCT. As his main accomplishment, he has been awarded more than 10 provincial- and ministerial-level industry awards, such as the first prize in natural science for the Ministry of Education, the second prize in scientific and technological progress in land and resources, and the second prize in science and technology of Jiangsu province.

Kun Yu

Kun Yu is a professor, the deputy director of the Institute of Germplasm Resources and Biotechnology, Jiangsu Academy of Agricultural Sciences, and the deputy director of the New Plant Variety Testing (Nanjing) Sub-center of the Ministry of Agriculture and Rural Development. His research areas include the response of nearshore-inland environmental change to human activities and climate change, agricultural information and agricultural big data, land use and revegetation change, and the remote sensing of resources, the environment, and agriculture. He has served as the deputy director of the Institute of Information Research of Jiangsu Academy of Agricultural Sciences, the deputy director of the International Cooperation Department, and the director of the Nanjing Branch of the Remote Sensing Application Center of the Ministry of Agriculture, and has also served as a member of the Standing Committee of the Agricultural Information Branch of the Chinese Society of Agriculture. He has presided over and participated in more than 30 national, ministerial, and provincial research projects and has published more than 40 research papers, of which 19 are SCI/EI-recorded papers.

Weiyang Zhang

Weiyang Zhang is an associate professor at the School of Geographic Sciences of East China Normal University, a researcher at the Key Research Base of the Ministry of Civil Affairs at East China Normal University, a Shanghai Morning Glory Scholar, and a Ph.D. graduate of Ghent University in Belgium. His research areas include theoretical and empirical studies of multi-scale urban networks, regional polycentric development, and Yangtze River Delta integration. He has presided over nearly 10 research projects, including those of the National Natural Science Foundation of China, the Shanghai Philosophy and Society Planning Project, and Shanghai Soft Science. He has published more than 40 papers in Urban Studies, Regional Studies, The Journal of Urban Technology, Cities, and other domestic and international journals, including more than 10 SSCI papers as the first/corresponding author; he has published one monograph in English and co-edited four books in English and Chinese. He also serves as a member of the Urban Agglomeration Committee of the China Society of Urban Economics and is an editorial board member of Space–Society–Economy, etc. He is a member of the American Association of Geography (AAG), the Regional Studies Association (RSA), and the China Geographic Society (CGS).

Jiawei Wu

Jiawei Wu is an assistant professor at the Nanjing Institute of Geography and Limnology, Chinese Academy of Sciences, and his research area covers industrial geography and regional sustainable development. He has presided over and participated in more than 10 national and provincial projects, such as projects for the National Natural Science Foundation of China, the Science and Technology Service Network Program of the Chinese Academy of Sciences, the research topics of the National Development and Reform Commission, etc. He has published more than 20 SCI/SSCI and CSCD retrieved papers.

Article

Changes in the Spatial Distribution of the Employed Population in the Yangtze River Delta Region since the 21st Century: An Analysis and Discussion Based on Census Data

Chen Chen

Institute of Urban and Demographic Studies, Shanghai Academy of Social Sciences, Shanghai 200020, China; chenchen@sass.org.cn

Abstract: Focusing on the Yangtze River Delta region, the spatial distribution and change characteristics of the employed population were assessed by selecting three time points: 2000, 2010 and 2020. Firstly, a correlation was established between population employment statistics and spatial units of administrative divisions to analyze the spatial distribution characteristics of the employed population in general and by industry; secondly, the changing characteristics of the spatial distribution of the employed population over time, including the migration of the centroid and density changes, were analyzed; thirdly, a systematic clustering approach was adopted to carry out a typological analysis of 41 cities in the Yangtze River Delta from three perspectives: industrial structure, time stage and spatial level. It was found that (1) regional differences within the Yangtze River Delta are still significant, but are narrowing; (2) different cities or regions show different characteristics of development stages, and late-developing regions can learn from early developing regions; (3) metropolitan areas are still the main areas of employment concentration, and the spatial distribution of employment in some cities is beginning to suburbanize.

Keywords: employed population; mean center; density change; cluster analysis; industry structure

Citation: Chen, C. Changes in the Spatial Distribution of the Employed Population in the Yangtze River Delta Region since the 21st Century: An Analysis and Discussion Based on Census Data. *Land* **2023**, *12*, 1249. https://doi.org/10.3390/land12061249

Academic Editors: Wei Sun, Zhaoyuan Yu, Kun Yu, Weiyang Zhang and Jiawei Wu

Received: 16 May 2023
Revised: 15 June 2023
Accepted: 16 June 2023
Published: 19 June 2023

Copyright: © 2023 by the author. Licensee MDPI, Basel, Switzerland. This article is an open access article distributed under the terms and conditions of the Creative Commons Attribution (CC BY) license (https://creativecommons.org/licenses/by/4.0/).

1. Introduction

Since entering the 21st century, urban agglomerations have become hotspots for population concentration and the growth of employment activities in China. According to census data, in 2020, the five provinces and three directly administered municipalities where the Beijing–Tianjin–Hebei Region, Yangtze River Delta, and Pearl River Delta are located accounted for one-third of China's employed population. The employed population growth in major cities within these regions is also generally higher than the national average. From 2000 to 2020, the employed population in the three cities of Beijing, Tianjin, and Shijiazhuang in the Beijing–Tianjin–Hebei region grew by 21.3%; in the four cities of Shanghai, Nanjing, Hangzhou, and Hefei in the Yangtze River Delta region, it grew by 61.2%; and in the nine cities of Guangzhou, Shenzhen, etc., in the Pearl River Delta region, it grew by 75.2%.

This study focuses on the Yangtze River Delta (YRD) region, a densely populated and economically active area encompassing Shanghai, Jiangsu, Zhejiang, and Anhui. The YRD region has a land area of about 360,000 km^2, accounting for less than 4% of the country. According to 2020 data, the region has a total population of 235.21 million people and a GDP of RMB 24.4694 trillion, accounting for 16.3% and 22.5% of national total, respectively. The region has created a large degree of economic scale with a relatively small land area, which can be attributed in large part to the labor creation of the vast employed population. Due to the high level of economic and social development in the YRD region, many cities in the region are able to provide more attractive employment opportunities and are among the priority areas for the majority of the employed population to consider for career development.

It is worth noting that there are significant differences in the spatial characteristics of urban and rural areas and the level of economic development in the YRD region, which will affect the spatial distribution characteristics of the employed population to a certain extent. Additionally, after the initial rapid development, urbanization and economic growth in the YRD region have entered a new stage, characterized by a higher level of development but a lower growth rate. This has also resulted in differences in the temporal characteristics of employed population changes. Therefore, the main objective of this study is to discover the spatial and temporal characteristics of employment in the YRD region since the 21st century, with the aim of providing research support for the region's spatially optimized development in the new era.

In terms of identifying the spatial distribution characteristics of the employed population, the amount of literature available is relatively more focused on the city and metropolitan area scales, with relatively few studies at the larger regional and national scales. Typical studies at the city and metropolitan area scales include, for example, the work of Richard Shearmur et al. (2002), who used Canadian statistics to analyze the spatial structure of employment in four metropolitan areas: Toronto, Montreal, Vancouver, and Ottawa-Hull [1]; William J. Coffey et al. (2002), who used workplace data to study the spatial variation in employment of several business services in the Montreal region of Canada during 1981–1996 [2]; Mehdi Alidad et al. (2018) analyzed the spatial distribution characteristics of employment in the Tehran metropolitan area [3]; Wang Bo et al. (2011) identified the spatial distribution characteristics of employment in different industries in Nanjing based on ArcGIS kernel density analysis [4]; Zhan Dongsheng et al. (2017) conducted an empirical analysis of the spatial structure of employment and residence in urban areas of Beijing from a sub-industry perspective based on business enterprise registration data and census data [5]; Zuo Wei et al. (2017) investigated and analyzed the commonalities and differences in the spatial distribution of the employment of migrant workers of different genders in the main urban area of Nanjing [6]; Li Pengfei (2019) analyzed the spatial layout characteristics of employment in Shenyang based on cell phone signaling data [7]; Sun Chen et al. (2016) analyzed the overall distribution characteristics of new jobs and the spatial distribution characteristics of specific industries, such as real estate, based on job posting information [8]; and Wang Hui et al. (2014) combined statistical and survey data to compare the characteristics of the overall employment space and the employment space of the mobile population in Nanjing [9,10].

Typical studies at regional and national scales include the work of: Xin Lao et al. (2013), who analyzed and compared the characteristics of employment density distribution in two regions, the Yangtze River Delta and the Pearl River Delta, using data from economic census [11]; Enrico Marelli (2004), who analyzed the employment distribution among the main production sectors in 145 regions of the European Union during 1983–1997 and found that differences in employment structure may be more pronounced within countries [12]; and Wang Zhenbo et al. (2007), who analyzed the spatial distribution characteristics of the employed population in China using 2000 census data, and summarized six types of employment spatial patterns based on the spatial distribution characteristics of employment by industry [13]. Most of the studies have been conducted using the attributes of industries, occupations, employment positions contained within the statistical data of the population census, economic census, and sample surveys, in combination with administrative division data, to analyze spatial distribution and change characteristics using the GIS platform. Some city-scale studies in recent years have started to use data such as cell phone signaling and network reviews. These are explorations of positive significance, but there are limitations to their application due to the limited individual attributes that can be reflected by such data.

The structural characteristics of the spatial distribution of employment comprise a hot topic of interest in such studies. It is common to conduct cross-regional and cross-temporal dimensional analysis based on employment-related data to summarize the spatio-temporal patterns of employment spatial distribution changes. Peter Gordon et al. (1986) and Kenneth A. Small et al. (1994) found a trend that was characteristic of population and employment polycentric dispersion in Los Angeles during the 1970s [14,15]; William J. Coffey et al. (2001) found an important role for business services and manufacturing in the polycentric process of Montreal [16]; Richard Shearmur et al. (2002) found three different patterns of development in the spatial structure of employment in four Canadian metropolitan areas [1]; William J. Coffey et al. (2002) found evidence of decentralization in a relative sense in Montreal [2]; Catherine Baumont et al. (2004) found significant monocentric characteristics in the spatial distribution of population and employment in the urban agglomeration of Dijon, France [17]; Klaus Desmet et al. (2005) found industrial differences in the characteristics of the changing spatial distribution of jobs in the United States, with a relative concentration of employment in the service sector [18]; Rachel Guillain (2006) analyzed the spatial distribution of employment in the Île-de-France region in 1978 and 1997 and found a process of suburbanization of employment in Paris and its surrounding regions [19]; Lijing Dong et al. (2008) found the obvious spatial suburbanization of employment in Shenyang [20]; Zeng Haihong et al. (2010) found that the spatial agglomeration of advanced service sectors in Shenzhen has been strengthened [21]; Liz Rodríguez-Gámez et al. (2012) analyzed the employment centers in Hermosillo, Mexico, and found that the monocentric model is still significant, but that the role of the central business district has changed [22]; Sun Bindong et al. (2014) analyzed the evolutionary characteristics of the spatial structure of employment in Shanghai in 1996, 2004 and 2008 and found that the spatial polycentricity of both employment and population was increasing [23]; Jae Ik Kim et al. (2014) found a spatial polycentric structure of employment in Seoul, with the emerging Gangnam sub-center becoming the largest employment center in Seoul [24]; and Chen Chen (2014) found a significant suburbanization of manufacturing employment in China between 2000 and 2010, with a clear tendency for the service sector to cluster in central cities [25]. Studies have been conducted to analyze the evolution of the spatial distribution of the employed population, both in general and in specific industry. The research results indicate that the spatial evolution of the employed population may differ significantly in different cities, regions, development stages, and industries.

The spatial characteristics of different regions have a crucial impact on the layout and development of employment due to their different development situations [26]. Several scholars have discussed various factors that affect the spatial and temporal distribution of employment, including industry, innovation, housing, public service support, and transportation. Regarding industrial factors, Xu Xianglong (2009) and Sun Tieshan et al. (2014) explored the influence of industrial structure on employment structure and distribution [27,28]. Regarding the factor of innovation, Michael Fritsch et al. (2011), Matthias Buerger (2012), and Chang Jifa et al. (2018) found that innovation activities have a general or localized promotional effect on employment [29–31]. In terms of residential factors, William Levernier et al. (1994) found that housing cost and quality are important determinants of population and employment distribution [32]. Sun Tieshan et al. (2015), Zhan Dongsheng et al. (2013), and Wang Bei et al. (2020) explored the spatial relationship between employment and residence in Beijing, highlighting characteristics such as industrial differences, circle differentiation, and agglomeration and dispersion [33–35]. Regarding the public service factor, Jia Yanfei et al. (2008) and Han Li et al. (2019) identified the key influence of public services on the spatial layout of employment [36,37]. Concerning transportation factors, Harry J. Holzer et al. (2003), Jiao Huafu et al. (2011), Stephen J. Appold (2015), and Wang De. et al. (2020) studied the significant effects of transportation costs and convenience on employment distribution, with public transportation exhibiting a more pronounced boosting effect in metropolitan regions [38–41]. Some scholars have tried to use some quantitative analysis methods to assess the coupling degree between the spatial distribution of population and regional development patterns to further propose policy recommendations in order to promote the enhancement of coordination between them [42–46].

Based on the purpose of this study and the literature analysis, three hypotheses are proposed for the spatial and temporal changes of the employed population in the YRD region, which will be the focus of the subsequent empirical study: (1) cities (regions) at different levels of economic development have significant differences in the industrial structure of the employed population; (2) the early employed population changes characteristics exhibited in relatively developed cities (regions) may occur in relatively backward cities (regions) after a certain period of time; and (3) the employed population (in general or within certain industry categories) in some large cities may show significant diffusion to the suburbs.

2. Materials and Methods

2.1. Data

The use of census data by administrative division to conduct spatial analysis of the distribution of the employed population is a common practice in existing studies [3,13,25]. The data required for this study include (1) census data on the employed population by industry and county, and (2) spatial data on administrative divisions at the county level. The time points of all data include 2000, 2010 and 2020. The spatial scope of this study includes one directly administered municipality (Shanghai) and three provinces (Jiangsu, Zhejiang, and Anhui) in the YRD region. This study was conducted in a total of 41 municipalities and prefecture-level cities in the region.

Data on the employed population were obtained from the Chinese population census by county [47–49]. This information contains statistics specifically focused on the employment of the population in each county. It should be noted that the statistics on employment are contained within the long-form data of the population census. Unlike the full data, the sample of the long-form data is about 10% of the full data, and the sampling rate is basically the same in all counties. There are two classifications of the employed population in the census data, namely, by occupation and by industry. This study uses data classified by industry. The specific classification differs slightly in the three years, but is generally consistent. The industry categories of the employed population included in the primary, secondary, and tertiary industries are shown in Table 1.

Table 1. Employed population categories (based on the 2020 census of China).

Industry Sectors	Industry Category of Employed Population
Primary Industry	Agriculture, forestry, animal husbandry and fishery
Secondary Industry	Mining; manufacturing; electricity, heat, gas and water production and supply; construction
Tertiary Industry	Wholesale and retail trade; transportation, storage and postal services; accommodation and catering; information transmission, software and information technology services; finance; real estate; rental and business services; scientific research and technical services; water, environment and public facilities management; residential services, repairs and other services; Education; health and social work; culture, sports and entertainment; public administration, social security and social organizations; international organizations

Regarding the data of county administrative divisions, this study was conducted based on the GIS database to sort out the data. The basic spatial units are municipal districts, county-level cities, and counties in the YRD region. This study considers changes in administrative divisions, such as county boundary adjustments and county mergers, during the period from 2000 to 2020.

2.2. Methods

This study uses ArcGIS 10.5 platform to carry out the spatial analysis of the data.

First, the association is established between population employment statistics and spatial units of administrative divisions, the basic characteristics of spatial distribution of employed population data are analyzed, and the changes in the centroids of the population are identified using spatial statistical analysis tools.

Secondly, density statistical analysis of the spatial distribution of the employed population is carried out to compare the variation in density distribution characteristics of different years. Since there a certain amount of adjustment of county administrative divisions takes place between different years, if the density statistics of different years are directly calculated as the difference, there will be a certain distortion of analysis results within the range of divisional adjustment. In this study, based on the attempted basis of existing studies, the analysis method of focal statistics is used to assess mean of the density statistics results in a certain range (Figure 1), which can achieve the effect of smoothing the density analysis results [25]. The benefits of smoothing the original density raster data for calculation processing include two aspects: one is to avoid abrupt changes in the values of density changes at the adjustment of administrative boundaries; the second is to consider that there is a certain radius activity range of human employment activities in space, that there is a certain interaction of the employed population on both sides of the administrative boundaries, and that the smoothing calculation of spatial values is closer to the reality. According to general experience, the daily employment activities of people generally do not exceed 20 km, and 20 km is selected as the search radius for focal statistical analysis in this study [25].

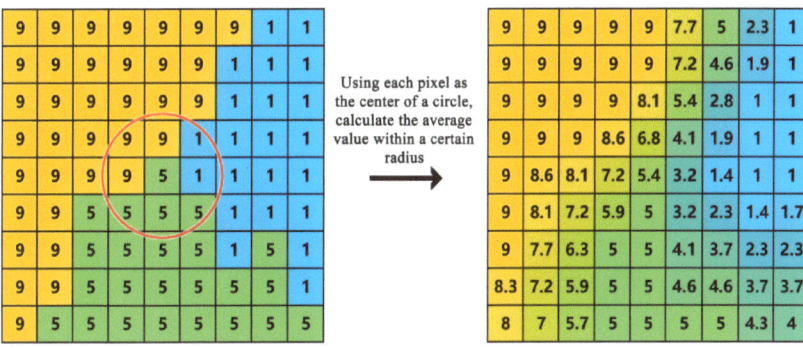

Figure 1. Principles of Focal Statistical Analysis (Source: author's own processing).

In addition, the data of each city are extracted and analyzed. The object of this study is cities at the prefecture level and above, whose spatial scope includes the concepts of central urban and municipal jurisdiction areas. The central urban area is generally the urban area where the city government is located, i.e., the urbanized area with the city center as the center of the circle, while the municipal jurisdiction area is all the areas within the administrative boundaries of the city, including the counties and county-level cities under its jurisdiction, in addition to the municipal district where the central urban area is located. There are significant differences in urbanization, industry, employment and other characteristics between the two spatial scopes, and so it is necessary to extract the relevant data from these two spatial scopes for each city separately for analysis. Using the spatial analysis tool in ArcGIS, the data of 41 city locations were extracted as central urban area data based on the focal statistical analysis results of the density raster. According to the administrative boundaries of 41 cities, the data of counties, county-level cities and municipal districts under their jurisdiction are aggregated as the data at the spatial level of municipalities. It should be noted that the boundaries of the prefectures and higher-level administrative areas of Shanghai, Jiangsu and Zhejiang were not adjusted during the period 2000–2020, but that some prefecture-level boundaries in Anhui province had scope for among themselves. For example, the counties of Chaohu City were assigned to Hefei City, Ma'anshan City, and Wuhu City; Shou County in Lu'an City was assigned to Huainan City; and Zongyang County in Anqing City was assigned to Tongling City. In order to ensure the consistency of spatial scope, the data of prefecture-level cities in Anhui Province at the three time points of 2000, 2010 and 2020 were uniformly counted according to the scope of 2020 when comparing the municipal data. Finally, clustering analysis was carried out. Systematic clustering in SPSS was used to classify and identify the data performance of 41 cities from the perspectives of industrial structure, time stage change, and spatial change.

3. Results
3.1. Changes in the Spatial Distribution Characteristics of the Employed Population
3.1.1. Spatial Distribution Characteristics of the Employed Population in 2020

- Resident population density

Determined from the spatial density distribution of the resident population (Figure 2), the central urban areas of each city are the areas with high resident population density, with Shanghai, Nanjing, Hangzhou and Hefei being particularly prominent. The contiguous areas with high resident population densities include the Taihu Plain area, consisting of Shanghai and southeastern Jiangsu and northern Zhejiang, which generally reaches more than 1000 people per square kilometer. These areas are also the core areas of the YRD. In addition, the contiguous areas with the second highest resident population density include the coastal areas around Wenzhou and Taizhou in Zhejiang, the area around Jinhua in central Zhejiang, the areas around Ma'anshan, Wuhu, Tongling and Anqing along the Yangtze River in Anhui, the area in northwestern Anhui and the area in northern Jiangsu, etc. The regions with lower resident population density include the south of Anhui Province and southwestern Zhejiang Province, and the topography of these regions is mainly mountainous.

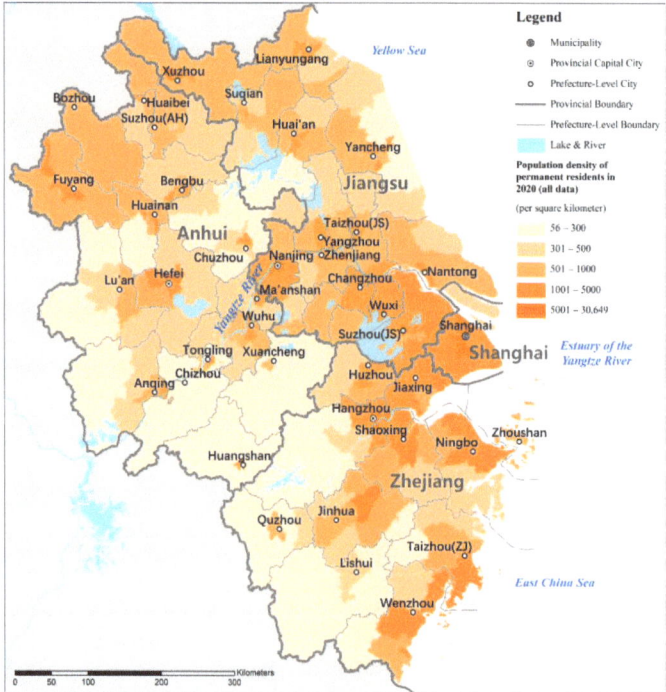

Figure 2. Population density of permanent residents in 2020 (all data) (Source: author's own processing).

- Employed population density

The spatial density distribution of the employed population (Figure 3, note that the employed population is derived from the long-form data of the census and is 10% of the total sample) shows a more clustered spatial density distribution compared to that of the resident population. The employed population is more concentrated in the Taihu Plain region consisting of Shanghai, southeastern Jiangsu, and northern Zhejiang, showing the attractiveness of the core region of the YRD for employment. Directly administered municipality and provincial capitals are the heartlands of employment concentration. In other regions, except for the central urban areas of prefecture-level cities where employed population is relatively concentrated, the density is generally low. In regions such as northern Jiangsu and northern Anhui, where the density of resident population is high, the density of employed population is significantly low, which shows that these regions are less attractive for employment.

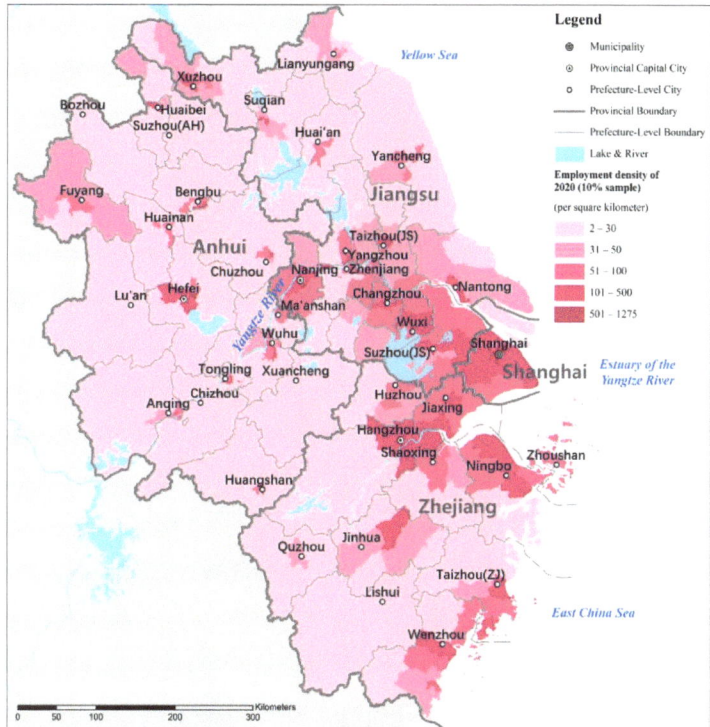

Figure 3. Employment density of 2020 (long-form data, 10% sample) (Source: author's own processing).

The spatial density analysis of the employed population is carried out separately in the primary, secondary and tertiary industries. The spatial density distribution of the primary industry is shown. The regions with a higher spatial density of the employed population in the primary industry include northern and eastern Jiangsu, and northern Anhui, indicating that more people are engaged in agriculture, forestry, animal husbandry and fishery industries in these regions (Figure 4). The regions with higher spatial density of the employed population in the secondary industry include Shanghai, south-central Jiangsu, northern Zhejiang and the eastern coastal areas, indicating that these regions have employ more people in manufacturing, construction and other industries (Figure 5). The spatial density distribution characteristics of the employed population in the tertiary industry are generally similar to those in the secondary industry, although the employed population in the tertiary industry is more centrally distributed in the centers and surrounding areas of major cities (Figure 6).

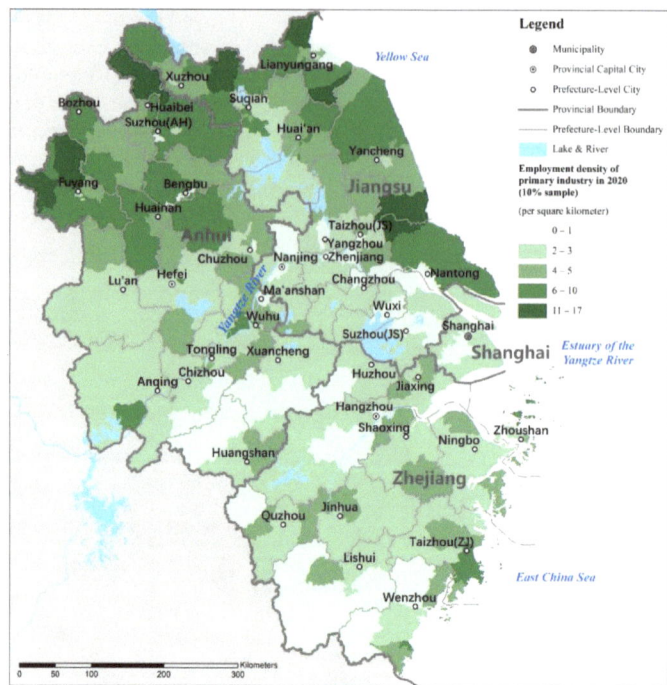

Figure 4. Employment density of primary industry in 2020 (long-form data, 10% sample) (Source: author's own processing).

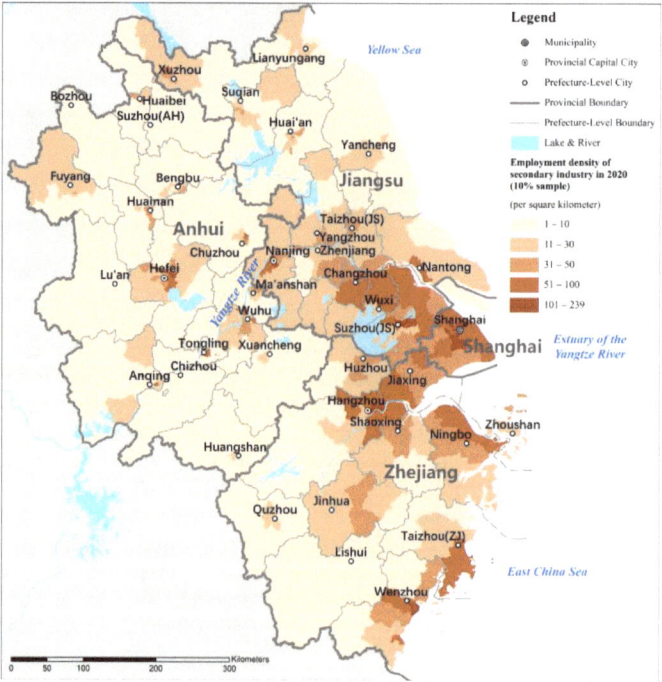

Figure 5. Employment density of secondary industry in 2020 (long-form data, 10% sample) (Source: author's own processing).

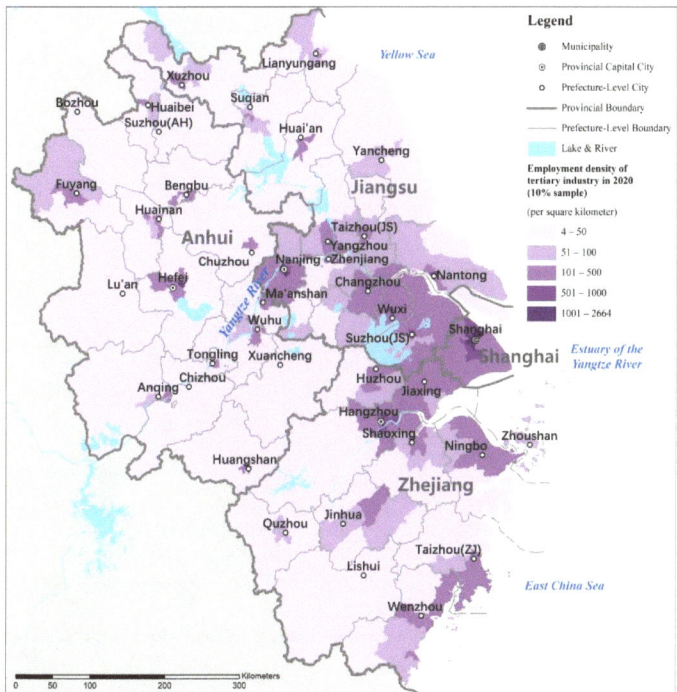

Figure 6. Employment density of tertiary industry in 2020 (long-form data, 10% sample) (Source: author's own processing).

3.1.2. Changes in the Centroids of the Spatial Distribution of the Employed Population from 2000 to 2020

Based on the spatial extent of the YRD, the centroids of the spatial distribution of the resident and employed populations at three time points in 2000, 2010, and 2020 were calculated, and the results are shown in Figure 7. In general, the centroids of both resident and employed populations are distributed in the central part of the YRD region, specifically in the southwestern part of Jiangsu Province. Specifically, the centroid of the employed population is more to the east than the centroid of the resident population, indicating that the eastern coastal region of the YRD is more attractive for employment.

Moreover, the changes in the position of the centroids are compared for the three years. It can be observed that the centroids of both the resident and employed populations shifted towards the southeast, with the centroids of the employed population moving a greater distance. This indicates that during the period from 2000 to 2020, there was a clear tendency for the population in the YRD region to move in the southeast direction. Comparing the two time periods, it can be found that the centroid of population moved less in the later decade (2010–2020) than in the earlier decade (2000–2010), which indicates that the trend of southeastward migration of resident and employed population in the YRD region has begun to slow down in recent years. This phenomenon indicates that the balanced development strategy of the YRD region has begun to show results. In recent years, some cities in the western and northern parts of the YRD region (e.g., Hefei) have developed faster and become more attractive methods of population clustering.

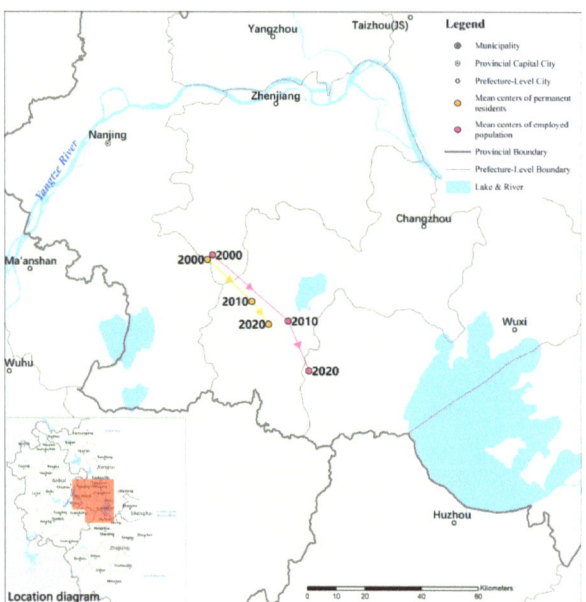

Figure 7. Mean centers of permanent residents and employed population from 2000 to 2020 (Source: author's own processing).

3.1.3. Changes in the Spatial Distribution Density of the Employed Population from 2000 to 2020

The results of the focal statistical analysis based on spatial density (the results for 2020 are shown in Figure 8) were used to carry out the analysis of the changes in the spatial distribution of the employed population during the period from 2000 to 2020.

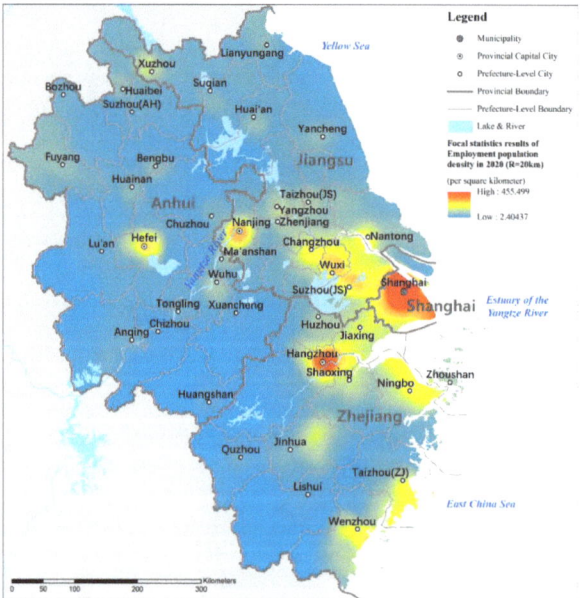

Figure 8. Focal statistics results of Employed population density in 2020 (R = 20 km) (Source: author's own processing).

- Changes in employed population density

In general, the employed population is mainly clustered in the core regions of the YRD, namely Shanghai, southern Jiangsu, and northern Zhejiang (Figure 9). In addition, the more obvious regions with increasing employed population density include Hefei in Anhui Province, Wenzhou-Taizhou coastal area in Zhejiang Province, and Jinhua surrounding area in Zhejiang Province. Regions with significant decreases in employed population density include northern Anhui, as well as northern and central Jiangsu. This feature indicates that relatively developed areas and areas around large cities have the advantage of attracting a concentration of employment.

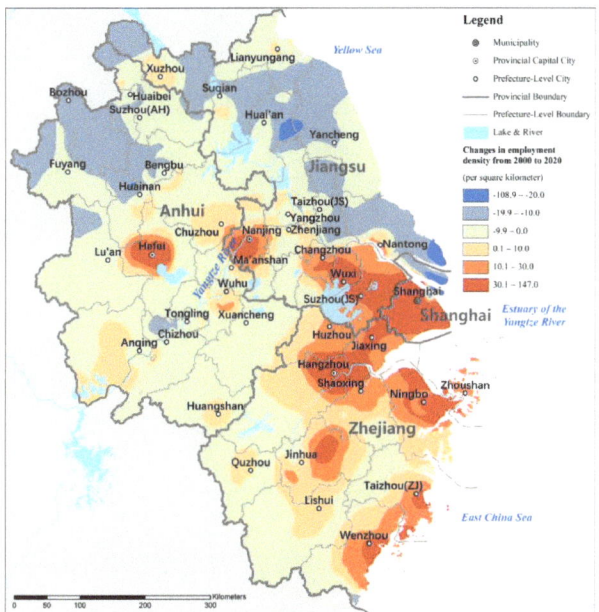

Figure 9. Changes in employment density from 2000 to 2020 (Source: author's own processing).

- Changes in density of employed population in different industrial sectors

The employed population in the primary sector is characterized by a decrease across almost the entire YRD region (Figure 10). This phenomenon indicates that the level of urbanization in the YRD region continues to increase. Specifically, the regions with the most significant decreases are northern Anhui and north-central Jiangsu. These regions are relatively late-developing regions in the YRD region, and more people employed in the primary industry have moved to the core region for employment in the secondary and tertiary industries in recent years. The change in the density of secondary industry employment shows a trend of significant increasing in some regions and significant decreasing in some regions (Figure 11). The regions that significantly increased include southeastern Jiangsu, northern Zhejiang, southeastern coastal Zhejiang, central Zhejiang, and Hefei, Anhui. The regions that significantly decreased include the central urban areas of Shanghai, Nanjing and Xuzhou. Among them, the degree of decrease in Shanghai was particularly prominent. This phenomenon indicates that the development of secondary industry shows different characteristics in different regions. In cities with earlier development such as Shanghai and Nanjing, their central areas have started to take the lead in deindustrialization. In recent years, the areas around major cities with better locational development conditions have become hot spots for the development of the secondary industry. The employment density of the tertiary industry has grown more significantly in the central areas of major cities and the surrounding areas (Figure 12). The four cities with the most prominent growth are Shanghai and the three provincial capitals: Nanjing, Hangzhou, and Hefei. This phenomenon indicates that large cities with higher administrative levels have outstanding advantages in attracting service sector employment.

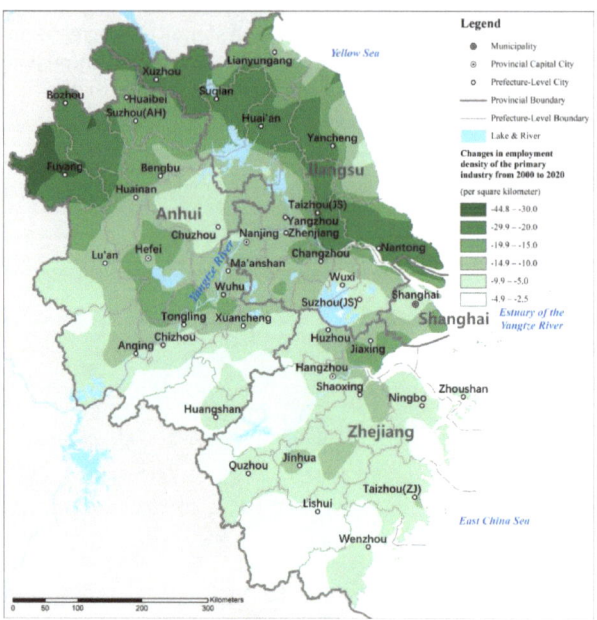

Figure 10. Changes in employment density of the primary industry from 2000 to 2020 (Source: author's own processing).

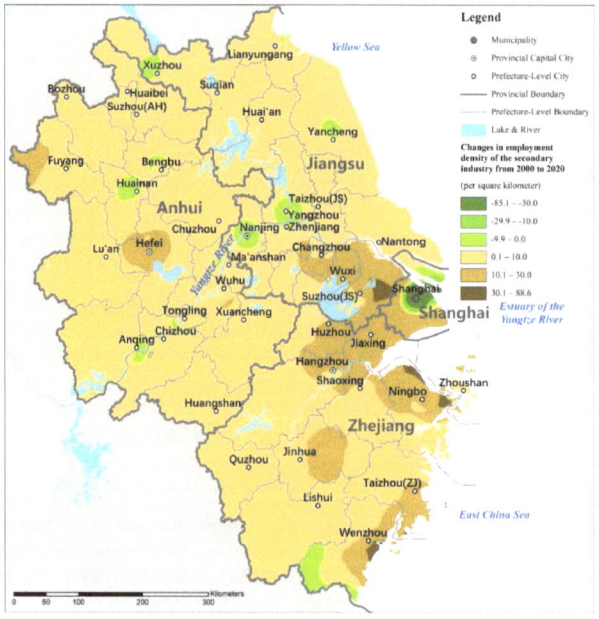

Figure 11. Changes in employment density of the secondary industry from 2000 to 2020 (Source: author's own processing).

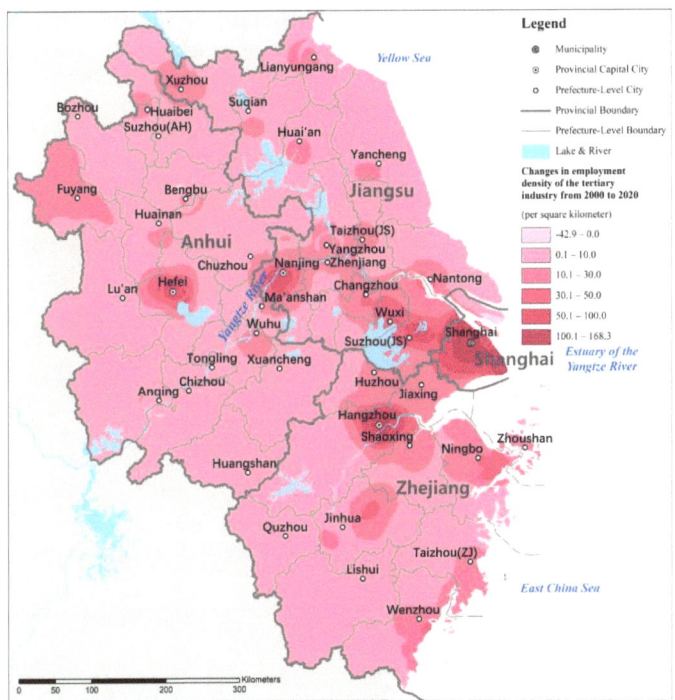

Figure 12. Changes in employment density of the tertiary industry from 2000 to 2020 (Source: author's own processing).

3.2. Cluster Analysis of Change Characteristics Based on 41 Cities

A cluster analysis was conducted on the characteristics of employed population changes in 41 cities in the YRD region. Systematic clustering in SPSS is used to carry out the analysis and discussion from three perspectives: industrial structure, time period, and spatial level. The clustering method was average linkage (between groups) method using Euclidean distance measurement. Based on the tree genealogy map obtained from the analysis (Figure 13), the classification and branching relationships of the 41 cities were identified, and the main categories were mapped into a visual map for subsequent analysis and interpretation.

3.2.1. Cluster by Industry Structure

First, the types of employed population structures in various industries in the 41 urban centers in 2020 are analyzed. The following indicators were selected as variables for analysis: (1) the proportion of employed population in primary industry; (2) the proportion of employed population in secondary industry; and (3) the proportion of employed population in tertiary industry. The results of the analysis are shown in Figure 13. Based on the tree genealogy chart, the 41 cities were classified into 5 categories (Figure 14). Category A includes one directly administered municipality of Shanghai, and three provincial capitals, Nanjing, Hangzhou and Hefei. These cities are the most important central cities in the YRD region, and the industrial structure characteristics of their employed population are significantly different from other cities, as shown by the extremely low proportion of the employed population in the primary industry (generally less than 2%), the low proportion in the secondary industry (about 20%, the lowest 4 of 41 cities), and the high proportion in the tertiary industry (more than 70%, the highest 4 of 41 cities). Among the cities in category B, category B-1 includes the 4 cities of Wenzhou, Taizhou, Jiaxing and Huzhou in Zhejiang, which are characterized by a slightly higher rate of employment in the secondary industry than in the tertiary industry; conversely, category B-2 includes 33 cities, which are characterized by a higher employment in the tertiary industry than in the secondary industry. Among the 33 B-2-1 category cities, Suzhou (Jiangsu) and 18 other B-2-1-1 category cities are characterized by a significantly higher proportion of tertiary industries than secondary industries, while the proportion of primary industries is extremely low; Bengbu and 10 other B-2-1-2 category cities are characterized by a significantly higher proportion of tertiary industries than secondary industries, but the proportion of primary industries is still of a certain scale (about 10%). Huaibei, Suzhou and Bozhou, the 3 northern Anhui cities of category B-2-2, have the highest proportion of primary industry among the 41 cities, reaching about 20%.

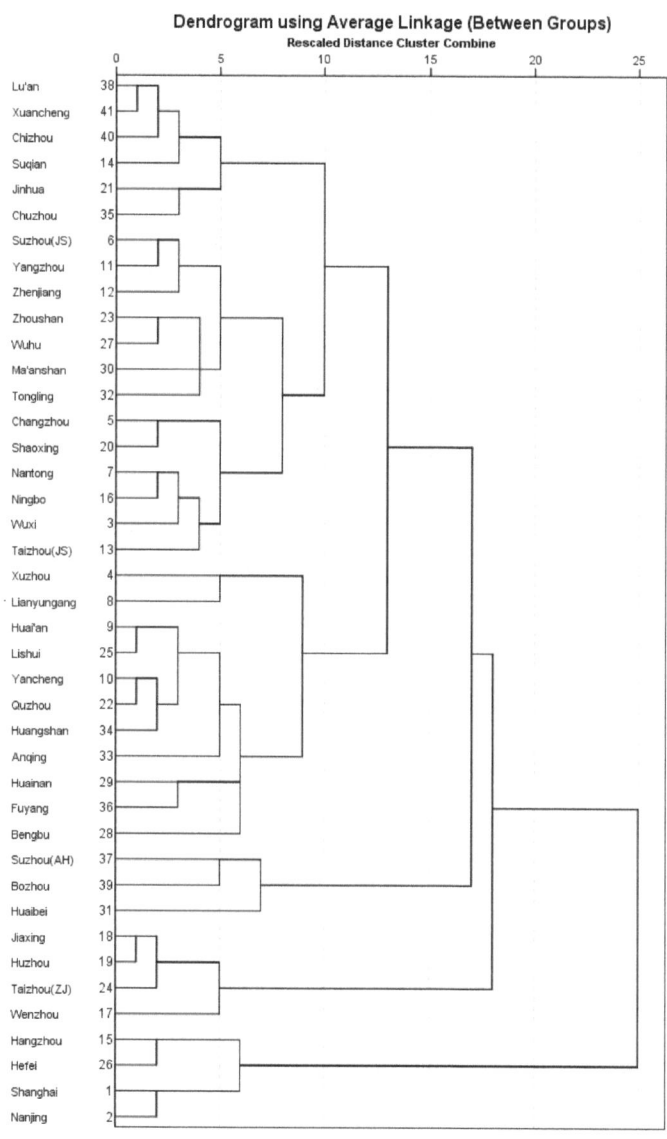

Figure 13. Dendrogram using average linkage of Cluster by industry structure of 2020.

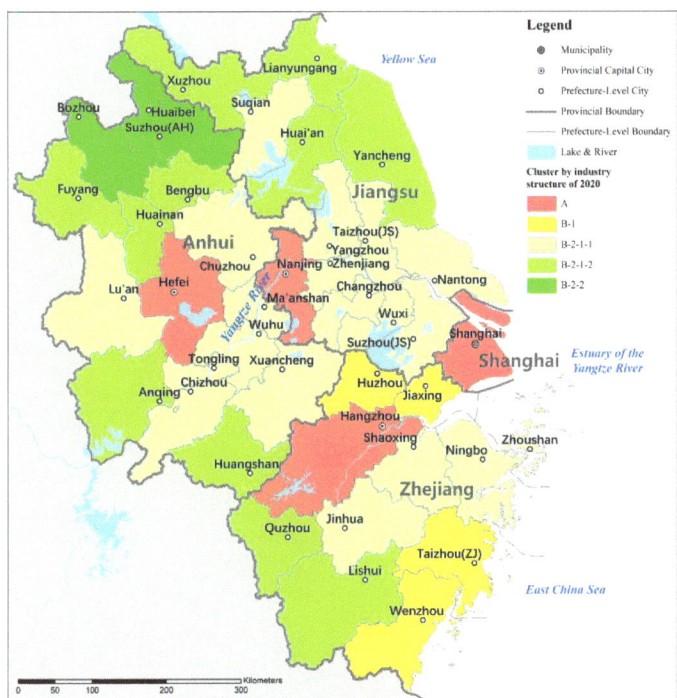

Figure 14. Map visualization results of industrial structure clustering analysis in 2020 (Source: author's own processing).

3.2.2. Cluster by Time Stage

The types of characteristics of changes in the employed population in the 41 urban centers were analyzed for two time periods: 2000–2010 (called the first decade) and 2010–2020 (called the second decade). First, the analysis of the characteristics of the overall change in the employed population was carried out by selecting the following indicators as variables: (1) the growth rate of the employed population from 2000 to 2010 and (2) the growth rate of the employed population from 2010 to 2020. The results of the analysis are shown in Figure 15, and the 41 cities are divided into 7 categories based on the tree spectrum diagram. Category A includes 2 cities, Suzhou (Jiangsu) and Wenzhou (Zhejiang), which are characterized by the extremely fast growth of the employed population (over 70%) in the first decade, but little change in the second decade. Among category B-1 cities, Hangzhou and Hefei of category B-1-1-1 have distinctive characteristics. This shows that they maintain fast growth in both decades, with growth rates of about 40%, reflecting the relatively stable employment growth dynamics of both cities. Huzhou and Chuzhou, 2 B1-1-2 category cities, are characterized by smaller increases in both decades, with the second decade seeing a slightly higher rate than the first decade decade. Shanghai, Ningbo and Wuxi, 10 B-1-2 category cities, are characterized by growth or flat growth in both decades, but with significantly lower growth rates in the second decade than the first decade. Among the B-2 cities, Jinhua and Anqing, 2 B-2-1-1 cities, show a significant decrease in employment in the first decade, but a significant increase in the second decade, reflecting a strong development momentum going forward. The change in employed population in the 20 B-2-1-2 category cities, including Xuzhou, Wuhu and Quzhou, was not significant in the two decades. The 3 B-2-2 category cities, Bozhou, Chizhou and Huai'an, showed little change in the first decade, but decreased significantly in the second decade.

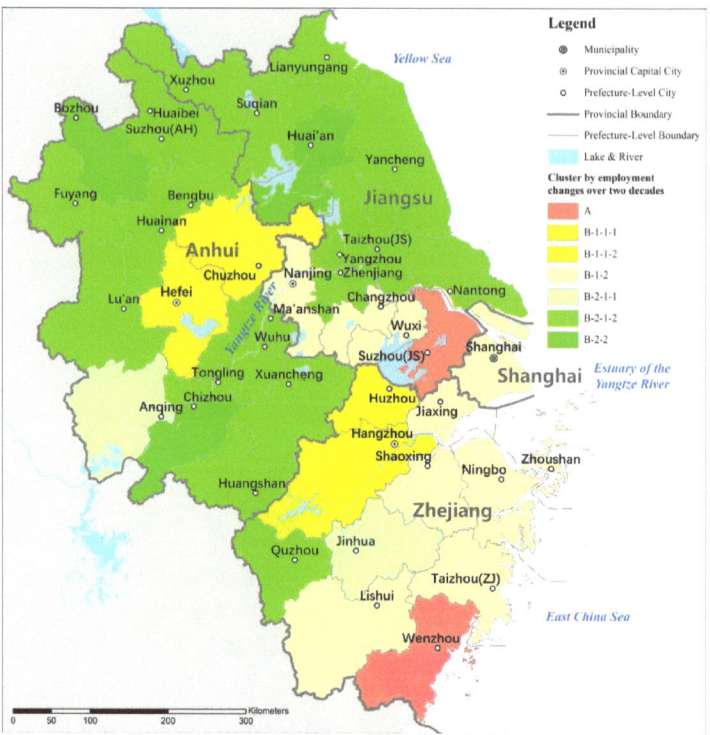

Figure 15. Map visualization results of employment changes clustering analysis over two decades of 2000–2010 and 2010–2020 (Source: author's own processing).

The analysis of the characteristics of the changes of the employed population in the three industries is carried out, and the following indicators are selected as variables for the analysis: (1) the growth rate of the employed population in the primary, secondary and tertiary industry from 2000 to 2010; and (2) the growth rate of the employed population in the primary, secondary and tertiary industry from 2010 to 2020. The results of the analysis are shown in Figure 16, and the 41 cities are divided into 6 categories based on the tree genealogy diagram. Category A includes 4 cities, among which only 1 city, Bozhou, is in category A-1. The city possesses outstanding characteristics, its secondary industry employed population has maintained a high growth rate during the two decades in question, and its total employment growth rate is ranked No. 1 among the 41 cities. Category A-2 includes 3 Anhui cities, Fuyang, Lu'an, and Suzhou, with similar characteristics to Bozhou. The employed population in the secondary industry grew faster, but the increase was significantly lower in the second decade. Among the cities in category B, category B-1-1 includes 1 city, Chuzhou, which is characterized by a significantly higher growth in both secondary and tertiary employment in the second decade than in the first decade, showing better development momentum. Category B-1-2 includes 2 cities, Jinhua and Xuancheng, which are characterized by a significantly higher increase in tertiary employment in the second decade; category B-2-1 includes 7 cities such as Lishui and Huangshan, which are characterized by a significantly higher increase in tertiary employment in the second decade; and category B-2-2 includes 27 cities, such as Bengbu, Changzhou and Quzhou, whose growth rate has been relatively stable in the two decades.

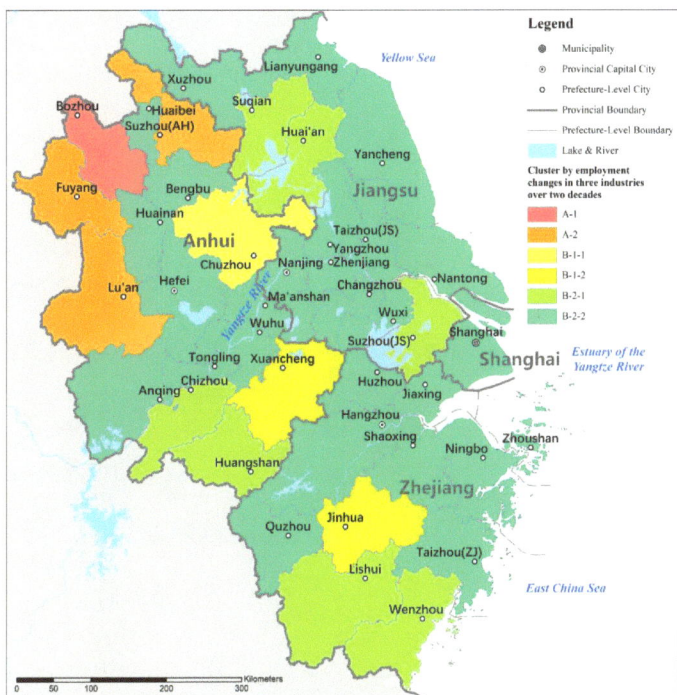

Figure 16. Map visualization results of employment industry structure changes clustering analysis over two decades of 2000–2010 and 2010–2020 (Source: author's own processing).

3.2.3. Cluster by Spatial Level

Finally, the characteristics of employed population changes at two spatial-level scales, central urban and municipal jurisdiction area, are analyzed for 41 cities. Firstly, the analysis of the characteristics of the distribution of employment in the three industries in 2020 was carried out, and the following indicators were selected as variables for the analysis: (1) the proportion of primary, secondary and tertiary industry employment in the central urban area; and (2) the proportion of primary, secondary and tertiary industry employment in the municipal jurisdiction area. The results of the analysis are shown in Figure 17, and the 41 cities are divided into 6 categories based on the tree spectrum diagram. Category A includes 2 cities, Suzhou (Jiangsu) and Wenzhou, whose share of tertiary industry employed population is the highest among the 41 cities, regardless of the central urban or the municipal jurisdiction area. Among category B cities, category B-1-1 includes 4 Zhejiang cities, Huzhou, Taizhou, Wenzhou and Jiaxing, whose share of tertiary industry employed population is higher than that of secondary industry, regardless of the central urban or the municipal jurisdiction area; category B-1-2 includes 9 cities, including Ningbo, Wuxi and Shaoxing, which are characterized by a higher proportion of tertiary employment than secondary employment in the central urban area, but with the opposite being true within the municipal jurisdiction area, where the proportion of secondary employment is higher than tertiary employment; and category B-2-1 includes 16 cities, such as Zhoushan, Huangshan and Suqian, which are characterized by a higher proportion of tertiary employment than secondary employment in both the central urban and municipal jurisdiction area. The proportion of the employed population in the tertiary industry is higher than that in the secondary industry in both the central urban and the municipal jurisdiction area; category B-2-2 includes 7 cities such as Xuzhou and Bengbu, which are characterized by the proportion of the employed population in the primary industry, reaching more than 20% within the municipal jurisdiction area, while the proportion of the employed population in the tertiary industry is generally higher than that in the secondary industry. Category C city is Suzhou in Anhui Province, which is characterized by a relatively similar proportion of people employed in primary, secondary and tertiary industries within the municipal jurisdiction area.

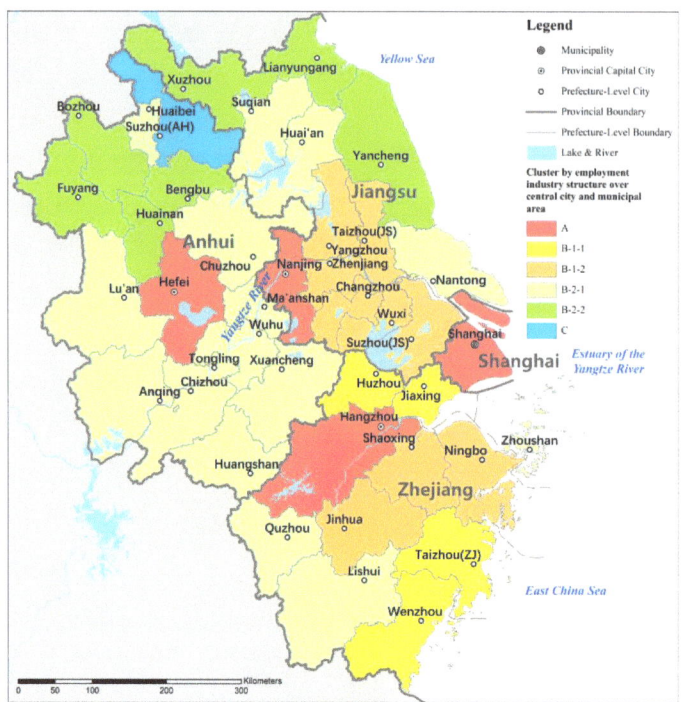

Figure 17. Map visualization results of employment industry structure clustering analysis over central urban and municipal jurisdiction area (Source: author's own processing).

An analysis combining spatial level and employed population change characteristics was carried out, and the following indicators were selected as variables for the analysis: (1) employed population growth rate in central urban area from 2000 to 2020; and (2) employed population growth rate in the municipal jurisdiction area from 2000 to 2020. The results of the analysis are shown in Figure 18. A total of 41 cities are classified into 6 categories based on the tree spectrum diagram. In total, 15 cities in category A have the common feature of employed population growth in both the central urban and the municipal jurisdiction area. Among them, Hangzhou and Hefei, which are the top 2 cities in terms of employment growth in the central urban area, are in the A-1-1 and A-1-2 categories, respectively, but Hefei is relatively less prominent in terms of employment growth in the municipal jurisdiction area; the A-2-1 category includes 5 cities, namely Suzhou (Jiangsu), Ningbo, Jiaxing, Wenzhou, and Wuxi, which are characterized by the second highest overall employment growth after Hangzhou and Hefei, while the overall employment growth in the central urban area is slightly higher than that in the municipal jurisdiction area. A-2-2 cities include Shanghai, Nanjing, Shaoxing and 5 other cities, whose employed population growth is generally lower than that of A-2-1, while the levels of employed population growth in the central urban and the municipal jurisdiction area a43 close to each other. In category B, there are 26 cities whose employed population in the central urban or municipal jurisdiction area is generally flat or decreasing. Among them, 12 cities in category B-1, including Wuhu, Quzhou, Nantong, etc., have seen a small increase in employed population in their central urban areas and a small decrease in employed population in their municipal jurisdiction areas. In total, 14 cities in category B-2, including

Chizhou, Taizhou (Jiangsu), Bozhou, etc., have seen a significant decrease in employed population in both their central urban and municipal jurisdiction areas.

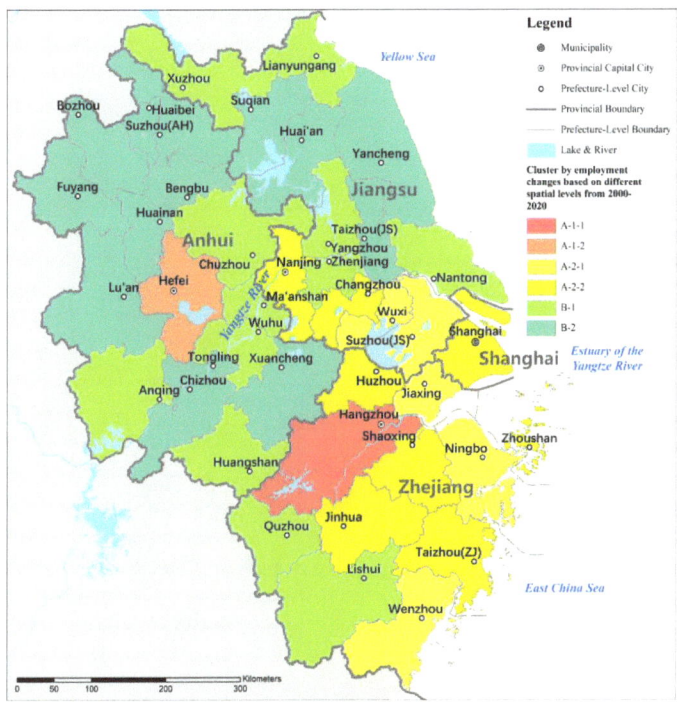

Figure 18. Map visualization results of employment changes clustering analysis based on different spatial levels from 2000 to 2020 (Source: author's own processing).

4. Discussion

The results of the above empirical analysis provide some evidence for the three hypotheses presented in the introduction of this paper, which will be discussed further in this subsection.

4.1. Regional Differences within the YRD Are Still Significant, but They Are Narrowing

On the whole, the economic development of the southeast region of the YRD is more developed compared to the northwest region. The relatively well-developed southeastern region roughly includes Shanghai, southern Jiangsu (cities such as Suzhou, Wuxi, Changzhou, and Nanjing), northern Zhejiang (cities such as Hangzhou, Jiaxing, and Shaoxing), and the southeast coast of Zhejiang (cities such as Ningbo, Taizhou, and Wenzhou). The spatial distribution of the employed population in these regions has the following typical characteristics: (1) a high proportion of the employed population working in secondary and tertiary industries; (2) the continuous agglomeration growth of the employed population in secondary and tertiary industries; and (3) employment has already grown faster in the first decade of the period 2000–2020. In contrast, the western and northern regions of the YRD are relatively underdeveloped, and these regions still maintain a relatively high proportion of employment in the primary sector, with limited ability to gather employment in the secondary and tertiary sectors and a clear tendency for employment to flow out to other regions. The differences in internal characteristics reflected in the YRD region are also present at scales, such as the national scale. For example, Enrico Marelli (2004) found that differences in employment structure are more pronounced inside countries, and Wang Zhenbo et al. (2007) found several different types of spatial patterns of employment in different regions of China [12,13]. However, the differences in characteristics within the Yangtze River Delta region are in fact gradually narrowing. The following evidence supports this view: (1) in the period 2000–2020, the distance that the centroid of the spatial distribution of the employed population moved from northwest to southeast was smaller in the second decade than in the first decade; (2) the regional differences in the structure of the employed population narrowed, the range of values of the proportion of the employed population in the secondary industry in 41 cities (municipal jurisdiction area) changed from [3.1%, 54.2%] in 2000 to [22.8%, 58.4%] in 2020, and the standard deviation of the proportion of the employed population in the primary, secondary and tertiary industries of 41 cities (municipal jurisdiction area) changed from 0.217, 0.152, 0.077 in 2000 to 0.094, 0.091, 0.070 in 2020; and (3) some cities in the northwestern YRD region have had a strong ability to attract employment in recent years, e.g., Hefei City, Anhui Province, ranked 1st and 2nd, respectively, among 41 cities in terms of employment growth in the central urban and municipal jurisdiction area during 2010–2020.

4.2. Different Cities or Regions Show Different Characteristics of Development Stages, and Late-Developing Regions Can Learn from Early Developing Regions

Due to differences in regional development, the development stages of different regions and cities show different characteristics in the spatial distribution of the employed population. According to the results of the previous analysis, the following outcomes are typical: (1) a high proportion of the employed population in the tertiary industry, a significant tendency towards agglomeration, and a significant spreading tendency of the secondary industry employment to the periphery, represented by Shanghai and Nanjing; (2) a high proportion of the employed population in the tertiary industry and a significant tendency of agglomeration of the employed population in both the secondary and tertiary industries, represented by Hangzhou and Hefei; (3) a significant growth of the employed population in the secondary industry in the first decade, but a significant decrease in the second decade, while the employment in the tertiary industry continues to cluster, represented by Suzhou, Jiangsu and Ningbo, Zhejiang; (4) the employment in the secondary industry grows rapidly in the first decade, but is basically stable in the second decade, while the proportion of employment in the tertiary industry begins to exceed that in the secondary industry, represented by Nantong and Wuhu; (5) the employment in primary industry decreases rapidly in the first decade, but the decrease in the second decade is significantly smaller, while the concentration of the population in secondary industry is more significant in the first decade, and the proportion of the employment in secondary industry is moderately low, represented by Yancheng and Lianyungang in Jiangsu; and (6) the proportion of the employment in primary industry is high, but significantly decreases in the second decade, while the proportion of the employed population in the secondary industry is low but the growth is rapid, represented by Bozhou and Suzhou in Anhui. From the stage characteristics of the industrial structure of the employed population and the growth of the agglomeration, it is obvious that these categories of cities are at different stages of development, with category 1 cities being at the highest stage of development in the region.

4.3. Metropolitan Areas Are Still the Main Areas of Employment Concentration, and the Spatial Distribution of Employment in Some Cities Is Beginning to Suburbanize

In general, the larger its size, the more attractive a city is for the concentration of employment. According to China's city size classification standards, urban area with populations of 10 million or more are considered megacities, populations of 5–10 million are supercities, populations of 3–5 million are type I large cities, populations of 1–3 million are type II large cities, populations of 0.5–1 million are medium cities, and populations of less than 500,000 are small cities. Based on the data of central urban areas, the employed population growth of the 41 cities between 2000 and 2020 was 22.0% for megacities, 68.5% for supercities, 66.4% for type I megacities, 14.4% for type II megacities, 4.0% for medium cities, and 8.2% for small cities. It is clear from this that the employed population in medium and small cities did not grow as much as in other larger cities. The most pronounced growth occurred in supercities and type I large cities, which have urban populations of between 3 and 10 million people. The megacities did not have the highest growth rate, which may be related to the high cost of living due to the shortage of urban space resources. It is worth noting that cities at higher administrative levels also performed better in terms of employment growth due to their advantages in terms of public services and other resources. Overall, the average growth of municipalities, provincial capitals or sub-provincial cities was 63.1%, while the average growth rate of ordinary prefecture-level cities was 12.3%. Looking specifically at the five type I large cities in the region, one of them, the provincial capital (Hefei), experienced a 101.4% increase in employed population, while the other four cities grew by an average of 57.6%. Many cities showed obvious characteristics of a circular structure in the spatial distribution variation in the employed population. In Shanghai and Nanjing, although the employed population grew in general, the growth of the central city is smaller than that of the municipal jurisdiction area, which indicates that the spatial suburbanization of employment was occurring. This phenomenon is similar to the trend of employment dispersion observed in Los Angeles by Kenneth A. Small et al. (1994), and the process of employment suburbanization found in Paris by Rachel Guillain (2006) [14,15,19]. If we look at the employed population in different categories, the secondary industry has shown a decrease in employment in central city areas and an increase in the peripheral areas in many cities; the tertiary industry is still generally clustered in the central city. Similar findings were derived from Klaus Desmet et al.'s (2005) analysis of differences in spatial variation in employment characteristics by industry in the United States and Zeng Haihong et al.'s (2010) analysis of spatial agglomeration characteristics of high-end service industries in Shenzhen [18,21].

5. Conclusions

Focusing on the YRD region, this study selected data from three time points, 2000, 2010 and 2020, to carry out a analysis of the spatial distribution and change characteristics of the employed population. Combined with spatial statistics and cluster analysis, the above characteristics were found to be similar or different in different regions, different types of cities, different time stages, and different industrial fields. This study concludes that: (1) regional differences within the YRD are still large but are narrowing; (2) different cities or regions exhibit different development stage characteristics, and early development regions can provide experience for later development regions; (3) the scale level and administrative level are important factors affecting the changing characteristics of the employed population, and the suburbanization of employment is occurring in large cities with better levels of development, although the characteristics of this vary by industry. These are the main findings of this study, some of which are useful in supporting the vision of the integrated development plan for the YRD region regarding balanced regional development.

Looking ahead, it is likely that the change in the employed population in the YRD region over the next decade or longer, compared to the previous two decades, will not be a simple linear fitting relationship. In other words, the trend of change in the previous 20 years cannot simply be used to predict the characteristics of future change. The main reasons for this include, but are not limited to, the following facts: (1) China's total population has peaked, the country will begin to experience negative population growth in 2022, and the YRD region will face the same challenge. (2) The population continues to age, while the number of births is declining significantly. The proportion of the population aged 0–19 in the YRD region decreased from 27.8% to 19.6%, and the proportion of the population aged 60 or above increased from 12.3% to 20.3% during 2000 to 2020). (3) Urbanization development tends to stabilize and, after the early rapid development, the urbanization rate in the YRD region reaches 70.8% in 2020, with limited room for future growth. (4) The pattern of large cities in the region is basically stable, existing small- and medium-sized cities have limited abilities to attract employment in future, and the pattern of the urban system might not undergo major changes in a short period of time. (5) Technological innovation changes the future employment mode and the development of artificial intelligence and other emerging industries may bring important impacts on the employment structure, employment scale, the layout of employment places, and the spatial relationship between employment and residence.

Based on the existing research, there are further directions for this research to deepen in the future. Firstly, scholars should refine the spatial units. The current study uses county-level spatial units, which can basically meet the research needs at the regional scale of the YRD. In the future, if data conditions allow, researchers should consider using more accurate spatial unit data (e.g., sub-district or township units) to obtain more accurate results of spatial variability characteristics. Secondly, scholars should refine the industrial categories. The current study divides the industries of the employed population into three major categories: primary industry, secondary industry and tertiary industry. These are the commonly used classifications, and this research has identified significant differences between the secondary and tertiary industries. It is worth noting, however, that there are significant differences between productive and consumer services in the tertiary sector. Within the productive service industry, there may also be significant differences between industries such as finance, information service, R&D, and goods transportation. Therefore, future research can be conducted by further subdividing the industry sectors. Thirdly, this study is an analysis of existing facts and provides limited support for future trend prediction. In the future, trend prediction can be carried out by introducing relevant variable factors in response to the latest situation of population development in order to improve the application value of this study for regional development strategic planning.

Funding: This research was funded by the Shanghai Philosophy and Social Science Planning Youth Project, grant number 2019ECK007.

Data Availability Statement: The census data used in this study can be found in the relevant publications and has already been cited in the text for illustration. The map data required for the study can be found in the atlases of the relevant years.

Conflicts of Interest: The author declares no conflict of interest.

References

1. Shearmur, R.; Coffey, W.J. A Tale of Four Cities: Intrametropolitan Employment Distribution in Toronto, Montreal, Vancouver, and Ottawa-Hull, 1981–1996. *Environ. Plan. A Econ. Space* **2002**, *34*, 575–598. [CrossRef]
2. Coffey, W.J.; Shearmur, R.G. Agglomeration and Dispersion of High-order Service Employment in the Montreal Metropolitan Region, 1981–1996. *Urban Stud.* **2002**, *39*, 359–378. [CrossRef]
3. Alidadi, M.; Dadashpoor, H. Beyond Monocentricity: Examining the Spatial Distribution of Employment in Tehran Metropolitan Region, Iran. *Int. J. Urban Sci.* **2018**, *22*, 38–58. [CrossRef]
4. Bo, W.; Feng, Z. An Analysis of Spatial Distribution of Urban Employment in Nanjing. *Hum. Geogr.* **2011**, *26*, 58–65.
5. Zhan, D.; Zhang, W.; Meng, B.; Dang, Y.; Liu, Q. Spatial Structure of Urban Residence and Employment in Beijing. *Sci. Geogr. Sin.* **2017**, *37*, 356–366.
6. Wei, Z.; Xiao, W.; Qiang, H. Analysis on the Differentiation of Employment Space of Migrant Workers from the Perspective of Gender: An Empirical Study on the Main City of Nanjing. *City Plan. Rev.* **2017**, *41*, 54–64.
7. Li, P. Detecting the Spatial Characteristics of Urban Employment Using Mobile Phone Signaling Data: A Case Study of Shenyang City. *Geomat. World* **2019**, *26*, 25–30.
8. Chen, S.; Feng, Z.; Enyu, C.; Yang, C. Spatial Distribution of New Employment Demand in Nanjing Based on the Data from Zhaopin.com. *Econ. Geogr.* **2016**, *36*, 83–90.
9. Hui, W.; Xiao, W.; Hao, Z. A Preliminary Analysis of the Correlation of Migrant Workers' Employment Spatial Agglomeration and Urban Employment Space: A Case of Nanjing Main City. *Hum. Geogr.* **2014**, *29*, 31–39.
10. Hui, W.; Xiao, W.; Zheng, H. A Preliminary Analysis of Spatial Distribution of Employment Spatial in Nanjing Main City. *Econ. Geogr.* **2014**, *34*, 115–123.
11. Lao, X.; Shen, W.; Wen, F. A Comparative Research on Employment Density Distribution between Yangtze River Delta and Pearl River Delta. *Urban Dev. Stud.* **2013**, *20*, 137–142.
12. Marelli, E. Evolution of Employment Structures and Regional Specialisation in the EU. *Econ. Syst.* **2004**, *28*, 35–59. [CrossRef]
13. Wang, Z.; Zhu, C. Employment Spatial Models and Regionalization of China. *Acta Geogr. Sin.* **2007**, *62*, 191–199.

14. Gordon, P.; Richardson, H.W.; Wong, H.L. The Distribution of Population and Employment in a Polycentric City: The Case of Los Angeles. *Environ. Plan. A* **1986**, *18*, 161–173. [CrossRef] [PubMed]
15. Small, K.A.; Song, S. Population and Employment Densities: Structure and Change. *J. Urban Econ.* **1994**, *36*, 292–313. [CrossRef] [PubMed]
16. Coffey, W.J.; Shearmur, R.G. Intrametropolitan Employment Distribution in Montreal, 1981–1996. *Urban Geogr.* **2001**, *22*, 106–129. [CrossRef]
17. Baumont, C.; Ertur, C.; Le Gallo, J. Spatial Analysis of Employment and Population Density: The Case of the Agglomeration of Dijon 1999. *Geogr. Anal.* **2004**, *36*, 146–176. [CrossRef]
18. Desmet, K.; Fafchamps, M. Changes in the Spatial Concentration of Employment across US Counties: A Sectoral Analysis 1972–2000. *J. Econ. Geogr.* **2005**, *5*, 261–284. [CrossRef]
19. Guillain, R.; Le Gallo, J.; Boiteux-Orain, C. Changes in Spatial and Sectoral Patterns of Employment in Ile-de-France, 1978–1997. *Urban Stud.* **2006**, *43*, 2075–2098. [CrossRef]
20. Dong, L.; Zhang, P. Spatial Differentiation of Employment Structure in Shenyang City from 1990s. *Hum. Geogr.* **2008**, *23*, 32–37.
21. Zeng, H.; Meng, X.; Li, G. Spatial Structure of Employment and Its Evolution in Shenzhen City: 2001–2004. *Hum. Geogr.* **2010**, *25*, 34–40.
22. Rodríguez-Gámez, L.; Dallerba, S. Spatial Distribution of Employment in Hermosillo, 1999–2004. *Urban Stud.* **2012**, *49*, 3663–3678. [CrossRef]
23. Bindong, S.; Xuhong, W. Spatial distribution and structure evolution of employment and population in Shanghai Metropolitan Area. *Acta Geogr. Sin.* **2014**, *69*, 747–758.
24. Kim, J.I.; Yeo, C.H.; Kwon, J. Spatial Change in Urban Employment Distribution in Seoul Metropolitan City: Clustering, Dispersion and General Dispersion. *Int. J. Urban Sci.* **2014**, *18*, 355–372. [CrossRef]
25. Chen, C. Changes of the Current Chinese Urban and Rural Employment Space and Extended Discussion. *City Plan. Rev.* **2014**, *38*, 72–78.
26. O'Regan, K.M.; Quigley, J.M. Where Youth Live: Economic Effects of Urban Space on Employment Prospects. *Urban Stud.* **1998**, *35*, 1187–1205. [CrossRef]
27. Xu, X. Research on the Evolution Characteristics and Interaction Efficiency of Industrial Structure and Employment Structure in Guangdong Province. *Acad. Res.* **2009**, 90–96.
28. Sun, T.; Qi, Y.; Liu, X. Changing Intra-Metropolitan Spatial Distribution of Employment with Economic Restructuring in Beijing Metropolitan Area. *Econ. Geogr.* **2014**, *34*, 97–104.
29. Fritsch, M.; Schindele, Y. The Contribution of New Businesses to Regional Employment—An Empirical Analysis. *Econ. Geogr.* **2011**, *87*, 153–180. [CrossRef]
30. Buerger, M.; Broekel, T.; Coad, A. Regional Dynamics of Innovation: Investigating the Co-evolution of Patents, Research and Development (R&D), and Employment. *Reg. Stud.* **2012**, *46*, 565–582.
31. Chang, J.; Cui, L. The Impact of Technological Innovation and Industrial Structure on Employment—Based on Threshold Effect and Spatial Effect. *J. Lanzhou Univ. Financ. Econ.* **2018**, *34*, 35–44.
32. Levernier, W.; Cushing, B. A New Look at the Determinants of the Intrametropolitan Distribution of Population and Employment. *Urban Stud.* **1994**, *31*, 1391–1405. [CrossRef]
33. Sun, T. Spatial mismatch between residences and jobs by sectors in Beijing and its explanations. *Geogr. Res.* **2015**, *34*, 351–363.
34. Zhan, D.; Meng, B. Spatial clustering analysis of residential and employment distribution in Beijing based on their social characteristics. *Acta Geogr. Sin.* **2013**, *68*, 1607–1618.
35. Wang, B.; Wang, L.; Liu, Y.; Yang, B.; Huang, X.; Yang, M. Characteristics of jobs-housing spatial distribution in Beijing based on mobile phone signaling data. *Prog. Geogr.* **2020**, *39*, 2028–2042. [CrossRef]
36. Jia, Y.; Zhen, F. Population and Employment Distribution Characters and the Relationships between them for Small Sized Cities: Case Study of Sihong, Jiangsu. *Mod. Urban Res.* **2008**, 54–62.
37. Li, H.; Wei, Y.D.; Wu, Y. Urban Amenity, Human Capital and Employment Distribution in Shanghai. *Habitat Int.* **2019**, *91*, 102025. [CrossRef]
38. Holzer, H.J.; Quigley, J.M.; Raphael, S. Public Transit and the Spatial Distribution of Minority Employment: Evidence from a Natural Experiment. *J. Policy Anal. Manag.* **2003**, *22*, 415–441. [CrossRef]
39. Jiao, H.; Hu, J. Spatial Match Between Employment and Housing in Wuhu City, Anhui Province of China. *Sci. Geogr. Sin.* **2011**, *31*, 788–793.
40. Appold, S.J. The Impact of Airports on US Urban Employment Distribution. *Environ. Plan. A Econ. Space* **2015**, *47*, 412–429. [CrossRef]
41. Wang, D.; Li, D.; Fu, Y.Z. Employment space of residential quarters in Shanghai: An exploration based on mobile signaling data. *Acta Geogr. Sin.* **2020**, *75*, 1585–1602.
42. Sun, P.J.; Xiu, C.L. Coupling Degree Assessment of the Man-Land Coupling System of the Mining City from the Vulnerability Perspective. *Areal Res. Dev.* **2010**, *29*, 75–79.
43. Ren, Y.W.; Cao, W.D.; Zhang, Y.; Su, H.F.; Wang, X.W. Temporal and Spatial Coupling Characteristics of Urbanization and Ecological Environment of Three Major Urban Agglomerations in the Yangtze River Economic Belt. *Resour. Environ. Yangtze Basin* **2019**, *28*, 2586–2600.

44. Hong, G.; Lan, C. Coupling Analysis of Urban Construction Land and Urban Population in the Yangtze River Delta. *Popul. Soc.* **2020**, *36*, 40–48.
45. Zhou, Y.; Huang, X.; Xu, G.; Li, J. The coupling and driving forces between urban land expansion and population growth in Yangtze River Delta. *Geogr. Res.* **2016**, *35*, 313–324.
46. Chen, Q.; Tang, J. The Spatial and Characteristics of Population Evolution and Its Dynamic Mechanism in City-and-Town Concentrated Areas in the Yangtze River Delta. *J. Zhejiang Univ. Technol. Soc. Sci.* **2009**, *8*, 27–33.
47. Population Census Office under the State Council; Department of Population and Social Science and Technology Statistics, National Bureau of Statistics. *Tabulation on the 2010 Population Census by County*; China Statistics Press: Beijing, China, 2003.
48. Population Census Office under the State Council; Department of Populationand Employment Statistics, National Bureau of Statistics. *Tabulation on the 2010 Population Census of the People's Republic of China by County*; China Statistics Press: Beijing, China, 2012.
49. Office of the Leading Group of the State Council for the Seventh National Population Census. *Tabulation on 2020 China Population Census by County*; China Statistics Press: Beijing, China, 2022.

Disclaimer/Publisher's Note: The statements, opinions and data contained in all publications are solely those of the individual author(s) and contributor(s) and not of MDPI and/or the editor(s). MDPI and/or the editor(s) disclaim responsibility for any injury to people or property resulting from any ideas, methods, instructions or products referred to in the content.

Article

The Impact of Intra-City and Inter-City Innovation Networks on City Economic Growth: A Case Study of the Yangtze River Delta in China

Xianzhong Cao, Bo Chen, Yi Guo and Zhenzhen Yi *

The Center for Modern Chinese City Studies, East China Normal University, Shanghai 200062, China; xzcao@geo.ecnu.edu.cn (X.C.); 51203902028@stu.ecnu.edu.cn (B.C.); 52203902001@stu.ecnu.edu.cn (Y.G.)
* Correspondence: zzyi@iud.ecnu.edu.cn

Abstract: Innovation networks promote regional innovation and economic growth. Using the patent data of cooperative inventions and the panel data of socio-economic statistics for 2010–2019, this study quantitatively analyzes the spatial structure evolution of intra-city and inter-city innovation networks for 41 cities in the Yangtze River Delta and their influence on economic growth. This study shows that these networks are increasingly connected and have a highly similar Z-shaped spatial structure. City economic growth is generally high, relatively stable, and mainly positively influenced by inter-city innovation networks. Intra-city innovation networks have no significant effect on economic growth; however, they are complementary to the inter-city ones.

Keywords: innovation networks; economic growth; spatio-temporal evolution; influence mechanism; Yangtze river delta

Citation: Cao, X.; Chen, B.; Guo, Y.; Yi, Z. The Impact of Intra-City and Inter-City Innovation Networks on City Economic Growth: A Case Study of the Yangtze River Delta in China. *Land* **2023**, *12*, 1463. https://doi.org/10.3390/land12071463

Academic Editor: Bernardino Romano

Received: 10 June 2023
Revised: 11 July 2023
Accepted: 20 July 2023
Published: 22 July 2023

Copyright: © 2023 by the authors. Licensee MDPI, Basel, Switzerland. This article is an open access article distributed under the terms and conditions of the Creative Commons Attribution (CC BY) license (https:// creativecommons.org/licenses/by/ 4.0/).

1. Introduction

Innovation network refers to the cooperation of innovation actors, such as government, enterprises, universities, research institutions, and intermediary service agencies in technology research and development [1]. Considering cities as boundaries, innovation networks can be divided into intra-city and inter-city ones. With the economic globalization, the innovation paradigm has changed from the traditional closed linear model to the modern open network one, and the influence of innovation networks on regional innovation and economic growth has gradually become the frontier scientific problem for economic geographers [2,3]. As the representative of developed regions in developing countries, The Yangtze River Delta has experienced rapid economic development and strong innovation potential, and the Chinese central government attaches great importance to the construction of a regional innovation community. On 20 December 2020, the Ministry of Science and Technology in China released the Development Plan for the Construction of Innovation Community in the Yangtze River Delta, which clearly pointed out that it was necessary to focus on high-tech industries, such as biomedicine, new materials, integrated circuits, equipment manufacturing, and to achieve cross-border cooperation with innovation actors, such as universities, research institutions, enterprises, and intermediaries in 41 cites, so as to promote the free flow of innovation elements. All these make it a typical region for studying intra-city and inter-city innovation networks; however, little research has focused on the relationship between innovation networks and economic growth, especially in terms of the former influencing the latter.

Economic geographers have studied the structure characteristics, spatial scale, influence mechanism, and effect of innovation networks. Among them, the research on innovation networks in different spatial scales, such as global, local, and global–local, has attracted the attention of many scholars; however, such research generally focuses on the analysis of the network structure rather than on the relationship mechanisms, and the influence of innovation networks of different spatial scales (intra-city and inter-city) on economic growth remains controversial [4]. Therefore, taking 41 cities in the Yangtze River Delta as examples, this paper analyzes the spatial and temporal evolution characteristics of intra-city and inter-city innovation networks and economic growth, trying to answer the following three questions: first, what are the characteristics of intra-city and inter-city innovation networks in the Yangtze River Delta and the differences between them? Second, what are the characteristics of the spatial and temporal evolution of economic growth in the Yangtze River Delta? Third, what is the impact of intra-city and inter-city innovation networks on economic growth in the Yangtze River Delta and the impact mechanism?

The remainder of this paper is organized as follows: the subsequent section presents the literature review. Section 3 describes the data and empirical variables used in analyzing the innovation networks and economic growth. With the help of ArcGIS 10.6 and MaxDEA, Section 4 analyzes the spatio-temporal evolution of innovation networks and economic growth in the Yangtze River Delta. Based on the regression model, Section 5 discusses the mechanism of innovation networks' impact on economic growth. The final section concludes and discusses the paper.

2. Literature Review
2.1. Relationship between Innovation Networks and Economic Growth

New knowledge based on technological innovation is recognized by scholars as the foundation of promoting economic growth [5,6]. On the basis of the traditional economic growth theory, the endogenous economic growth theory emphasizes the key role of knowledge in driving productivity and economic growth [7]. It also points out that economic growth and innovation networks are interrelated, as knowledge creation, accumulation, and transfer can effectively explain the differences in city economic growth levels [8]. The innovation network is an important way to promote regional growth [9]. The purpose of innovation networks is to improve the innovation ability and performance of innovation actors, and then to transform the research and development of innovative products into economic benefits. Innovation can be regarded as one of the necessary conditions for regional economic growth and development. The existing research has also shown that the knowledge flow in innovation networks determines technological innovation ability and the level of regional economic growth [10].

Innovation networks strengthen the knowledge flow inside and outside the region and is a key capital investment in the process of regional economic growth [11]. However, there is still controversy about the relationship between innovation networks of different spatial scales and economic growth. Local knowledge creation and global knowledge acquisition interact with regional economic growth [12], and we can see knowledge integrators with high competitiveness innovate by integrating global and local knowledge [13]. Crespo mainly analyzed the influence of the structural attributes of local knowledge networks on the promotion of regional competitiveness [14], while Breschi et al. pointed out that non-local innovation networks are more conducive to regional economic development [15]. There are considerable differences in the influence of different spatial scales of knowledge on regional development; the course of regional economy and innovation not only depends on localized production and knowledge creation, but also needs to combine the "local buzz" and "global pipelines" [16–18]. At the same time, it should be emphasized that there are costs for innovation actors to cooperate with local and non-local innovators. Esposito et al. found that too much local interaction would lead to the disappearance of the boundaries of innovation actors and the decline in regional technological innovation, when innovation actors engage in non-local interaction, the costs may also exceed the benefits [12]. Bianchi et al. observed that acting as interregional broker cities, especially connecting Latin American cities to the rest of the world, negatively affected patent outcomes [19].

2.2. Influence of Innovation Networks on City Economic Growth

City economic growth is the ultimate embodiment of the economic effect of innovation networks, and innovation is the necessary condition to maintain stable economic growth and development. Existing research results show that knowledge flow in innovation networks determines innovation ability and economic growth [10]. The position and rights of actors in innovation networks are the important factors that affect its knowledge acquisition and control the knowledge flow. Bianchi et al. found that cities in the center of innovation networks had more innovative activities, while as regional "gatekeepers", accessing and using external knowledge required costs and internal capabilities [19]. Moreover, cities acting as regional brokers have a negative influence on their own innovation capability due to the effects of "information overload" and "mobilization failure" [20]. On the contrary, Le Gallo et al. used collaborative patent data in the biomedical industry to find that gatekeepers who directly acquired non-local knowledge could improve innovation performance and benefit the entire region [21]. The influence of proximity among innovation agents in innovation networks has also attracted scholars' attention, including geographical, cognitive, social, institutional, and cultural proximity [22]; proximity significantly promotes the formation of innovation networks in an ITISA and contributes to the improvement of innovation ability [23]. In addition, the selection of knowledge sources in innovation networks is also important for regional economic growth. Balland et al. found that what matters is not being connected to other regions per se, but being connected to regions that provide complementary capabilities [24]. Regional knowledge absorption capacity moderates the impact of innovation networks on regional economic growth. The lack of regional knowledge absorption and utilization capacity leads to the evolution of local knowledge into path lock [25], which eventually reduces the power of regional economic growth. Some scholars pointed out that the characteristics of knowledge determine the network value, which in turn affects the economic growth [11]. The structural attribute of innovation networks modulates the effect of the knowledge flow, which determines the innovation capability and performance of regions or enterprises. Regional development is restricted by the innovation capability of the respective regions or enterprises, and areas occupying the core position in innovation networks are more likely to bring about regional economic growth [26].

Generally speaking, compared to the role of traditional economic factors in city economic growth, knowledge flow and innovation have become the key driving factors for city economic growth and productivity improvement [27]. However, the related research needs to study in more depth the relationship mechanism; the existing research pays more attention to the relationship between technological innovation and economic growth, or regards the innovation network as a whole but pays little attention to the relationship between intra-regional and inter-regional innovation networks and economic growth. Therefore, this study analyzed the differences in the spatial and temporal evolution of economic growth from the perspective of innovation networks and compared the influence of intra-city and inter-city networks, exploring their influence mechanisms.

3. Methods and Data

3.1. Research Sample

The Yangtze River Delta is one of the most dynamic, open, and innovative regions in China. In 2020, the GDP of the Yangtze River Delta was CNY 3851 billion, accounting for 24.1 percent of the whole country's GDP. High-quality and regional integration have become the core solutions for the Yangtze River Delta. On 30 December 2020, the Ministry of Science and Technology in China issued the Development Plan for the Construction of Science and Technology Innovation Community in the Yangtze River Delta, further clarifying the advantages of regional innovation resources, optimizing the regional innovation layout and collaborative innovation ecology, enhancing the regional collaborative innovation capability, and supporting the high-quality development and regional integration of the Yangtze River Delta. According to the Regional Integration Development Plan of the

Yangtze River Delta issued by the government in December 2019, the research sample of this paper included the whole area of three provinces and one municipality in the Yangtze River Delta (Shanghai, Jiangsu, Zhejiang, and Anhui provinces, covering 258,000 square kilometers), with 41 cities.

3.2. Calculation of Innovation Networks in the Yangtze River Delta

Innovation can be regarded as one of the necessary conditions for regional economic growth and development. However, with the increasing complexity and uncertainty of knowledge innovation, an actor's innovation can no longer be satisfied only by its own resources. The purpose of the innovation network is to improve the innovation ability and performance of innovation actors, and then to transform the research and development of innovative products into economic benefits. Essentially, innovation networks are the connections between different innovation actors, including government, enterprises, universities, research institutions, and intermediary service agencies, while the urban innovation network classifies the cooperation links to the urban level.

In this study, we constructed multi-level innovation networks, including intra-city and inter-city innovation networks. Collaboration among the innovators within a city formed the intra-city innovation networks. Each node represents an innovator and the edges represent their collaborative relations. Similarly, collaborations among the cities formed the inter-city innovation networks. Each node represents a city and the edges represent collaborative relations among cities. Considering that different innovation actors within a city belong to the same city, we used the number of collaborations between innovation actors to measure the intensity of collaborative innovations within the city. Moreover, we used the social network analysis (SNA) to analyze innovation networks. Degree centrality is one of the indicators to measure the centrality of the nodes in the network, and its calculation formula is as follows:

$$DC = \sum_{j=1}^{n} X_{ij} \quad (1)$$

where n is the number of cities in the network, X_{ij} is the level of cooperation between i and j. The higher the centrality, the better the position of a city in sharing, integrating, and utilizing complementary and heterogeneous resources.

The patent literature is the largest source of technical information in the world. The report by the World Intellectual Property Organization (WIPO) shows that about 90–95 percent of global R&D outputs is contained in patents, and the rest is embodied in scientific literature, such as papers and publications [28]. The patent literature offers the advantages of openness, timeliness, detailed content, and easy comparisons between industrial technologies or different spaces, and has become an important data source for studying knowledge production and innovation activities [29]. A joint patent application is an interactive innovation process based on knowledge sharing between organizations and resource integration embedded in social networks. Economic geographers have widely recognized that joint patent application data can be researched on innovation networks and knowledge spillovers [30]; the invention patent represents the original technology, which can better reflect the technological innovation achievements. The patent data are sourced from the incoPat database and processed as follows: (i) time selection—because it takes 18 months for patent applications to be published, our research limited the patent application dates to the period from 1 January 2010 to 31 December 2019, and the locations to Shanghai, Anhui, Zhejiang, and Jiangsu province. The patents with 1 number of applicants were filtered, and individuals were filtered in the applicant category to obtain 256,934 joint patent applications in the Yangtze River Delta. (ii) Matching the innovation actors and geographical location—we matched the geographical positions of the applying actors by using the enterprise database of Tianyancha (https://www.tianyancha.com/search, accessed on 20 June 2021) and cross-processed the data of three or more actors, thus obtaining 400,201 linkages of city innovation networks. (iii) The classification of innovation networks—we divided the innovation networks within the Yangtze River Delta into intra-city and inter-city ones, and finally obtained 161,766 linkages of intra-city and 86,487 linkages of inter-city innovation networks (Figure 1).

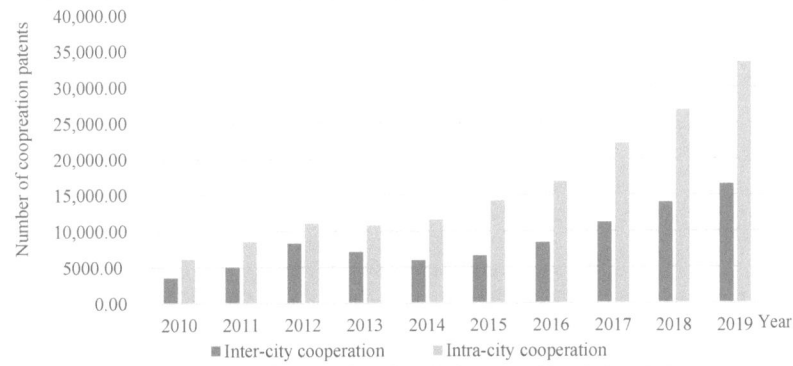

Figure 1. Diagram of the intra-city and inter-city innovation cooperation networks in the Yangtze River Delta from 2010 to 2019.

3.3. Calculation Method of Economic Growth in the Yangtze River Delta

3.3.1. Total Factor Productivity of DEA Malmquist Index

The total factor productivity (TFP), which is generally accepted by scholars, is used as an index to measure city economic growth, while the TFP of the DEA Malmquist index is mostly used as the measurement method [31]. Based on the relative efficiency, Data Envelopment Analysis (DEA) evaluates its relative effectiveness by comparing the degree of deviation of (decision-making unit: DMU) from the DEA frontier, which is a non-parametric method. The advantage of this method is that it does not need any specific function form or distribution assumption. It is effective in dealing with the efficiency problems of multi-input and multi-output DMUs and is suitable for the efficiency evaluation of urban complex economic systems. Therefore, based on the method of the DEA Malmquist index, this paper analyzed the dynamic characteristics of economic growth in the Yangtze River Delta.

The Malmquist productivity index is based on the DEA model, which uses the ratio of the distance function to calculate the input–output efficiency. The Malmquist productivity index given to output index variables is:

$$M_0^t = \frac{D_0^t(x_{t+1}, y_{t+1})}{D_0^t(x_t, y_t)}; \quad M_0^{t+1} = \frac{D_0^{t+1}(x_{t+1}, y_{t+1})}{D_0^{t+1}(x_t, y_t)} \quad (2)$$

$D_0^t(x_t, y_t)$ and $D_0^{t+1}(x_{t+1}, y_{t+1})$ are the output distance functions over the same periods of t and t + 1; $D_0^t(x_{t+1}, y_{t+1})$ and $D_0^{t+1}(x_t, y_t)$ are the output distance functions over the mixed periods of t and t + 1.

Fare et al. calculated the Malmquist index of the fixed 4e direction output with the geometric average of the two Malmquist productivity indices [32], which can be written in the following equivalent form:

$$M_0(x_t, y_t, x_{t+1}, y_{t+1}) = \left[\frac{D_0^t(x_{t+1}, y_{t+1})}{D_0^t(x_t, y_t)} * \frac{D_0^{t+1}(x_{t+1}, y_{t+1})}{D_0^{t+1}(x_t, y_t)}\right]^{1/2} \quad (3)$$

Further decomposition:

$$M_0(x_t, y_t, x_{t+1}, y_{t+1}) = \frac{D_0^{t+1}(x_{t+1}, y_{t+1})}{D_0^t(x_t, y_t)} * \left[\frac{D_0^t(x_{t+1}, y_{t+1})}{D_0^{t+1}(x_{t+1}, y_{t+1})} * \frac{D_0^t(x_t, y_t)}{D_0^{t+1}(x_t, y_t)}\right]^{1/2} \quad (4)$$

Formula (3) is a variant of Formula (2), which was used to express the separation of technical change and technical efficiency change, that is, MI = EF * TC. Furthermore, the change in technical efficiency can be further divided into pure technical efficiency change (PEC) and scale efficiency change (SEC), that is, MI = PEC * SEC * TC.

3.3.2. The Input–Output Index System

This paper studied the TFP of cities as a decision-making unit, which is based on the (Cobb–Douglas: C–D) production function model in economic growth theory. The essence of national and regional economic development is the input of capital, land, labor, and other production factors. According to the neoclassical economic growth theory and the understanding of economic growth in cities and urban agglomerations [33], this study selected capital, land, and labor as the input indicators and urban economic aggregates as the output indicators. The data sources and processing methods of each variable were described in continuation(Table 1):

Table 1. Selection and explanation of the growth input and output indexes in the Yangtze River Delta.

Indicator Type	Secondary Indicator	Measure Indicators
Input indicator	Capital	Social fixed-assets investment
	Land	Built-up area of each city
	Labor	The sum of employees in urban units and private and individual employees
Output indicator	The economic output	GDP

First, input variables:

(i) capital variables. Scholars often use the perpetual inventory method of social fixed-assets investment to estimate the capital stock. The formula is: $K_{it} = K_{it-1}(1-\delta) + I_{it}/p_t$, where δ is the depreciation rate; p_t is the fixed assets the investment price index. For the capital stock in the base year, this study used the material capital stock of three provinces and one municipality in the Yangtze River Delta in 2000, and calculated the material capital stock of cities in the Yangtze River Delta in the following way. The depreciation rate δ was 9.6 percent, and the fixed-assets investment price index referred to China's fixed-assets investment price published annually by the National Bureau of Statistics (https://data.stats.gov.cn/easyquery.htm?cn=C01, accessed on 21 June 2021); the data of new fixed-assets investments in each city was obtained from the China Urban Statistical Yearbook.

$$\frac{\text{The material capital stock of cities in 2000}}{\text{The material capital stock of provinces in 2000}} = \frac{\text{GDP of cities in 2000}}{\text{GDP of provinces in 2000}} \quad (5)$$

(ii) Labor variables. The number of employees in each city was obtained and the specific indicators were characterized by the sum of employees in urban units and private and individual employees.

(iii) Land variables. We selected the built-up area of each city to represent the input of urban land elements.

Second, output variables: the total annual GDP of each city was selected to represent the economic output of the city.

4. Results Analysis

4.1. Characteristics of Innovation Networks in the Yangtze River Delta Region

Based on the joint patent applications of the cities in the Yangtze River Delta from 2010 to 2019, ArcGIS 10.6 software was used to describe the spatiotemporal evolution of innovation networks within and between these cities(Figure 2).

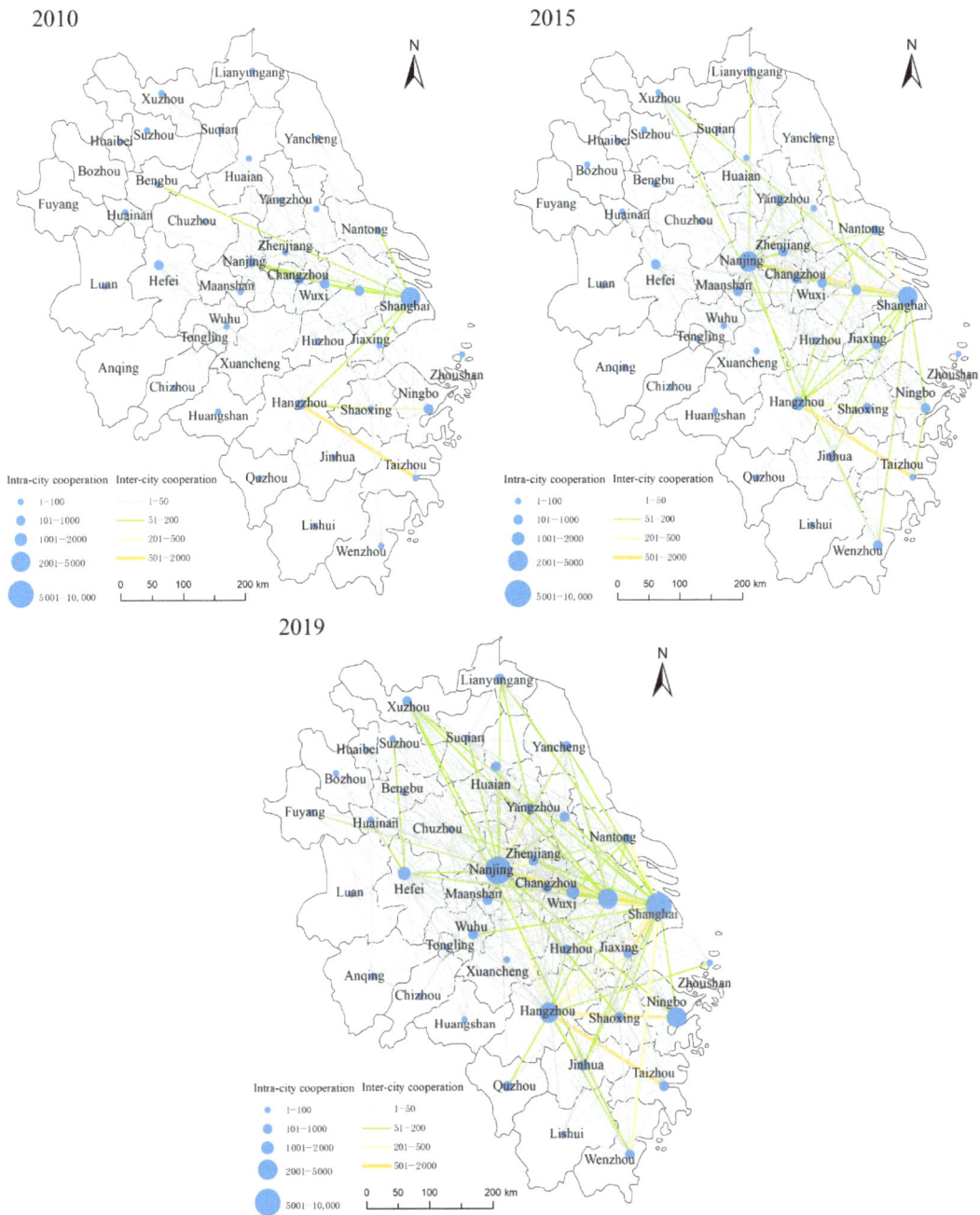

Figure 2. Illustration of the intra-city and inter-city innovation networks in the Yangtze River Delta from 2010 to 2019.

The intra-city innovation networks of the Yangtze River Delta present a "core—periphery" structure, and the core changed from a single center in Shanghai to multi-centers "Nanjing–Shanghai–Hangzhou–Ningbo". This result also accords with the studies of Rombach et al. [34] and Zhang et al. [35], and refers to their measurements of network structure. Specifically, the highlands of the intra-city innovation cooperation are mainly concentrated in the core cities of Shanghai, Jiangsu, and Zhejiang. Anhui, southwest of Zhejiang and north of Jiangsu, has become a depression for intra-city cooperation. In 2010, Shanghai became the single core of the Yangtze River Delta with 2736 intra-city innovation collaborations. Hangzhou was far behind, with only 902 intra-city innovation collaborations; the difference between the two was more than three-times greater. In 2019, Shanghai and Nanjing became the cores with 7191 and 6488 intra-city innovation collaborations, respectively. In addition, Hangzhou, Ningbo, and Suzhou each had 4007, 3176, and 3088 intra-city innovation collaborations as the sub-core. Only Hefei in Anhui entered the top ten with 1686 intra-city collaborations. From the perspective of growth and changes, from 2010 to 2019, intra-city innovation cooperation in 41 cities in the Yangtze River Delta was on the rise, with the core cities maintaining a growth rate of 3 to 10 times; Nanjing experienced an increase of nearly 10 times in the past 10 years, totaling 5915 intra-city innovative collaborations, while the peripheral cities, such as Bozhou, Fuyang, Suqian, or Lishui, due to the relatively small base of innovation cooperations within the city in 2010, experienced a significant increase in 10 years. However, there is still a big gap between the absolute amount of growth and the core cities.

The inter-city innovation networks of the Yangtze River Delta also present a similar "core–periphery" structure, while being more compact. Specifically, cities with strong innovation capabilities, such as Hangzhou, Shanghai, and Nanjing, have become the core of the inter-city innovation networks, and other cities in Anhui province, except Hefei, with less inter-city cooperation, have become the edge of the network. This spatial distribution is highly similar to the spatial pattern of inter-city technology flow in the Yangtze River Delta based on the patent transfer network observed in the existing studies. The inter-city innovation cooperation pairs also increased from 152 in 2010 to 361 in 2019. The scale of the main body of innovation cooperation between cities is also increasing accordingly; from the number of "partner cities", it can be observed that the average scale of the top-ten city pairs in 2010 was 227, which increased nearly 3 times, to 667, in 2019 (Table 2). The centrality of the inter-city innovation networks also increased from 3.55% in 2010 to 10.03% in 2019, an increase of nearly 3 times, which fully shows that these networks are becoming more closely connected; at the same time, the core is also improving. The status of core cities in inter-city innovation networks is becoming more and more important.

Table 2. Comparison of the top-ten innovation cooperation scales within and between cities in the Yangtze River Delta in 2010 and 2019.

2010		2019	
Partner Cities	Cooperation Scale (Pieces)	Partner Cities	Cooperation Scale (Pieces)
Hangzhou–Taizhou	1149	Hangzhou–Ningbo	1327
Hangzhou–Ningbo	213	Shanghai–Suzhou	1118
Hangzhou–Shanghai	179	Hangzhou–Taizhou	785
Shanghai–Suzhou	178	Jiaxing–Shanghai	610
Jiaxing–Shanghai	125	Nanjing–Suzhou	594
Nanjing–Shanghai	117	Nanjing–Shanghai	562
Shanghai–Wuxi	98	Hangzhou–Jinhua	443
Suzhou–Wuxi	79	Nanjing–Yangzhou	421
Nantong–Shanghai	77	Hangzhou–Shaoxing	414
Hangzhou–Shaoxing	59	Hangzhou–Shanghai	394

Overall, the spatial structure of intra-city and inter-city innovation networks in the Yangtze River Delta is relatively consistent; over time, the "Z-shaped" core–periphery structure is gradually appearing. Shanghai, Nanjing, and Hangzhou became the core nodes of intra-city and inter-city innovation networks in the region; Suzhou, Ningbo, Hefei, and Wuxi became the sub-core nodes, which is consistent with the spatial structure of the "local—cross-border" innovation network in the Yangtze River Delta cities based on cooperative patent data.

4.2. Temporal and Spatial Evolution Characteristics of Economic Growth in the Yangtze River Delta

Based on the endogenous growth model, urban capital stock, labor force, and land were selected as the input indicators, and urban economic aggregates as the output indicators. We used MaxDEA software to calculate the TFP of the 41 cities from 2010 to 2019 and to describe the spatiotemporal evolution of urban economic growth.

Figure 3 shows the changes in the average TFP of cities in Shanghai, Jiangsu, Zhejiang, and Anhui from 2010 to 2019. We found that Shanghai had the highest total factor productivity, followed by Jiangsu and Zhejiang, and Anhui was the lowest; Shanghai, Jiangsu, and Zhejiang were higher than the regional average. Anhui became a depression for economic development in the Yangtze River Delta region. From the perspective of dynamic trends, the overall TFP of the Yangtze River Delta had relatively stable changes; while the TFP of Shanghai fluctuated greatly, showing an overall upward trend, Anhui and Zhejiang generally showed a trend of rising first and then decreasing. Jiangsu's TFP showed a slight upward trend in fluctuations.

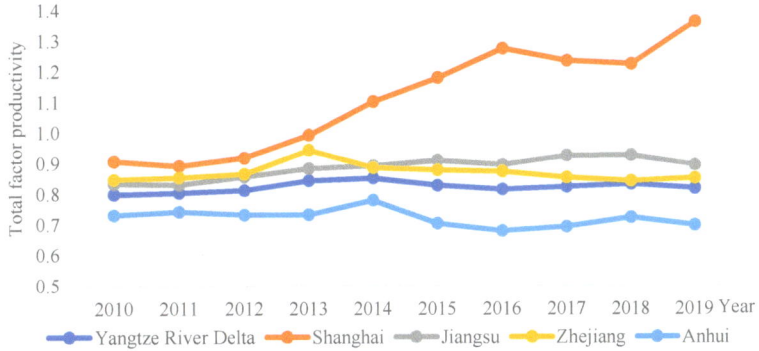

Figure 3. Calculated value of total factor productivity of cities in the Yangtze River Delta from 2010 to 2019.

Figure 4 shows the dynamic evolution of TFP in 41 cities in the Yangtze River Delta in 2010, 2015, and 2019 and the average value in 2010–2019. In general, the TFP level of these cities was high; however, the spatial differences were significant. The high-value areas were mainly distributed in the coastal areas and along the Yangtze River. The Shanghai metropolitan area has become the leader of regional economic development. The low-value areas were mainly distributed in the northern part of the Yangtze River Delta and the western and southern parts of Anhui. Specifically, in 2010, there were 4 cities with TFPs exceeding 1, namely, Suzhou (1.1793), Taizhou (1.1028), Jinhua (1.0037), and Maanshan (1.0702), among which Suzhou was the highest; in 2015, there were 6 cities with TFPs exceeding 1, namely, Shanghai (1.1791), Suzhou (1.0942), Zhenjiang (1.1064), Jiaxing (1.1046), Jinhua (1.2102), and Tongling (1.0964), among which Jinhua was the highest, followed by Shanghai; and in 2019, the number of cities with TFPs exceeding 1 increased to 8 cities, namely, Shanghai (1.3616), Wuxi (1.0856), Suzhou (1.1352), Nantong (1.0606), Jiaxing (1.0004), Jinhua (1.0492), Lishui (1.0241), and Bengbu (1.1338), among which Shanghai was the highest at 1.362. According to the average TFPs of cities in the Yangtze River Delta from 2010 to 2019, it can be observed that cities with better economic development are mainly distributed along the coast and along the Yangtze River. The cities with TFPs exceeding 1 were Shanghai, Suzhou, Taizhou, Jiaxing, and Jinhua; Suzhou was the highest (1.133), followed by Shanghai (1.108). Cities with relatively poor economic development were mainly distributed in the northern part of the Yangtze River Delta and the western and southern parts of Anhui. Huangshan, Huaibei, Xuancheng, Lu'an, Chizhou, and Chuzhou have become economic development depressions in the Yangtze River Delta region, with an average TFP of about 0.6.

Figure 4. The spatial dynamic evolution of total factor productivity in cities in the Yangtze River Delta from 2010 to 2019.

5. Influence Mechanism

5.1. Model Construction and Variable Selection

We constructed the full-sample panel data of 41 cities in the Yangtze River Delta from 2010 to 2019 and analyzed the impact of intra-city and inter-city innovation networks on urban economic growth. Compared to the traditional research using regional economic aggregates to measure city economic growth, calculating the TFP based on input and output variables can more comprehensively reflect the level of urban economic growth. We used the urban TFP as a dependent variable, which represents urban economic growth, while the core independent variables included intra-city and inter-city innovation networks, measured by the scale of intra-city and inter-city cooperative invention patents, respectively. The data were sourced from the incoPat patent database.

Referring to the relevant studies [36], we selected control variables that had an important impact on urban economic growth, including economic development level, degree of openness, city size, population density, and fiscal autonomy of local governments. To reduce the effect of heteroskedasticity, we logged all variables. Table 3 shows the descriptive statistics of the variables.

Table 3. Descriptive statistics of related variables.

Variables	Symbol	Description	Mean	SD	Minimum	Maximum
Total factor productivity	TFP	Super-efficient total factor productivity	0.821	0.181	0.421	1.865
Inter-city cooperation	Inter	Number of cooperative patents between cities (piece)	421.82	825.55	0	5463
Intra-city cooperation	Intra	Number of cooperative patents in the city (piece)	394.55	1010.1	0	7191
Economic development	GDP	GDP per capita (CNY/person)	67,051.6	32,818.7	9068	199,017
Degree of openness	FDI	Foreign direct investment/GDP (%)	0.463	0.368	0.029	2.853
City size	Size	Population of the city (10,000 people)	207.76	225.40	29	1469
Population density	Dens	Population of the city/urban area (10,000 people/square kilometer)	1.2762	0.5983	0.5630	3.8535
Financial autonomy	GOV	Public budget revenue/public budget expenditure (%)	0.641	0.232	0.069	1.116

The model was constructed as follows:

$$\text{LnTFP}_{ct} = \alpha + \beta_1 \text{LnInter}_{ct} + \beta_2 \text{LnIntra}_{ct} + \beta_3 \text{LnInter}_{ct} * \text{LnIntra}_{ct}$$
$$+ \beta_4 \text{LnInter}_{ct}^2 + \beta_5 \text{LnIntra}_{ct}^2 + \beta_6 \text{LnGDP}_{ct} + \beta_7 \text{LnFDI}_{ct}$$
$$+ \beta_8 \text{LnSize}_{ct} + \beta_9 \text{LnDens}_{ct} + \beta_{10} \text{LnGOV}_{ct} + \mu_c + \varepsilon$$

where c and t represent the city and year, respectively; LnTFP_{ct} is the logarithm of the TFP of city c in year t; LnIntra_{ct} and LnInter_{ct} are the logarithms of the intra-city and inter-city innovation network scales, respectively, in city c in year t; LnIntra_{ct}^2, LnInter_{ct}^2, $\text{LnInter}_{ct} * \text{LnIntra}_{ct}$ represent the logarithmic forms of the square and interaction terms of the intra-city and inter-city innovation network scales in city c year t; LnGDP_{ct}, LnFDI_{ct}, LnSize_{ct}, LnDens_{ct}, LnGOV_{ct}, respectively, represent the logarithmic form of the control variable in c city t year; μ_c represents the fixed city effect; and α and ε represent the constant and random-error terms.

5.2. Analysis of Model Regression Results

Using the full sample data of the Yangtze River Delta urban innovation networks from 2010 to 2019 to draw a scatter diagram of the intra-city and inter-city innovation networks (Figure 5), we found that there was a high correlation between the two, which also confirmed the above conclusion on their spatial structure similarities. At the same time, using Stata software to conduct a correlation analysis, we found that the correlation coefficient between the two was as high as 0.8437. In order to avoid the model being affected by the correlation between the independent variables, the regression analysis of the influence of intra-city and inter-city innovation networks on regional TFP was conducted.

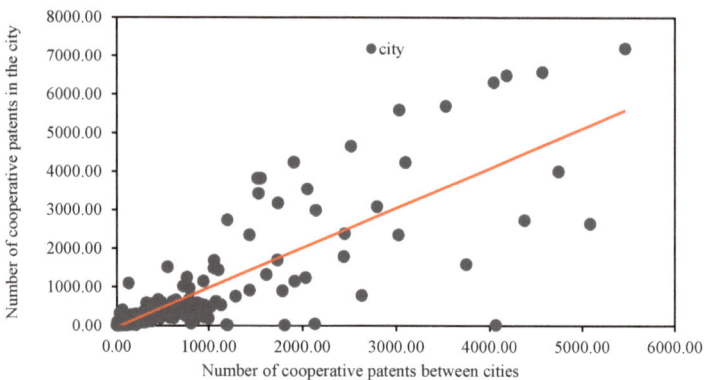

Figure 5. Scatter diagram of the distribution of the intra-city and inter-city innovation networks in the Yangtze River Delta from 2010 to 2019.

The results of multiple regression using panel data are shown in Table 4. Model 1 mainly depicts the influence of control variables on urban TFPs. The economic development level index is significantly positive, indicating that cities with higher economic development levels have higher TFPs. The index of openness is significantly negative, indicating that foreign direct investment (FDI) has not significantly promoted the economic growth of cities in the Yangtze River Delta. This may be because the FDI introduced by China may be mainly concentrated in low-tech fields, and not only did this not produce a high technology spillover effect, it also produced a strong crowding-out effect. The index of city size is significantly negative, indicating that cities in the Yangtze River Delta have diseconomies of scale; as cities expand, large-scale urban diseases, such as traffic congestion and pollution, cause external diseconomies, and smaller cities cannot obtain economies of scale. The index of population density is significantly positive, indicating that the agglomeration economy brought about by labor agglomerations in these cities has a positive effect on urban development, which further verifies the relevant research conclusions of Huang and He using population density to measure the agglomeration economy [37]. The index of financial autonomy is positive, but does not reach statistical significance, indicating that the political and economic development strategies adopted by local governments have a stable impact on urban economic efficiency; in general, the more active the local government's fiscal policy, the more likely it is for the city's economic efficiency to increase.

The core independent variables of models 2 and 3 are the primary terms of the scales of inter-city and intra-city innovation networks, respectively, and the core independent variables of models 4 and 5 are the secondary terms of their scales, respectively. It can be seen from the results that the first-order coefficient of the inter-city innovation networks is positive; however, it does not reach statistical significance; after adding the square term of the scale of the inter-city innovation networks, the coefficient of the square term is significantly positive, indicating that the inter-city innovation network has a non-linear positive impact on urban economic growth. There may be a positive U-shaped relationship between inter-city innovation networks and urban economic growth.

In the low-level stage of inter-city innovation networks, the positive effect on urban economic growth gradually diminishes, because most of the cities in the low-level stage are underdeveloped. The attractiveness and absorptive capacity of these cities for non-local knowledge are at a low level [38]; therefore, it is difficult to build high-level inter-city innovation networks, and its role in promoting urban economic growth is also weak. Conversely, at the high-level stage, inter-city innovation networks can significantly promote urban economic growth, and the larger the scale is, the stronger the promotion effect. This is because core cities have a large number of heterogeneous innovation actors, and they are widely used in a broad range of the economic and technological fields conducting innovation activities and, thus, their demands, attractiveness, and absorptive capacity for non-local knowledge are high. Therefore, most of the inter-city innovation networks constructed are of high quality and have better innovation effects, which can bring vitality to urban economic development and promote growth.

Table 4. Regression results of urban innovation networks and regional growth in the Yangtze River Delta urban agglomeration from 2010 to 2019.

	Model (1)	Model (2)	Model (3)	Model (4)	Model (5)	Model (6)
	LnTFP	LnTFP	LnTFP	LnTFP	LnTFP	LnTFP
LnInter		0.0294		−0.314 **		−0.0760 *
		(0.0434)		(0.131)		(0.0443)
LnIntra			−0.0628		−0.0488	−0.278 ***
			(0.0409)		(0.124)	(0.0749)
LnInter*LnIntra						0.471 ***
						(0.11)
LnInter^2				0.478 ***		
				(0.174)		
LnIntra^2					−0.0231	
					(0.162)	
LnGDP	0.0763 ***	0.0756 ***	0.0790 ***	0.0796 ***	0.0789 ***	0.0803 ***
	(0.0171)	(0.0172)	(0.0178)	(0.0174)	(0.0179)	(0.0179)
LnFDI	−0.0243 ***	−0.0238 ***	−0.0249 ***	−0.0247 ***	−0.0250 ***	−0.0244 ***
	(0.00637)	(0.00642)	(0.00646)	(0.00639)	(0.00649)	(0.00649)
LnSize	−0.122 ***	−0.124 ***	−0.122 ***	−0.108 ***	−0.122 ***	−0.116 ***
	(0.0282)	(0.0282)	(0.0283)	(0.0297)	(0.0285)	(0.0287)
LnGOV	0.0171	0.0169	0.0168	0.0125	0.017	0.0143
	(0.0179)	(0.0181)	(0.0178)	(0.0201)	(0.0178)	(0.0191)
LnDens	0.145 ***	0.146 ***	0.146 ***	0.133 ***	0.146 ***	0.144 ***
	(0.04)	(0.0398)	(0.0404)	(0.0402)	(0.0407)	(0.0402)
constant	0.573 ***	0.582 ***	0.580 ***	0.443 ***	0.583 ***	0.536 ***
	(0.142)	(0.144)	(0.141)	(0.161)	(0.144)	(0.144)
Region fixed effect	Yes	Yes	Yes	Yes	Yes	Yes
Observations	410	410	410	410	410	410
R-squared	0.688	0.688	0.688	0.694	0.688	0.695

* $p < 0.10$; ** $p < 0.05$; *** $p < 0.01$.

For intra-city innovation networks, the coefficients of the primary and quadratic terms are both negative and do not reach statistical significance, indicating that, according to the existing data, these networks have no significant impact on the urban economic growth in the Yangtze River Delta, while the quadratic term is negative, which indicates that there may be an inverted U-shaped relationship between intra-city innovation networks and urban economic growth. Intra-city innovation networks represented by cooperative patents play a certain role in promoting regional economic development; however, too many of them may bring about a redundancy of knowledge, which leads to the development dilemma of path locking in the region.

Model 6 takes the interaction term of intra-city and inter-city innovation networks as the core independent variable, exploring the mechanism of interaction between them and urban economic growth. The results show that the interaction term is significantly positive, indicating that they have complementary roles in promoting growth. Studies have shown that cities with dense local and non-local innovation networks tend to achieve better development performances [39]; the dense local network creates a good innovation environment and also strengthens the region's ability to absorb non-local knowledge to transform it and apply it locally [40]. At the same time, inter-city innovation networks provide innovation vitality and expand external knowledge sources for the development of intra-city innovation networks, thus avoiding industrial and technological stagnations or lock-ins due to locally contained interactions and over-embedding within a regionally rigid inward-looking system [41]. The interaction of intra-city and inter-city innovation networks provides an inexhaustible impetus for urban economic growth.

To verify the robustness of the benchmark model, a robustness test was performed with the urban TFP lagging one period as the dependent variable, and the results are shown in Table 5. From the regression results, the coefficients and symbols of the core explanatory variables are consistent with the benchmark regression results in Table 4, which further verifies the robustness of the model results presented in Table 4.

Table 5. Robustness test of the regression results of urban innovation networks and regional growth in the Yangtze River Delta urban agglomeration from 2010 to 2019.

	Model (7)	Model (8)	Model (9)	Model (10)	Model (11)
	L.LnTFP	L.LnTFP	L.LnTFP	L.LnTFP	L.LnTFP
LnInter	0.0542		−0.310 **		−0.0453
	(0.0469)		(0.137)		(0.0521)
LnIntra		−0.0533		−0.0186	−0.253 **
		(0.0523)		(0.164)	(0.0894)
LnInter*LnIntra					0.420 ***
					(0.123)
LnInter^2			0.505 ***		
			(0.171)		
LnIntra^2				−0.0581	
				(0.217)	
Control variables	Yes	Yes	Yes	Yes	Yes
constant	0.538 ***	0.514 **	0.371	0.524 **	0.497 *
	(0.218)	(0.215)	(0.230)	(0.213)	(0.218)
Region fixed effect	Yes	Yes	Yes	Yes	Yes
Observations	369	369	369	369	369
R-squared	0.684	0.684	0.690	0.684	0.690

* $p < 0.10$; ** $p < 0.05$; *** $p < 0.01$.

To sum up, the empirical analysis shows that there are different mechanisms of action between intra-city and inter-city innovation networks and urban economic growth. The relevant mechanisms are summarized in Figure 6.

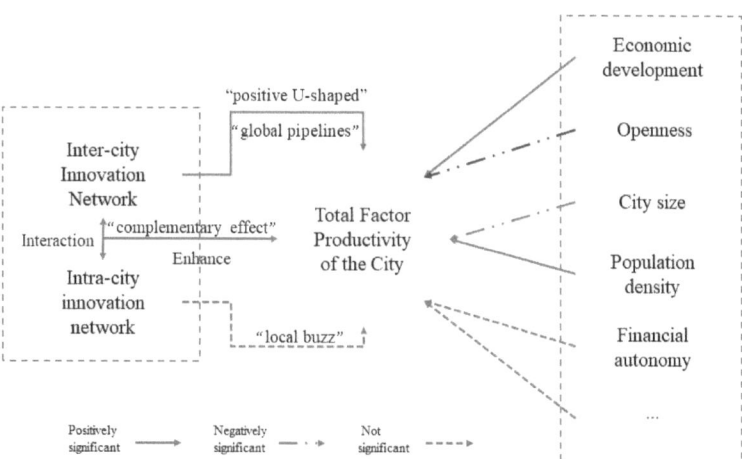

Figure 6. Diagram of the interaction mechanism between urban innovation networks and regional growth in the Yangtze River Delta.

6. Discussion and Conclusions

Based on the panel data of cooperative application for invention patents and the socio-economic statistics of cities in the Yangtze River Delta from 2010 to 2019, the input–output index system of economic growth was established in this study. With the methods of TFP and multiple regression analysis, this paper explored the dynamic characteristics, spatial evolution, and influencing mechanism of intra-city and inter-city innovation networks on economic growth in the Yangtze River Delta.

First, the connection density of intra-city and inter-city innovation networks in this region increased, and the spatial structure presented a highly similar core–edge structure. Specifically, the Z-shaped spatial structure of the networks with Hefei–Nanjing–Shanghai–Hangzhou–Ningbo as the core became clearer. Comparatively speaking, most of Anhui, western Zhejiang, and northern Jiangsu became the edge of the innovation networks in the Yangtze River Delta. This spatial distribution was in agreement with the distribution of the economic structure and innovation ability among cities in this area. Due to the implementation of policies, such as regional integration and industrial transfer, the intra-city and inter-city innovation networks presented a trend of core-driven marginal development towards the overall regional coordinated development.

Second, the economic growth of the 41 cities in the Yangtze River Delta was generally in a relatively stable state, and few cities were characterized by fluctuations. Specifically, the level of economic growth in Shanghai was the highest and increasing, followed by Jiangsu and Zhejiang, while Anhui became the depression of economic development in the region. From the perspective of spatial distribution, unlike the Z-shaped core–edge structure of intra-city and inter-city innovation networks, the core cities were distributed along the coast and along the golden waterway of the Yangtze River; the metropolitan area with Shanghai as the core became the leader of regional economic development.

Third, there was strong heterogeneity in the influence of innovation networks on economic growth. This influence presented a positive non-linear relationship and occurred when the inter-city innovation networks reached a certain scale, which could significantly promote city economic growth; however, the influence was not significant, but the interaction between them was significantly positive, which indicated that intra-city and inter-city innovation networks had a strong complementary effect on promoting economic growth. Intensive intra-city cooperation can acquire local tacit knowledge that is difficult for inter-city networks to acquire, while the inter-city collaboration can acquire external innovation knowledge that is difficult for intra-city innovation networks to acquire.

Theoretically, there was heterogeneity in the action mechanism of the networks based on cooperative invention patents on economic growth. We suggested strengthening the research on the influence mechanisms, such as network capital-oriented, social capital-oriented, formal, and informal, to provide a theoretical reference for analyzing the source of economic growth from the perspective of innovation networks. Practically, our suggestions are as follows: first, adhere to the six-in-one policy of government and industry-university research to promote the cooperation of regional science and technology innovation in the Yangtze River Delta, and provide a power source for the science and technology innovation community and regional high-quality integration; second, insist on performing good industrial transfer in central cities and undertaking industrial transfers in marginal areas, and promote the coordinated development of marginal areas with industrial transfer and industrial cooperation in core cities; third, to support the construction of intra-city and inter-city innovation networks, such as industrial technology innovation alliances, innovation enclaves, industrial technology transfer, and transformation platforms, to enhance inter-regional knowledge flow, regional innovation capabilities, and promote regional economic development.

Author Contributions: Conceptualization, X.C. and Z.Y.; methodology, B.C. and Y.G.; software, B.C.; validation, B.C., Y.G. and X.C.; formal analysis, X.C.; data curation, B.C.; writing—original draft preparation, X.C. and B.C.; writing—review and editing, X.C. and Z.Y.; visualization, Y.G.; project administration, X.C.; funding acquisition, X.C. All authors have read and agreed to the published version of the manuscript.

Funding: This research was funded by National Natural Science Foundation of China (42171184 and 42130510).

Data Availability Statement: The data presented in this study are available on request from the corresponding author.

Conflicts of Interest: The authors declare no conflict of interest.

References

1. Gluckler, J. How controversial innovation succeeds in the periphery? A network perspective of BASF Argentina. *J. Econ. Geogr.* **2014**, *14*, 903–927. [CrossRef]
2. Fernandes, C.; Farinha, L.; Ferreira, J.J.; Asheim, B.; Rutten, R. Regional innovation systems: What can we learn from 25 years of scientific achievements? *Reg. Stud.* **2020**, *55*, 377–389. [CrossRef]
3. Bathelt, H.; Cantwell, J.A.; Mudambi, R. Overcoming frictions in transnational knowledge flows: Challenges of connecting, sense-making and integrating. *J. Econ. Geogr.* **2018**, *18*, 1001–1022. [CrossRef]
4. Denney, S.; Southin, T.; Wolfe, D.A. Entrepreneurs and cluster evolution: The transformation of Toronto's ICT cluster. *Reg. Stud.* **2021**, *55*, 196–207. [CrossRef]
5. Antonelli, C. *Localised Technological Change: Towards the Economics of Complexity*; Routledge: London, UK, 2011.
6. Mahmood, N.; Zhao, Y.; Lou, Q.; Geng, J. Role of environmental regulations and eco-innovation in energy structure transition for green growth: Evidence from OECD. *Technol. Forecast. Soc. Chang.* **2022**, *183*, 121890. [CrossRef]
7. Aghion, P.; Howitt, P. *Endogenous Growth Theory*; MIT Press: Cambridge, MA, USA, 1998.
8. Roberts, M.; Setterfield, M. Endogenous regional growth: A critical survey. In *Handbook of Alternative Theories of Economic Growth*; Setterfield, M., Ed.; Edward Elgar: Cheltenham, UK, 2010.
9. Capello, R.; Caragliu, A.; Fratesi, U. Breaking down the border: Physical, institutional and cultural obstacles. *Econ. Geogr.* **2018**, *94*, 485–513. [CrossRef]
10. Ramadani, V.; Gërguri, S.; Rexhepi, S. Innovation and economic development: The case of FYR of macedonia. *J. Balk. Near East. Stud.* **2013**, *15*, 324–345. [CrossRef]
11. Huggins, R.; Thompson, P. Networks and regional economic growth: A spatial analysis of knowledge ties. *Environ. Plan. A* **2017**, *49*, 1247–1265. [CrossRef]
12. Esposito, C.R.; Rigby, D.L. Buzz and pipelines: The costs and benefits of local and nonlocal interaction. *J. Econ. Geogr.* **2019**, *19*, 753–773. [CrossRef]
13. Buciuni, G.; Pisano, G. Knowledge integrators and the survival of manufacturing clusters. *J. Econ. Geogr.* **2018**, *18*, 1069–1089. [CrossRef]
14. Crespo, J.; Suire, R.; Vicente, J. Lock-in or lock-out? How structural properties of knowledge networks affect regional resilience. *J. Econ. Geogr.* **2014**, *14*, 199–219. [CrossRef]
15. Breschi, S.; Lissoni, F.; Miguelez, E. Foreign-origin inventors in the USA: Testing for diaspora and brain gain effects. *J. Econ. Geogr.* **2017**, *17*, 1009–1038. [CrossRef]
16. Bathelt, H.; Malmberg, A.; Maskell, P. Clusters and knowledge: Local buzz, global pipelines and the process of knowledge creation. *Prog. Hum. Geogr.* **2004**, *28*, 31–56. [CrossRef]
17. Storper, M. Separate worlds? Explaining the current wave of regional economic polarization. *J. Econ. Geogr.* **2018**, *18*, 247–270. [CrossRef]
18. Galaso, P.; Kovářík, J. Collaboration networks, geography and innovation: Local and national embeddedness. *Pap. Reg. Sci.* **2021**, *100*, 349–377. [CrossRef]
19. Bianchi, C.; Galaso, P.; Palomeque, S. The trade-offs of brokerage in inter-city innovation networks. *Reg. Stud.* **2021**, *19*, 36–64. [CrossRef]
20. Operti, E.; Kumar, A. Too much of a good thing? Network brokerage within and between regions and innovation performance. *Reg. Stud.* **2023**, *57*, 300–316. [CrossRef]
21. Le Gallo, J.L.; Plunket, A. Regional gatekeepers, inventor networks and inventive performance: Spatial and organizational channels. *Res. Policy* **2020**, *49*, 103981. [CrossRef]
22. Boschma, R. Proximity and innovation: A critical assessment. *Reg. Stud.* **2005**, *39*, 61–74. [CrossRef]
23. Cao, X.; Zeng, G.; Ye, L. The structure and proximity mechanism of formal innovation networks: Evidence from Shanghai high-tech ITISAs. *Growth Chang.* **2019**, *50*, 569–586. [CrossRef]
24. Balland, P.A.; Boschma, R. Complementary interregional linkages and Smart Specialisation: An empirical study on European regions. *Reg. Stud.* **2021**, *55*, 1059–1070. [CrossRef]

25. Boschma, R. Relatedness as driver of regional diversification: A research agenda. *Reg. Stud.* **2017**, *51*, 351–364. [CrossRef]
26. Cicerone, G.; McCann, P.; Venhorst, V.A. Promoting regional growth and innovation: Relatedness, revealed comparative advantage and the product space. *J. Econ. Geogr.* **2020**, *20*, 293–316. [CrossRef]
27. Huggins, R.; Thompson, P. A Network-based view of regional growth. *J. Econ. Geogr.* **2014**, *14*, 511–545. [CrossRef]
28. Prabhakaran, T.; Lathabai, H.H.; Changat, M. Detection of paradigm shifts and emerging fields using scientific network: A case study of information technology for engineering. *Technol. Forecast. Soc. Chang.* **2015**, *91*, 124–145. [CrossRef]
29. Ter Wal, A.L.J. Cluster emergence and network evolution: A longitudinal analysis of the inventor network in sophia-antipolis. *Reg. Stud.* **2013**, *47*, 651–668. [CrossRef]
30. Patra, S.K.; Muchie, M. An assessment of south african technological capability using patent data from WIPO patentscope database. *Afr. J. Sci. Technol. Innov. Dev.* **2022**, *14*, 333–340. [CrossRef]
31. Wang, S.; Wang, J.; Wei, C.; Wang, X.; Fan, F. Collaborative innovation efficiency: From within cities to between cities—Empirical analysis based on innovative cities in China. *Growth Chang.* **2021**, *52*, 1330–1360. [CrossRef]
32. Fare, R.; Grosskopf, S.; Norris, M.; Zhang, Z. Productivity growth, technical progress, and efficiency changes in industrialized countries. *Am. Econ. Rev.* **1994**, *84*, 66–83.
33. McCombie, J.S.; Spreafico, M.R. Kaldor's 'technical progress function' and verdoorn's law revisited. *Camb. J. Econ.* **2015**, *40*, 1117–1136. [CrossRef]
34. Rombach, M.P.; Porter, M.A.; Fowler, J.H.; Mucha, P.J. Core-periphery structure in networks. *Siam J. Appl. Math.* **2014**, *74*, 167–190. [CrossRef]
35. Zhang, W.J.; Thill, J.C. Mesoscale structures in world city networks. *Ann. Am. Assoc. Geogr.* **2019**, *109*, 887–908. [CrossRef]
36. Zhang, W.; Derudder, B.; Wang, J.; Witlox, F. An analysis of the determinants of the multiplex urban networks in the Yangtze River delta. *Tijdschr. Voor Econ. En Soc. Geogr.* **2020**, *111*, 117–133. [CrossRef]
37. Huang, Z.J.; He, C.F. Industrial Innovation Investments and the Quality of Urban Economic Growth in China. *China Soft Sci.* **2013**, *3*, 89–100.
38. Trippl, M.; Grillitsch, M.; Isaksen, A. Exogenous sources of regional industrial change: Attraction and absorption of non-local knowledge for new path development. *Prog. Hum. Geogr.* **2018**, *42*, 687–705. [CrossRef]
39. Wen, H.; Zhang, Q.G.; Zhu, S.Z.; Huang, Y.Y. Inter- and intra-city networks: How networks are shaping China's film industry. *Reg. Stud.* **2021**, *55*, 533–545. [CrossRef]
40. Vale, M.; Carvalho, L. Knowledge networks and processes of anchoring in portuguese biotechnology. *Reg. Stud.* **2013**, *47*, 1018–1033. [CrossRef]
41. Boschma, R.; Iammarino, S. Related variety, trade linkages, and regional growth in Italy. *Econ. Geogr.* **2009**, *85*, 289–311. [CrossRef]

Disclaimer/Publisher's Note: The statements, opinions and data contained in all publications are solely those of the individual author(s) and contributor(s) and not of MDPI and/or the editor(s). MDPI and/or the editor(s) disclaim responsibility for any injury to people or property resulting from any ideas, methods, instructions or products referred to in the content.

Article

Reputation, Network, and Performance: Exploring the Diffusion Mechanism of Local Governments' Behavior during Inter-Governmental Environmental Cooperation

Yihang Zhao [1], Jing Xiong [2,3,4,*] and De Hu [5,*]

1. Institute of Urban and Demographic Studies, Shanghai Academy of Social Sciences, Shanghai 200020, China; zhaoyihang94@163.com
2. School of International and Public Affairs, Shanghai Jiao Tong University, Shanghai 200030, China
3. China Institute for Urban Governance, Shanghai Jiao Tong University, Shanghai 200030, China
4. Institute of Eco-Chongming, East China Normal University, Shanghai 202162, China
5. School of Urban and Regional Science, East China Normal University, Shanghai 200241, China
* Correspondence: bearnear@163.com (J.X.); dhu@re.ecnu.edu.cn (D.H.)

Citation: Zhao, Y.; Xiong, J.; Hu, D. Reputation, Network, and Performance: Exploring the Diffusion Mechanism of Local Governments' Behavior during Inter-Governmental Environmental Cooperation. *Land* **2023**, *12*, 1466. https://doi.org/10.3390/land12071466

Academic Editors: Wei Sun, Zhaoyuan Yu, Kun Yu, Weiyang Zhang and Jiawei Wu

Received: 7 June 2023
Revised: 13 July 2023
Accepted: 20 July 2023
Published: 23 July 2023

Copyright: © 2023 by the authors. Licensee MDPI, Basel, Switzerland. This article is an open access article distributed under the terms and conditions of the Creative Commons Attribution (CC BY) license (https://creativecommons.org/licenses/by/4.0/).

Abstract: The selective behavior of local governments during regional environmental cooperation could generate a diffusion effect through the black box of reputation mechanism. This study incorporates the reputation mechanism, social capital, and environmental governance performance into a unified analysis framework, empirically testing the moderating effect of the implementation rate of environmental cooperative projects (indicating reputation) on the relationship between two types of social capital and environmental governance performance among cities in the Yangtze River Delta (YRD) and Beijing–Tianjin–Hebei (BTH) regions. The inter-governmental environmental cooperation news and policies are collected by Data Capture technology as a dataset, and a set of social-economic data is also adopted. The spatial econometric regression results show that an increase in reputation could both strengthen the leadership and coordination ability (bridging social capital) of the central cities in the YRD and BTH regions, thus improving their environmental governance performance. However, the bonding social capital path could only significantly work in the BTH region, which unexpectedly increases pollutant emission through excessive internal cohesion. The results indicate that a "community of entangled interest" should be constructed among cities within urban agglomerations, which requires local governments to weaken the concept of their administrative boundary. At the same time, in order to avoid excessive internal condensation, a clear division of rights and responsibilities is also necessary during continuous inter-governmental environmental cooperation. We believe that these findings could provide empirical evidence for local governments to avoid failing to the traps of "agglomeration shadow".

Keywords: inter-governmental environmental cooperation; local government behavior; regional social capital; diffusion effect

1. Introduction

As is known, reputation can promote individuals' cooperation by the means of indirect reciprocity in social networks [1]. It has been proved that raising individuals' awareness of reputation can promote more people to participate in cooperation among the public goods game [2]. When applying this to the organization level, in the process of partner selection, enterprises usually consider reputation and trust as important criteria, which means that mutual trust could improve this cooperation performance based on the resources they possess [3,4].

Raising reputation is also of vital importance for promoting inter-governmental cooperation. Although it has been recognized that regional cooperation could effectively enhance the regional competitiveness in China [5], problems such as "Industrial isomorphism" and

"Broken roads" still appear at the junction of administrative boundaries, which sometimes hinder regional development [6]. Specifically, local governments sometimes leave these previous collaborative arrangements, which might cause damage to the reputation of these local governments and destroy their relationship of mutual trust [7].

In the field of environmental governance, although inter-governmental cooperation could effectively break through the restrictions posed by administrative divisions to address cross-regional pollution problems [8], collaborative frictions still appear among differential local governments [9], which could be influenced by the conflict between top-down pressure and local self-interest [10], as well as local favoritism behaviors during regional cooperation [11]. Specifically manifested as conspicuous collaboration risks and transaction costs, these might result in a campaign-style environmental governance effect [12].

Cross-regional environmental governance is usually a costly, time-consuming, and conflict-ridden process with an uncertain outcome [13]. It might bring out some direct and indirect negative effects when stakeholders renege and perform passive cooperation. Firstly, a direct effect will occur due to the dropout of collaborative resources, which might directly result in the failure of the regional cooperation project on environmental governance. Secondly, a defect of trust could also damage the cooperative performance throughout the diffusion effect of local governmental reputation [14], which could be treated as an indirect effect and lead to an unsustainable performance. The remaining research has defined the disconnected and differential phenomenon between the willingness and implementation of related cooperative projects in the process of inter-governmental collaboration as "selective cooperation" [10], which is especially evident in the field of environmental cooperative governance (Figure 1). In other words, effective implementation in the field of environmental cooperation has great potential for improvement. When local governments have a higher implementing rate towards regional cooperative projects, this could also develop a diffusion effect to improve their performance through the reputation mechanism.

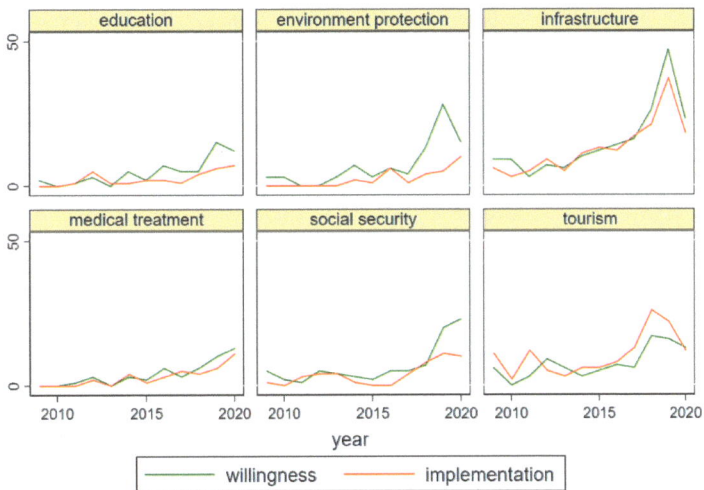

Figure 1. The time-varying trend of the inter-governmental cooperation policies and news text amounts in YRD and BTH regions. (Notes: We have collected the inter-governmental cooperation policies and news texts during 2009–2020 in the field of education, environmental protection, infrastructure, medical treatment, social security and tourism. It could be observed that the amount gap between cooperative implementation and cooperative willingness is the hugest during regional environmental cooperation).

This paper attempts to adopt the "reputation mechanism" to describe the black box process of inter-governmental strategic interaction, and the indictor "landing rate of environmental cooperative projects" is used to reflect local governments' reputation, which could perform the effect of signal transmission and cause the interaction and spillover effects of local government behaviors among regional cities. At the same time, it is still unclear how reputation adjusts the transmission of the micro-structure of a regional cooperation network, thus affecting the cooperative performance. Referring to the remaining studies, this paper combines social capital with the related network theory, and measures this social capital with the method of a social network analysis [15,16]. The quantitative strategy is to identify the regulatory effect of reputation on the impact path of different social capitals on environmental cooperative performance, and then summarize the influencing mechanism of local government behavior strategies in regional environmental cooperation.

2. Research Hypothesis and Theoretical Framework

In the process of regional environmental cooperation, its cooperative performance could be affected by the regional social capital connecting among cities. Social capital in regional research could be defined as the structural and cognitive resources formed by the local governments within an urban agglomeration during a long-term interaction. It mainly includes regional social trust, regional network, regional norms, and regional identification [17]. In this study, the social linking network is established by the local governments within an urban agglomeration through achieving a consensus of cooperation or promoting the actual implementation of cooperation, which could be believed as the regional social capital formed by their long-term contacts [18]. The stakeholders (local governments) located in this social linking network could achieve collective resources shared by the other participants within the network, and promote cooperation with others to improve their own behavioral performance through their wide communications and interactions [19]. With the development of the social network analysis method, more and more studies have focused on the relationship between the differential types of social capital embedded in network and governance performance. Specifically, regional social capital could be divided into bridging social capital and bonding social capital [20,21].

Bridging social capital corresponds to a sparse and open network structure, which has a related lower risk of cooperation among the network stakeholders. There are mainly coordination game problems among multiple subjects of the network members, and effective information sharing and transmission are needed to ensure the promotion of cooperation [22]. In the field of regional environmental governance, bridging social capital could reflect the ability of a central city connecting other cities who participate in the regional environmental cooperation. It could share and transmit related environmental governance information (such as the determination of each local government to control environmental pollution and the cost–benefit relationship of inter-governmental environmental cooperation) among the cities in an urban agglomeration during the process of regional environmental governance.

When the risk of cooperation increases, inter-governmental cooperation will be more dependent on bonding social capital, which features a dense and closed network structure. Bonding social capital is mainly used to solve the cooperation game problems among local governments [23]. It has a strong network relationship and its network members are closely connected, which could promote the generation of effective rules and mutual trust [24]. In the field of inter-governmental environmental cooperation, bonding social capital could control the defection risk and strengthen the executive force of local governments. For example, frequent interaction within small cooperative groups increases the trust among the local governments of cities and forms related legal norms, which could increase the cohesion of these cooperative groups.

To sum up, bridging social capital has a micro-structure that could benefit information bridging and transmission, while bonding social capital has a small group structure that features close ranks. The strength of related bridging and bonding social capital could have an impact on the performance of environmental cooperation. We propose Hypothesis 1:

Hypothesis 1. *Bridging social capital and bonding social capital could effectively improve environmental governance performance among urban agglomeration cities.*

In the process of inter-governmental environmental cooperation, the reputation of local governments is mainly reflected by the degree of their cooperation enforcement. Referring to the selective cooperation theory [25], influenced by self-interest and a top-down institutional arrangement, local governments usually perform the selective enforcement feature towards regional environmental cooperation agreements, which presents the disconnected and differential phenomenon between the willingness and implementation of related environmental cooperative projects. According to the social impact model, the behavior strategies of local governments within a cooperation network are usually influenced by the behaviors of other actors, which could also be treated as the spillover effect from the reputation of these other actors, and the influencing path is indirect [26].

In order to facilitate understanding, we simplify the analysis model (Figure 2): there is a central city in a small group within an urban agglomeration (Central city A). When the reputation (the implementation rate of cooperation projects) of this central city is increased, it will send a positive signal (spillover effect) to the other members embedded in the regional social network, including the members within the small group that have previously reached a cooperative relationship, and also including cities outside the small group that have not yet reached a cooperative relationship (City F and G) [2,27]. Reputation is the basis for this central city to strengthen the influence of its social capital [28,29]. An increase in reputation could promote more surrounding cities to learn and imitate the cooperation strategies of the central city, expand the range of cities connected to them, and enhance their influence and coordination ability. At the same time, it could also enhance the trust and regulatory constraints between the central city and the other cities that have been connected before, enhancing the cohesion and execution of a small group.

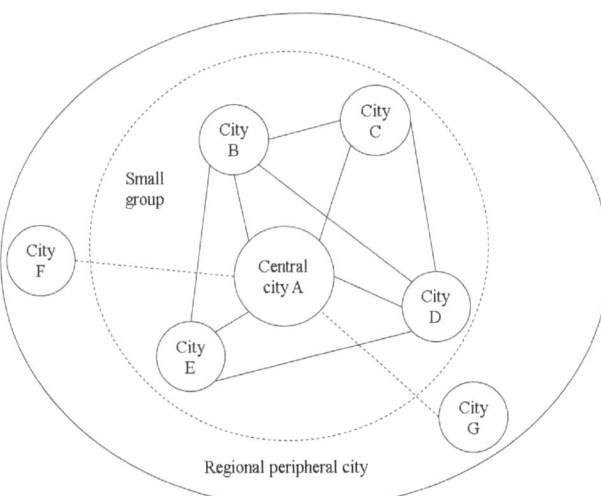

Figure 2. The simplified regional analytic model.

Reputation will firstly affect the role of the coordination ability of regional social capital. If the reputation of the local governments embedded in a regional social network increases, it will enhance their influence and leadership within the social network, and further play a coordinating role (external diffusion) by bridging more cities' local governments [30], which enhances the positive effect of bridging social capital on promoting cooperation performance. Through the more convenient transmission and sharing of real regional environmental governance information and resources, it could be easier for the central city to figure out the determination of chief officials from other cities to control pollution and the cost–benefit relationship of the cross-regional environmental cooperation in the various cities within the region, which is helpful for coordinating their behavior preferences and improving the efficiency of local governments in participating in inter-governmental environmental cooperation [31,32]. At the same time, the frequent exchange and coordination of information and resources could contribute to the function of the market mechanism, which could promote the convergence of resource allocation and suppress the occurrence of pollution shelters among stakeholders. Therefore, we propose Hypothesis 2:

Hypothesis 2. *The promotion of reputation could affect the function path of regional bridging social capital. That is, it could enhance the coordination ability of the central city and connect with more peripheral cities, so as to improve the environmental governance performance among urban agglomeration cities.*

Reputation could also affect the function process of the cooperation game behavior of local governments. When the reputation of central cities increases, it could also enhance their trust degrees in the social network. Other cities associated in a small group are more inclined to form close cooperative relations and enhance the stability of inter-governmental environmental cooperation projects (internal cohesion), which even fully breaks through the limitation of administrative boundaries. That is, this strengthens the function intensity of bonding social capital by the means of reaching commitments, releasing binding environmental laws and regulations, or forming compact relationships through mutual trust [33]. Stakeholders (local governments) can be encouraged to adopt the strategy of withdrawing from environmental cooperation to achieve higher returns for their own interest [34]. When the local government of a central city is worried about betrayal, effective supervision is needed to ensure the credible commitment of the network participants. All in all, reputation could enhance the function density of bonding social capital by forming closer small group relationships with other local governments [35]. We put forward Hypothesis 3:

Hypothesis 3. *The promotion of reputation could affect the function path of regional bonding social capital. That is, it could enhance the degree of mutual trust and cohesion of local governments within an environmental cooperation small group, so as to improve the environmental governance performance among urban agglomeration cities.*

According to the theory analysis above, we advance the following analytic framework (Figure 3):

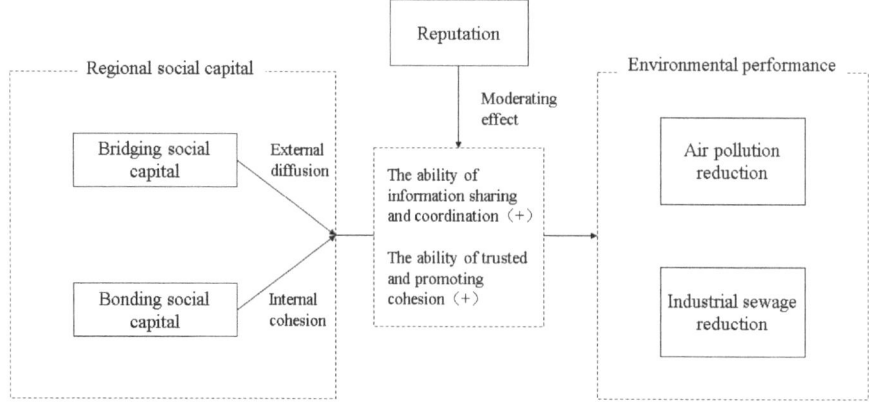

Figure 3. The theoretical analytic framework.

3. Variable Selection and Methodology

3.1. Variable Selection

3.1.1. Explained Variable

In order to fully measure the regional pollutant emission conditions, this paper uses per capita $PM_{2.5}$ emissions and industrial sewage emissions to indicate the atmosphere and water pollution conditions, respectively. A reduction in $PM_{2.5}$ emissions and industrial sewage emissions could effectively reflect the environmental collaborative governance effect, and they are adopted to evaluate the local atmosphere and water pollution governance performance by the Ministry of Ecology and Environment of China.

3.1.2. Core Explanatory Variables

The core explanatory variables are indictors of the various social capital formed by the regional environmental cooperative network. According to the Social Capital Theory, social capital could play the role of guiding cooperation, employment, and building trust among stakeholders. Combining the Social Capital theory, collaborative network structure, and cooperative performance, it could be further developed and adopted in the field of regional governance [36]. Specifically, bridging social capital plays the role of information sharing and coordination, while bonding social capital plays the role of trust consolidating and condensing within small groups in cooperative networks.

Bridging social capital could be indicated by the average degree of centrality within an environmental cooperative network, which is calculated by the number of direct connections among the central participant and other participants [15] (In Figure 2, only 4 cities directly link with the central city A, and the degree centrality of City A is 4). In the field of inter-governmental environmental cooperation, if a city could widely bridge to other cities (with higher bridging social capital), it could be easier to fully figure out the determination of other local governments to control environmental pollution and the cost–benefit relationship of the inter-governmental environmental cooperation. It could contribute to improving the environmental cooperation performance by leveraging its communication and coordination abilities within the cooperative network. Bonding social capital could be indicated by the average clustering coefficient within an environmental cooperative network, which is calculated by the proportion of links among the ego's partners that exist over the total number that could exist (The calculation formula is: $Cluster = n/C_m^2$) [22] (n represents the links among the ego's partners that exist, while m represents the city numbers directly linked with the central city A. In Figure 2, the average clustering coefficient of Central city A is $5/C_4^2 = 5/6$). Local governments within a small group have a mutually inclusive preference for regional environmental cooperation. Frequent interaction within this small group could promote forming legal norms among the local governments of cities and leverage their mutual trust and cohesion abilities, which is beneficial for promoting effective environmental cooperation among local governments and improving environmental governance performance. The panel data of these two indictors could be obtained through Ucinet 6.0 [37].

3.1.3. Moderating Variable

The moderating variable in this study is the reputation of local governments in regional environmental cooperation. The behavior selection of local governments determines their reputation during inter-governmental environmental cooperation. Referring to the indictor "project landing rate" in related research [38,39], this study adopts the indictor "implementation rate of environmental cooperative projects" ($Rate_{it}$) to reflect the selective cooperation behavior of local governments. This selective cooperation behavior could play a moderating effect through the black box of reputation mechanism. Formula (1) represents the implementation rate of environmental cooperative projects for city i in year j

$$Rate_{i,t} = \frac{\left(\sum_1^n Implement_{i,t,n} \times W_n + \sum_1^m Implement_{i,t+1,m} \times W_m\right) \times 0.5}{\sum_1^l Willing_{i,t,l} \times W_l + 1} \quad (1)$$

Referring to the quantitative operation of policy function intensity [40,41], this study measures the willingness and implementation intensity of inter-governmental environmental cooperation considering two dimensions, with policy (news) number and power. $Implement_{i,t,n}$ represents a total of n cooperative policies (news) implemented by city i in year t. $Willingness_{i,t,n}$ represents a total of n cooperative policies (news) that city i expresses its willingness for in year t. W_n represents the weight of the policies' (news) power, which gradually decreases according to the intervention level from central government ($W = 3$), provincial government ($W = 2$), to municipal government ($W = 1$). In order to avoid measurement errors, the arithmetic mean of the current year and lagging year is adopted to calculate the implementation frequency of inter-governmental cooperation (According to an analysis from the data of inter-governmental cooperation news, it could be speculated that the cycle from reaching cooperative willingness to promoting cooperative implementation is usually 1–2 years for most regional cooperation projects). At the same time, we also add 1 to the denominators in Equation (1) to avoid the situation where a denominator is 0.

3.1.4. Control Variables

A huge difference appears in the social and economic development levels among the cities in the YRD and BTH regions. In order to more accurately figure out the relationship among selective cooperation, regional social capital, and cooperative performance, this study adopts the indictors Total Population (pop), Gross Domestic Product (GDP), Proportion of the Secondary Industry to GDP (second), and Environmental Protection Expenditure (exp) as control variables, which could, respectively, reflect the current situation of each city in the fields of population, economics, industrial structure, and local government financial capacity. The study also logarithmizes all the control variables to reduce the volatility of the control variables over time and alleviate the heteroscedasticity in the model [42].

3.2. Methodology

According to the theoretical analysis, the black box of the reputation mechanism appears as the performance interaction among various actors (local governments). That is, there are strong spatial corrections among various factors of the cities in the regional cooperation network. At the same time, this study does not focus on discussing the results of the spillover effect analysis. The spatial econometric model is adopted for a main regression analysis. After a series of model selection tests and the consideration of spatial correlation errors, this study chooses the Spatial Autoregressive Model (SAR) to eliminate the impact of spatial autocorrelation and other factors on the results when performing an OLS regression [43]. The model is designed as shown in Formulas (2) and (3).

$$PM2.5_{it} = \alpha_0 + \rho W \times PM2.5_{it} + \alpha_1 Degree_{it} + \alpha_2 Cluster_{it} + \alpha_3 Rate_{it} + \alpha_4 (Rate_{it} \times Degree_{it}) + \alpha_5 (Rate_{it} \times Cluster_{it}) + \alpha_6 pop_{it} + \alpha_7 GDP_{it} + \alpha_8 second_{it} + \alpha_9 Exp_{it} + \mu_{it} + \epsilon_{it} \quad (2)$$

$$Sewage_{it} = \alpha_0 + \rho W \times Sewage_{it} + \alpha_1 Degree_{it} + \alpha_2 Cluster_{it} + \alpha_3 Rate_{it} + \alpha_4 (Rate_{it} \times Degree_{it}) + \alpha_5 (Rate_{it} \times Cluster_{it}) + \alpha_6 pop_{it} + \alpha_7 GDP_{it} + \alpha_8 second_{it} + \alpha_9 Exp_{it} + \mu_{it} + \epsilon_{it} \quad (3)$$

Among them, $PM2.5_{it}$ and $Sewage_{it}$ represent the per capita emissions of $PM_{2.5}$ and the industrial sewage in year t of city i, respectively. W represents the spatial weight matrix. The study adopts a spatial inverse distance matrix to conduct a spatial econometric analysis, and uses a spatial adjacency matrix in the robustness test. Coefficient ρ represents the spatial interaction relationship between local pollutant emissions and adjacent pollutant emissions. $Degree_{it}$, $Clustering_{it}$, $Rate_{it}$, Pop_{it}, GDP_{it}, $Second_{it}$, and exp_{it} represent the average degree of centrality, average clustering coefficient, implementation rate of environmental cooperative projects, total population, gross domestic product, proportion of the secondary industry to GDP, and the environmental protection expenditure in year t of city i, respectively. The interaction term $\text{Rate}_{it} \times \text{Degree}_{it}$ could indicate the moderating effect of the implementation rate on the relationship between the regional bridging social capital and environmental governance performance. The interaction term $\text{Rate}_{it} \times \text{Clustering}_{it}$ could measure the moderating effect of the implementation rate on the relationship between the regional bonding social capital and environmental governance performance. Since the role of regional social capital depends on the performance interaction and mutual imitative learning effect with neighboring cities [44], the core goal of this study is to analyze the regression coefficients of these two interaction terms (coefficients α_4, α_5). μ_{it} is the individual fixed effect, while ϵ_{it} is the random error term.

3.3. Data Source and Descriptive Statistics

As the two representative urban agglomerations in China, the Yangtze River Delta (YRD) and Beijing–Tianjin–Hebei (BTH) regions increasingly receive attention from scholars, especially after both being selected for national development strategies in China. The research target of this study is the cities' local governments in the YRD and BTH regions. Specifically, the inter-governmental cooperation in the YRD region is mostly driven by the self-interest of local governments, which makes it easier to form a win-win outcome. While the BTH region includes Beijing and Tianjin, two municipalities directly under the Central Government of China, it also covers some undeveloped cities in the Hebei province. In order to ensure the interest of capital development, the sustainable development of many cities in Hebei might not be balanced under top-down intervention. This provides an obvious regional heterogeneity for this study to conduct a comparative analysis.

The city-level social and economic data above were sourced from the China City Statistical Yearbook, China Urban Construction Statistical Yearbook and China Civil Affairs' Statistical Yearbook, ranging from 2010 to 2021. The variables relating to GDP and fiscal expenditure were converted to the level of 2009 constant prices. The pollutant emission data could be collected using several approaches: the city-level industrial sewage emission could be collected from the China City Statistical Yearbook, and the ground–based $PM_{2.5}$ concentrations could be gained from the website of the Atmospheric Composition Analysis Group. Compared with the Globe Annual $PM_{2.5}$ Grids at Columbia University, it could reflect the $PM_{2.5}$ concentration after 2016, in which the city-level $PM_{2.5}$ concentration was measured based on the V4.GL.03 geophysical satellite [45].

The implementation rate of inter-governmental environmental cooperation projects is adopted to reflect the behavior strategy of the local governments participating in regional environmental collaborative governance. Referring to the current mainstream literature, the regional environmental cooperation intensity could be indicated by the comprehensive indicator combining the policy (news) number and power of regional environmental cooperation [46]. The Newspaper database of the China Digital Literary Library (2009–2020) was adopted as the fundamental database. On the one hand, we collected the news information and policy documents about inter-governmental environmental cooperation in a wide range. The website information capture technology based on Python environment was adopted. On the other hand, news and policy documents of environmental cooperation were clustered according to their cooperative stages and promotion administrative levels. We took the LDA (Latent Dirichlet Allocation) topic analysis model [47] of machine learning to solve the problem of semantic mining in the text clustering by considering the relationships among words, topics, and texts. The inter-governmental environmental cooperation networks in the YRD and BTH regions are shown in Figures 4 and 5. It could be observed that there was a huge gap in the cooperative network density between the willingness and implementation stages, in both the YRD and BTH regions. It is necessary to explore the effect of inter-governmental behaviors on environmental cooperation performance.

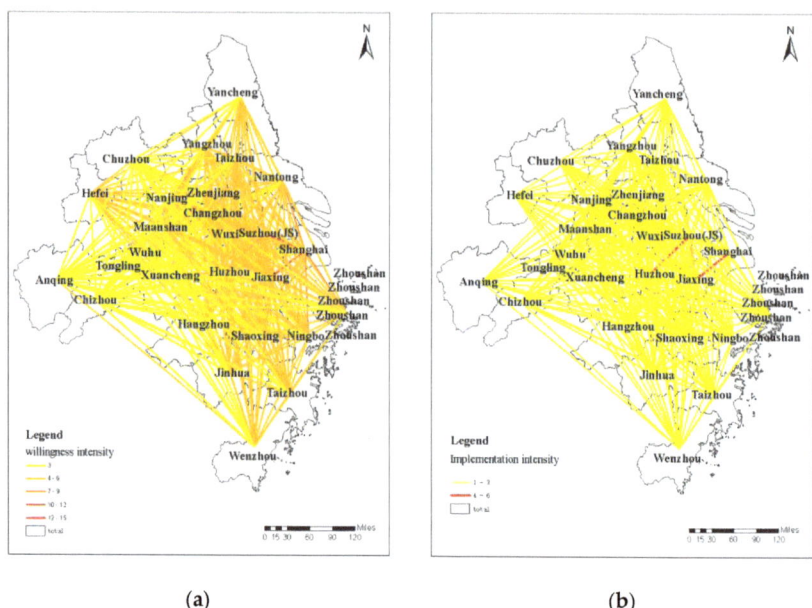

(a) (b)

Figure 4. The inter-governmental environmental cooperation network in YRD. (**a**) Willingness network; (**b**) Implementation network. (Note: the willingness network density is 4.3020, while implementation network density is 1.7721).

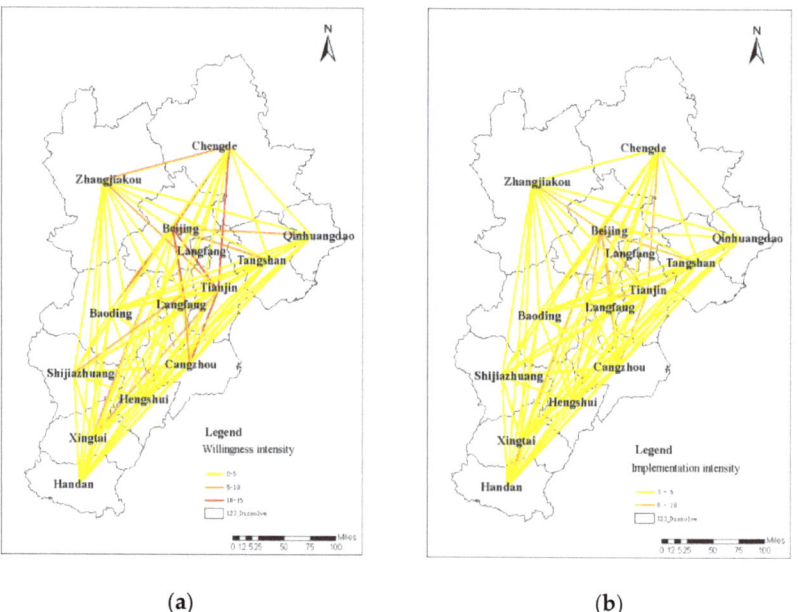

(a) (b)

Figure 5. The inter-governmental environmental cooperation network in BTH. (**a**) Willingness network; (**b**) Implementation network. (Note: the willingness network density is 7.411, while implementation network density is 3.949).

According to the database above, a panel data set covering 11 years, from 2009 to 2019 (according to Formula (1), one-year lagging data are used to calculate the implementation rate of environmental cooperation projects. As a result of this, the ultimate panel data only cover 11 years), is formed. There are 13 cities in the BTH region, while there are 27 cities in the YRD region, so the panel data set covers 440 samples. The descriptive statistics are shown in Table 1.

Table 1. Descriptive statistics.

Variable	Observations		Mean		Std.Dev		Min		Max	
	YRD	BTH	YRD	BTH	YRD	BTH	YRD	BTH	YRD	BTH
$PM_{2.5}$	297	143	48.052	57.588	12.807	25.247	18	13	71.739	110.121
$lnPM_{2.5}$	297	143	3.832	3.927	0.298	0.549	2.891	2.565	4.273	4.702
Sewage	297	143	14,604.471	8833.378	14,538.510	6317.984	486	615	80,468	31,058
lnSewage	297	143	9.148	8.786	0.983	0.857	6.186	6.422	11.296	10.344
degree	297	143	5.529	7.748	10.169	9.655	0	0	94	54
cluster	297	143	0.828	1.548	1.446	1.887	0	0	9	7
rate	297	143	0.733	0.803	0.830	0.699	0	0	8	3.75
degree×rate	297	143	4.720	6.378	10.538	11.021	0	0	92.932	83.464
cluster×rate	297	143	0.699	1.324	2.054	2.485	0	0	20.8	18.75
lnpop	297	143	6.055	6.531	0.629	0.467	4.301	5.66	7.293	7.244
lngdp	297	143	1.262	17.219	0.969	0.941	14.714	15.691	19.639	19.573
lnsecond	297	143	3.898	3.781	0.153	0.259	3.286	2.876	4.314	4.096
lnexp	297	143	15.227	15.401	0.903	0.982	13.095	13.715	18.27	18.192

4. Empirical Analysis

4.1. Spatial Autocorrelation Analysis

Before estimating the spatial econometric regression coefficients, a spatial autocorrelation analysis should be used to reveal the temporal and spatial characteristics of the air pollutant emissions and industrial sewage pollutant emissions, which could be indicated by global Moran's index. Using the software Arcgis 10.4, the coordinates of the 13 cities in BTH and 27 cities in YRD were extracted, and we could then transfer them into the inverse distance matrix with Stata 15.0. The result of the spatial autocorrelation analysis is shown in Table 2.

Table 2. The global Moran's index.

Year	YRD				BTH			
	$PM_{2.5}$		Industrial Sewage		$PM_{2.5}$		Industrial Sewage	
	Moran's I	Prob.	Moran's I	Prob.	Moran's I	Prob.	Moran's I	Prob.
2009	0.226 ***	0.000	0.063 ***	0.001	0.051 **	0.029	0.139	0.202
2010	0.205 ***	0.000	0.054 ***	0.002	0.086 ***	0.007	0.123	0.178
2011	0.220 ***	0.000	0.091 ***	0.000	0.099 ***	0.004	0.080 *	0.059
2012	0.163 ***	0.000	0.074 ***	0.000	0.102 ***	0.003	0.095 *	0.086
2013	0.213 ***	0.000	0.049 ***	0.003	0.106 ***	0.003	0.072 *	0.089
2014	0.209 ***	0.000	0.061 ***	0.001	0.081 ***	0.008	0.089 *	0.098
2015	0.216 ***	0.000	0.065 ***	0.000	0.079 ***	0.009	0.094 *	0.079
2016	0.221 ***	0.000	0.064 ***	0.000	0.089 ***	0.006	0.116	0.108
2017	0.228 ***	0.000	0.071 ***	0.000	0.109 ***	0.002	0.140 *	0.079
2018	0.242 ***	0.000	0.067 ***	0.000	0.093 ***	0.005	0.162	0.134
2019	0.240 ***	0.000	0.076 ***	0.000	0.094 ***	0.005	0.155 *	0.084

(Note: * indicates $p < 0.10$, ** indicates $p < 0.05$, and *** indicates $p < 0.01$).

As shown in Table 2, it could be observed that most Moran's I of the dependent variables were significantly positive from 2009 to 2019, indicating a positive spatial correlation among the cities in the YRD and BTH regions in regard to their $PM_{2.5}$ and industrial sewage emissions, and the spatial agglomerative effect was significant. As a result of this, it is reasonable to adopt spatial econometric models instead of OLS regression in this study, which could effectively weaken the impact of the spatial mobility of the dependent variables.

4.2. Spatial Regression Analysis

In the benchmark analysis, the study quantitatively explored the influencing mechanism of local governments' reputation (the implementation rate of environmental cooperative projects) on the relationship between regional social capital and environmental cooperation performance. Tables 3 and 4 reflect the city-level empirical results of the Spatial Autocorrelation regression (SAR) in the YRD and BTH regions, respectively. Among them, the dependent variable in model (1), (2) is the emissions of $PM_{2.5}$, while the dependent variable in model (3), (4) is the emissions of industrial sewage. Models (2) and (4) are regressions, adding a series of social and economic control variables. The Spatial Autocorrelation regression results are shown below.

Table 3. The result of spatial autocorrelation regression in YRD region.

Variables	$PM_{2.5}$		lnsewage	
	(1)	(2)	(3)	(4)
Degree	−0.0175 ***	−0.0292 **	−0.0062 ***	−0.0056 **
	(−2.80)	(−2.08)	(−2.65)	(−2.36)
Cluster	−0.296 **	−0.258 *	−0.0058 **	−0.0035 **
	(−2.06)	(−1.77)	(−2.35)	(−2.02)
Rate	0.309	0.206	0.0608	0.0570 *
	(1.19)	(0.77)	(1.04)	(1.65)
Degree × Rate	−0.0049 **	−0.0079 ***	−0.0044 **	−0.0061 ***
	(−2.14)	(−3.23)	(−2.11)	(−3.49)
Cluster × Rate	0.294	0.276	−0.0056	−0.0082
	(1.55)	(0.98)	(−1.42)	(−1.62)
lnpop		−2.078		0.525 ***
		(−1.58)		(3.71)
lngdp		0.122		0.302
		(1.07)		(1.30)
lnsecond		3.747 **		0.627 **
		(2.611)		(2.16)
lnexp		−0.686 **		−0.270
		(−2.56)		(−1.52)
ρ	0.928 ***	0.889 ***	0.679 ***	0.601 ***
	(58.00)	(33.74)	(9.87)	(6.19)
Log likelihood	−755.1827	−749.419	−95.7470	−85.9291
Adjusted-R^2	0.1974	0.3179	0.4456	0.5244
observations	297	297	297	297

(Note: * indicates $p < 0.10$, ** indicates $p < 0.05$, and *** indicates $p < 0.01$).

Table 4. The result of spatial autocorrelation regression in BTH region.

Variables	$PM_{2.5}$		lnsewage	
	(1)	(2)	(3)	(4)
Degree	−0.0442 **	−0.0303 *	−0.00290 **	−0.00316 **
	(−2.42)	(−1.83)	(−2.05)	(−2.49)
Cluster	0.00976	0.0105	0.0247	0.0251
	(1.03)	(1.17)	(1.07)	(1.12)
Rate	0.984 **	1.122 **	0.0129 *	0.0147 *
	(2.04)	(2.18)	(1.82)	(1.65)
Degree × Rate	−0.329 ***	−0.441 **	−0.0302 ***	−0.0291 *
	(−3.01)	(−2.32)	(−3.19)	(−1.95)
Cluster × Rate	0.092 *	0.115 *	0.0011 **	0.00298 ***
	(1.71)	(1.85)	(2.20)	(2.57)

Table 4. *Cont.*

Variables	PM$_{2.5}$		lnsewage	
	(1)	(2)	(3)	(4)
lnpop		0.124		−0.356 ***
		(1.01)		(−3.84)
lngdp		3.117		−0.468 ***
		(0.41)		(−3.69)
lnsecond		1.629 **		0.318 **
		(2.37)		(2.53)
lnexp		−4.336 *		−0.362 **
		(−1.88)		(−2.36)
ρ	−0.881 ***	−0.827 ***	−1.385 ***	−1.285 ***
	(−3.61)	(−3.78)	(−4.92)	(−5.11)
Log likelihood	−483.4208	−477.2383	−139.8291	−81.9675
Adjusted-R^2	0.3482	0.4050	0.1788	0.4315
observations	143	143	143	143

(Note: * indicates $p < 0.10$, ** indicates $p < 0.05$, and *** indicates $p < 0.01$).

4.2.1. The Empirical Result of YRD Region

According to the SAR result with the sample of the 27 cities in the YRD region in Table 3, since the R-squared of the regressions adding the control variables ($R^2 = 0.3179/0.5244$) is higher than the regressions without the control variables ($R^2 = 0.1974/0.4456$), the regression coefficients in models (2) and (4) are mostly considered. Firstly, the spatial autoregressive coefficient ρ is positive and significant at the 1% level, which indicates that the PM$_{2.5}$ concentration and industrial sewage emissions show a strong spatial agglomeration with the feature of "high-high, low-low" distribution. This is consistent with the reality. Due to the huge difference in the industrial structure and environmental governance ability among the cities in YRD, the haze and industrial sewage pollution are more serious located in the north of Jiangsu and south of Zhejiang and Anhui, while the air and water resource quality are better in the south of Jiangsu, Shanghai, and north of Zhejiang.

Secondly, we analyze the coefficients of the independent variables indicating the bridging social capital (Degree) and bonding social capital (Cluster). The results show that these two types of social capital could both effectively increase the environmental governance performance to reduce the concentration of PM$_{2.5}$ and industrial sewage emissions. Hypothesis 1 has been fully validated in the YRD region. At the same time, the coefficients of the implementation rate of environmental cooperation projects (Rate) are not significant, which indicates that the reputation mechanism of local government could not work with a direct effect.

Thirdly, the study mainly focuses on the interaction effect of local governments' behavior strategies and these two types of social capital. The results show that the interaction term of average-degree centrality (bridging social capital) and implementation rate could significantly reduce the concentration of PM$_{2.5}$ and industrial sewage emissions (Degree×Rate, $β = −0.0079$ ***/0.0061 ***). It could be assumed that an increase in reputation could improve the leadership and influence of the central city within a regional cooperative network, facilitating its bridging and coordination role, which could promote the exchange of environmental governance information, thus improving the environmental control performance. However, the interaction term of the average clustering coefficient (bonding social capital) and implementation rate could not significantly affect the environmental governance performance. It could be supposed that the cohesive "small group" type of inter-governmental environmental cooperation is not common among cities in the YRD region. They tend to form open and widely-connected networks instead of a closed network structure. Hypothesis 2 could be fully validated, while Hypothesis 3 is not applicative in the YRD region.

4.2.2. The Empirical Result of BTH Region

The spatial econometric regression results of the 13 cities in the BTH region have an obvious difference compared to the sample of the YRD region. The results of model (2) and (4) are mainly adopted to analyze the regression coefficients (higher R-squared values). Firstly, the spatial autoregressive coefficient ρ is significantly negative at the 1% level, indicating a "high-low" spatial negative spillover feature of the dependent variables in the BTH region. This could be assumed to be due to the "siphon" phenomenon of the environmental governance ability of local governments among adjacent cities.

Second, we explore the relationship between these two types of regional social capital and pollutant emission reduction. It could be observed that the average-degree centrality has a significantly negative effect on the $PM_{2.5}$ concentration (β = −0.0303 *) and industrial sewage emissions (β = −0.00316 **). However, the coefficients of *Cluster* are not significant. These results indicate that the information sharing and coordination effect (bridging social capital) of a regional environmental cooperation network work to promote environmental governance performance. However, the impact of promoting trust and internal cohesion (bonding social capital) is masked during the process of inter-governmental environmental cooperation. All in all, Hypothesis 1 has been partly validated in the BTH region

Thirdly, we deeply explore the function mechanism of local governments' selective behavior during regional environmental cooperation. Similar to the result in the YRD region, the interaction term of average-degree centrality and the implementation rate could also significantly reduce the concentration of $PM_{2.5}$ and industrial sewage emissions (Degree × Rate, β = −0.441 **/−0.0291 *). However, the coefficients of the interaction term of the average clustering coefficient and implementation rate show the opposite results (Cluster × Rate, β = 0.115 */0.00298 ***), which significantly increase the pollutant emissions. On the one hand, an increase in reputation could enhance the coordination ability of the central city and share environmental governance information with more peripheral cities, so as to improve their environmental governance performance. On the other hand, an increase in reputation could also promote the ability of the internal cohesion of local governments within a small group, which further enhances the path dependence effect of their environmental governance preferences. Specifically, the central city will lead the regional environmental cooperation process, from decision-making to implementation, while peripheral cities will gradually lose their interest in participating in regional cooperation. This will do harm to environmental governance outcomes. Through a comparative analysis of the coefficient values, the absolute values of the pollutant emission reduction (Degree × Rate) are higher than the absolute values of the pollutant emission growth (Cluster × Rate), which indicates that the reputation mechanism could realize an overall pollutant reduction effect. According to the empirical results above, Hypothesis 1 has been totally proved, while Hypothesis 2 cannot be validated in the BTH region.

4.2.3. Robustness Check

Referring to the remaining literature [48,49], this study conducts a robustness test by replacing the spatial weight matrix based on spatial econometric regression. The specific approach is to replace the spatial inverse distance matrix with a spatial adjacency matrix. Since the construction principles of both types of matrices are based on the spatial location conditions of the city unit's centroid or boundary, the regression results should not deviate too much from the original results. The robustness test results of the SAR with a spatial adjacency matrix are shown in Table 5.

Table 5. The spatial autocorrelation regression results of YRD and BTH regions (spatial adjacency matrix).

Variables	YRD		BTH	
	$PM_{2.5}$ (1)	lnsewage (2)	$PM_{2.5}$ (3)	lnsewage (4)
Degree	−0.00945 **	−0.00434 **	−0.0411 ***	−0.00328 *
	(−2.44)	(−2.09)	(−2.56)	(−1.70)
Cluster	−0.158 *	−0.00418 **	0.060	0.0281
	(−1.81)	(−2.26)	(1.24)	(0.66)
Rate	0.142	0.0555 *	0.893 *	0.00823
	(0.89)	(1.86)	(1.77)	(1.24)
Degree × Rate	−0.0274 **	−0.0039 *	−0.206 *	−0.0062 **
	(−2.30)	(−1.93)	(−1.89)	(−2.52)
Cluster × Rate	0.0906	−0.00464	0.0836 **	0.00303 *
	(1.32)	(−0.36)	(2.32)	(1.82)
lnpop	−1.301	0.427 ***	0.780	−0.286 ***
	(−1.52)	(2.68)	(1.08)	(−3.91)
lngdp	−1.028	−0.279	−3.001	−0.315 ***
	(−0.87)	(−1.27)	(−0.53)	(−3.45)
lnsecond	1.214 *	0.815 ***	6.249 **	0.209 **
	(1.77)	(2.83)	(2.09)	(2.32)
lnexp	−0.567 ***	−0.0044 **	−0.322 **	−0.329 ***
	(−2.61)	(−2.03)	(−2.09)	(−2.78)
ρ	0.901 ***	0.144 **	−0.885 ***	−0.605 ***
	(52.88)	(2.20)	(−32.23)	(−6.22)
Log likelihood	−571.0521	−14.7855	−439.4345	−79.3500
Adjusted-R^2	0.2819	0.3381	0.2520	0.4331
observations	297	297	143	143

(Note: * indicates $p < 0.10$, ** indicates $p < 0.05$, and *** indicates $p < 0.01$).

As shown in Table 5, when a spatial adjacency matrix is adopted for a spatial autocorrelation regression analysis, it could be found that the coefficients of the core independent variables are similar to the results in Tables 3 and 4, regardless of the coefficient values and their significance. The results of the robustness test verify our benchmark regression results, indicating that our empirical conclusions are credible.

5. Conclusions and Discussion
5.1. Conclusions

Regional social capital could be formed from a regional cooperative network during the process of inter-governmental cooperation, and regional policies have effects through different types of social capital paths impacting the cooperative performance. This study incorporated the reputation mechanism, social capital, and environmental governance performance into a unified analysis framework and deeply explored the spatial diffusion process of local governments' behavior. The empirical strategy was to test the moderating effect of the implementation rate of environmental cooperative projects on the relationship between two types of social capital and the environmental governance performance among the cities in the YRD and BTH regions.

According to the empirical results from the YRD region, the reputation of local government works mainly through the bridging social capital path. An increase in reputation could improve the leadership and coordination ability of the central city among a cooperative network, which could promote the exchange of environmental governance information, thus improving the environmental governance performance. However, since the closed "small group" type of inter-governmental environmental cooperation is not common in the YRD region, the performance promotion path through internal cohesion is not feasible. Local governments in the YRD region tend to form open and widely-connected network structures.

According to the empirical results from the BTH region, the interaction effect of reputation and bridging social capital is similar to the result in the YRD region. However, the moderating effect of reputation through bonding social capital shows the opposite results compared to bridging social capital, which significantly increases the pollutant emissions. This indicates that excessive internal cohesion might strengthen the authority of the central city and enhance the path dependence of local governments' environmental governance preferences, further deviating from the original intention of inter-governmental environmental cooperation. The above internal cohesion function of reputation might weaken the environmental governance performance through cross-regional cooperation.

5.2. Discussion

Two reasons could be assumed to explain the differential regional cooperation patterns in the YRD and BTH regions. Firstly, the strategic positioning of YRD and BTH are different. The regional development goal of YRD is realizing inter-city "integration", which is determined to build the "strongest and most active economic growth pole" and a window of "all-round opening up" in China. However, the BTH region has been given the development goal of "regional collaboration", and its inter-governmental cooperation usually relies on a top-down arrangement to allocate production factors. Secondly, a difference appears concerning intergovernmental relations and their power structures in the YRD and BTH regions. In the YRD region, Shanghai, Jiangsu, Zhejiang, and Anhui are all provincial units and their intergovernmental relations are equal and independent, while in the BTH region, the relationship between Beijing and Tianjin (or Hebei) is more like a central–local relationship, since Beijing is the capital of China. As a result, YRD's urban agglomeration usually appears as an open and widely-connected cooperative network structure, while an authority-driven and enclosed cooperative pattern is common in BTH's urban agglomeration, which could generate excessive internal cohesion and the results might deviate from the original intention of the inter-governmental cooperation.

The empirical conclusions above could be beneficial for us to understand the effect of local governments' selective behavior on regional governance performance through the reputation mechanism. In order to improve the implementation rate of environmental cooperative projects, a "community of entangled interest" should be constructed among local governments [50], which requires these local governments to weaken the concept of administrative boundaries, promoting the integrity of environmental governance among the cities within an urban agglomeration. That is, local governments should put themselves in the position of other stakeholders. At the same time, excessive internal cohesion and fully breaking down administrative divisions are also undesirable during cross-regional environmental governance. Without a clear definition of rights and responsibilities, local governments will have little incentive to participate in environmental governance [51].

The information-sharing mechanism based on horizontal local governments and the punishment mechanism based on vertical power pressure should be adopted to ensure the continuous operation of inter-governmental environmental cooperation, which is useful for suppressing free-riding behavior [52]. At the same time, local governments should carry out cross-border linkage according to actual situations based on a clear division of rights and responsibilities. A representative case is the "United River Chief Policy Pilot" in the YRD integration demonstration zone. It breaks up the administrative border to grant joint enforcement powers to the environmental protection departments of its adjacent cities. Meanwhile, it also clearly defined the rights and responsibilities of different local governments. It achieved an outstanding environmental governance performance (People's Daily, 9 June 2023. Referring to: http://js.people.com.cn/n2/2023/0609/c360300-40450051.html, accessed on 20 July 2023). Within the current institutional context, local governments remain responsible for local environmental performance, and effective environmental regulation measures should also be adopted to achieve sustainable, instead of campaign-style, environmental governance performance when facing collective action failure.

Author Contributions: Conceptualization, Y.Z.; materials and methods, Y.Z. and J.X.; formal analysis, Y.Z. and D.H.; writing—original draft preparation, Y.Z.; writing—review and editing, Y.Z.; J.X. and D.H.; supervision, J.X. and D.H.; funding acquisition, J.X. All authors have read and agreed to the published version of the manuscript.

Funding: This paper was funded by the National Social Science Foundation of China (22BZZ059).

Data Availability Statement: Not applicable.

Acknowledgments: We want to thank the anonymous referees for their constructive suggestions on the earlier draft of our paper, upon which we have improved the content.

Conflicts of Interest: The authors declare no conflict of interest.

References

1. Fu, F.; Hauert, C.; Nowak, M.A.; Wang, L. Reputation-based partner choice promotes cooperation in social networks. *Phys. Rev. E* **2008**, *78*, 026117. [CrossRef] [PubMed]
2. Wang, L.; Chen, T.; Wu, Z. Promoting cooperation by reputation scoring mechanism based on historical donations in public goods game. *Appl. Math. Comput.* **2021**, *390*, 125605. [CrossRef]
3. Franco, M.; Haase, H. The role of reputation in the business cooperation process: Multiple case studies in small and medium-sized enterprises. *J. Strategy Manag.* **2021**, *14*, 82–95. [CrossRef]
4. Goldberg, A.I.; Cohen, G.; Fiegenbaum, A. Reputation building: Small business strategies for successful venture development. *J. Small Bus. Manag.* **2003**, *41*, 168–186. [CrossRef]
5. Li, Y.; Wu, F. Understanding city-regionalism in China: Regional cooperation in the Yangtze River Delta. *Reg. Stud.* **2017**, *3*, 313–324. [CrossRef]
6. Yang, L.; Chen, W.; Wu, F.; Li, Y.; Sun, W. State-guided city regionalism: The development of metro transit in the city region of Nanjing. *Territ. Politics Gov.* **2021**, 1–21. [CrossRef]
7. Chen, X.; Sullivan, A.A. Should I Stay or Should I Go? Why Participants Leave Collaborative Governance Arrangements. *J. Public Adm. Res. Theory* **2022**, *33*, 246–261. [CrossRef]
8. Ye, C.; Chen, R.; Chen, M.; Ye, X. A new framework of regional collaborative governance for PM2.5. *Stoch. Environ. Res. Risk Assess.* **2019**, *33*, 1109–1116. [CrossRef]
9. Zhao, C.; Wang, X.; Cheung, P.T.; Xu, J. Influence of External Authorities on Collaborative Frictions. *Public Adm. Rev.* **2023**, *83*, 603–622. [CrossRef]
10. Zhao, Y.; Wang, Y. *Why Do They Say One Thing and Do Another? Exploring the Factors Influencing the Strategy Selection of Local Governments in the Process of Regional Public Service Cooperation*; Working Paper. 2023; Unpublish.
11. Xing, P.; Xing, H. Blood is thicker than water: Local favouritism and inter-local collaborative governance. *Policy Stud.* **2022**, 1–19. [CrossRef]
12. Wang, Y.; Zhao, Y. Is collaborative governance effective for air pollution prevention? A case study on the Yangtze river delta region of China. *J. Environ. Manag.* **2021**, *292*, 112709. [CrossRef]
13. Huxham, C.; Vangen, S. *Managing to Collaborate: The Theory and Practice of Collaborative Advantage*; Routledge: London, UK, 2013.
14. Havakhor, T.; Soror, A.A.; Sabherwal, R. Diffusion of knowledge in social media networks: Effects of reputation mechanisms and distribution of knowledge roles. *Inf. Syst. J.* **2018**, *28*, 104–141. [CrossRef]
15. Yi, H. Network structure and governance performance: What makes a difference? *Public Adm. Rev.* **2018**, *78*, 195–205. [CrossRef]
16. Huang, C.; Chen, W.; Yi, H. Collaborative networks and environmental governance performance: A social influence model. *Public Manag. Rev.* **2021**, *23*, 1878–1899. [CrossRef]
17. Gao, X. Research on regional cooperation from the perspective of regional social capital. *Contemp. World Social.* **2013**, *5*, 123–126. (In Chinese)
18. Shi, J.; Tang, D. Research on regional innovation networks based on social capital theory. *Sci. Manag. Res.* **2007**, *5*, 10–13. (In Chinese)
19. Ferris, G.R.; Perrewe, P.L.; Douglas, C. Social effectiveness in organizations: Construct validity and research directions. *J. Leadersh. Organ. Stud.* **2002**, *9*, 49–63. [CrossRef]
20. Andrew, S.A.; Carr, J.B. Mitigating uncertainty and risk in planning for regional preparedness: The role of bonding and bridging relationships. *Urban Stud.* **2013**, *50*, 709–724. [CrossRef]
21. Feiock, R.C.; Lee, I.W.; Park, H.J. Administrators' and elected officials' collaboration networks: Selecting partners to reduce risk in economic development. *Public Adm. Rev.* **2012**, *72*, S58–S68. [CrossRef]
22. Berardo, R.; Scholz, J.T. Self-organizing policy networks: Risk, partner selection, and cooperation in estuaries. *Am. J. Political Sci.* **2010**, *54*, 632–649. [CrossRef]
23. Lee, I.W.; Feiock, R.C.; Lee, Y. Competitors and cooperators: A micro-level analysis of regional economic development collaboration networks. *Public Adm. Rev.* **2012**, *72*, 253–262. [CrossRef]
24. Coleman, J.S. Social capital in the creation of human capital. *Am. J. Sociol.* **1988**, *94*, S95–S120. [CrossRef]

25. Zhao, Y.; Wang, Y. Selective cooperation: The inter-governmental public service supply in regional governance of Yangtze River Delta. *J. Shanghai Adm. Inst.* **2022**, *23*, 27–37. (In Chinese)
26. Lopez-Pintado, D.; Watts, D.J. Social influence, binary decisions and collective dynamics. *Ration. Soc.* **2008**, *20*, 399–443. [CrossRef]
27. Jiao, Y.; Chen, T.; Chen, Q. The impact of expressing willingness to cooperate on cooperation in public goods game. *Chaos Solitons Fractals* **2020**, *140*, 110258. [CrossRef]
28. Liu, Y.; Chen, T. Sustainable cooperation based on reputation and habituation in the public goods game. *Biosystems* **2017**, *160*, 33–38. [CrossRef]
29. Liu, Y.; Chen, T.; Wang, Y. Sustainable cooperation in Village Opera based on the public goods game. *Chaos Solitons Fractals* **2017**, *103*, 213–219. [CrossRef]
30. Wang, Z.; Chen, T.; Wang, Y. Leadership by example promotes the emergence of cooperation in public goods game. *Chaos Solitons Fractals* **2017**, *101*, 100–105. [CrossRef]
31. Provan, K.G.; Sebastian, J.G. Networks within networks: Service link overlap, organizational cliques, and network effectiveness. *Acad. Manag. J.* **1998**, *41*, 453–463. [CrossRef]
32. Meier, K.J.; O'Toole, L.J., Jr. Managerial strategies and behavior in networks: A model with evidence from US public education. *J. Public Adm. Res. Theory* **2001**, *11*, 271–294. [CrossRef]
33. Burt, R.S. *Brokerage and Closure: An Introduction to Social Capital*; Oxford University Press: Oxford, UK, 2005.
34. Scholz, J.T.; Wang, C.L. Learning to cooperate: Learning networks and the problem of altruism. *Am. J. Political Sci.* **2009**, *53*, 572–587. [CrossRef]
35. Shrestha, M.K.; Feiock, R.C. Transaction cost, exchange embeddedness, and interlocal cooperation in local public goods supply. *Political Res. Q.* **2011**, *64*, 573–587. [CrossRef]
36. Yi, H.; Scholz, J.T. Policy networks in complex governance subsystems: Observing and comparing hyperlink, media, and partnership networks. *Policy Stud. J.* **2016**, *44*, 248–279. [CrossRef]
37. Borgatti, S.P.; Everett, M.G.; Freeman, L.C. Freeman. In *UCINET for Windows: Software for Social Network Analysis*; Analytic Technologies: Harvard, MA, USA, 2002.
38. Tan, J.; Zhao, J.Z. The rise of public–private partnerships in China: An effective financing approach for infrastructure investment? *Public Adm. Rev.* **2019**, *79*, 514–518. [CrossRef]
39. Tan, J.; Zhao, J.Z. Explaining the adoption rate of public-private partnerships in Chinese provinces: A transaction cost perspective. *Public Manag. Rev.* **2021**, *23*, 590–609. [CrossRef]
40. Peng, J.; Zhong, W.; Sun, W. Policy measurement, policy collaborative evolution and economic performance: An empirical study based on innovation policies. *J. Manag. World* **2008**, *180*, 25–36. (In Chinese)
41. Sun, W.; Peng, J.; Huang, Y. Evolution of technology policies in China: A comparative analysis between central and local levels. *J. Sci. Technol. Policy China* **2011**, *2*, 238–254. [CrossRef]
42. He, L.; Zhang, X. The distribution effect of urbanization: Theoretical deduction and evidence from China. *Habitat Int.* **2022**, *123*, 102544. [CrossRef]
43. Ord, K. Estimation methods for models of spatial interaction. *J. Am. Stat. Assoc.* **1975**, *70*, 120–126. [CrossRef]
44. Young, H.P. Innovation diffusion in heterogeneous populations: Contagion, social influence, and social learning. *Am. Econ. Rev.* **2009**, *99*, 1899–1924. [CrossRef]
45. Hammer, M.S.; van Donkelaar, A.; Li, C.; Lyapustin, A.; Sayer, A.M.; Hsu, N.C.; Levy, R.C.; Garay, M.J.; Kalashnikova, O.V.; Kahn, R.A.; et al. Global estimates and long-term trends of fine particulate matter concentrations (1998–2018). *Environ. Sci. Technol.* **2020**, *54*, 7879–7890. [CrossRef] [PubMed]
46. Du, H.; Guo, Y.; Lin, Z.; Qiu, Y.; Xiao, X. Effects of the joint prevention and control of atmospheric pollution policy on air pollutants-A quantitative analysis of Chinese policy texts. *J. Environ. Manag.* **2021**, *300*, 113721. [CrossRef] [PubMed]
47. Kar, M.; Nunes, S.; Ribeiro, C. Summarization of changes in dynamic text collections using Latent Dirichlet Allocation model. *Inf. Process. Manag.* **2015**, *51*, 809–833. [CrossRef]
48. Ma, D.; Zhang, J.; Wang, Z.; Sun, D. Spatio-temporal evolution and influencing factors of open economy development in the Yangtze River Delta area. *Land* **2022**, *11*, 1813. [CrossRef]
49. Zhao, Y.; Liang, C.; Zhang, X. Positive or negative externalities? Exploring the spatial spillover and industrial agglomeration threshold effects of environmental regulation on haze pollution in China. *Environ. Dev. Sustain.* **2021**, *23*, 11335–11356. [CrossRef]
50. Wang, H.; Ran, B. Network governance and collaborative governance: A thematic analysis on their similarities, differences, and entanglements. *Public Manag. Rev.* **2022**, *25*, 1187–1211. [CrossRef]
51. Xiong, J. The administrative division's logic of regional urban integration: Governing with administrative division and reforming administrative division with governance. *J. Shanghai Adm. Inst.* **2017**, *23*, 65–73. (In Chinese)
52. Song, M.; Lai, Y.; Zhang, Y.; Li, L.; Wang, E. From Neighbors to Partners: A quantum game model for analyzing collaborative environmental governance in China. *Expert Syst. Appl.* **2022**, *210*, 118248. [CrossRef]

Disclaimer/Publisher's Note: The statements, opinions and data contained in all publications are solely those of the individual author(s) and contributor(s) and not of MDPI and/or the editor(s). MDPI and/or the editor(s) disclaim responsibility for any injury to people or property resulting from any ideas, methods, instructions or products referred to in the content.

Article

Does the Opening of High-Speed Railway Improve High-Quality Economic Development in the Yangtze River Delta, China?

Chiming Guan [1,2,*], Liuying Chen [1] and Danyang Li [3]

[1] School of Economics & Management, Southeast University, Nanjing 211102, China; chenliuying@seu.edu.cn
[2] National School of Development and Policy, Southeast University, Nanjing 210096, China
[3] Department of Geography and Tourism, Katholieke Universiteit Leuven, 3000 Leuven, Belgium; danyang.li@kuleuven.be
* Correspondence: gchm@seu.edu.cn; Tel.: +86-13814048880

Citation: Guan, C.; Chen, L.; Li, D. Does the Opening of High-Speed Railway Improve High-Quality Economic Development in the Yangtze River Delta, China? *Land* **2023**, *12*, 1629. https://doi.org/10.3390/land12081629

Academic Editors: Wei Sun, Zhaoyuan Yu, Kun Yu, Weiyang Zhang and Jiawei Wu

Received: 4 July 2023
Revised: 12 August 2023
Accepted: 16 August 2023
Published: 18 August 2023

Copyright: © 2023 by the authors. Licensee MDPI, Basel, Switzerland. This article is an open access article distributed under the terms and conditions of the Creative Commons Attribution (CC BY) license (https:// creativecommons.org/licenses/by/ 4.0/).

Abstract: The Yangtze River Delta (YRD) is the area with the densest high-speed railway (HSR) network in China, and it leads the high-quality economic development (HQED) in the country. HSR plays an important role in regional development. However, research on the impact of the HSR on HQED is notably limited. Theoretically, this study develops an analytical framework for the mechanism of the HSR's influence on HQED. Empirically, it calculates the HQED index and then investigates the impact of the HSR on HQED and the regional discrepancies across cities in the YRD, based on data from 2011 to 2019 using the difference-in-differences model. The results show: (1) The mechanism lies in that the HSR improves urban accessibility, accelerates the flow of the production factors, and enhances the allocation efficiency of the input factors. (2) The distribution of the HQED level presents an obvious circular pattern, with Shanghai and Suzhou at the centre, showing the prominent principle of distance decay. (3) Both the regression model and the robustness tests show that the HSR significantly promotes HQED in the YRD. Additionally, the economic development, foreign capital spent, financial level, industry advancement, and living standard are conducive to HQED. (4) The results of the heterogeneity test reveal that the HSR has an obviously varied impact on HQED in cities depending on their size and location. The HSR has a significant promotional effect on HQED in cities with a large population and those far away from a provincial city.

Keywords: high-speed railway (HSR); high-quality economic development (HQED); the Yangtze River Delta (YRD); impact mechanism

1. Introduction

Since sustainable development, which addresses development issues in society, the economy, and the environment in an integrated manner [1], was proposed in the 1980s, relative concepts and practices have attracted attention globally [2]. Economic development quality was proposed at the end of the 20th century as a complementary part of sustainable development [3]. Thomas put forward that in order to obtain and ensure broad-based and long-term growth, combining actions on equality, quality, and sustainability with those for growth should be taken [4]. Boyle and Simms believe that economic growth, the sustainable development of humans and nature, and living standards should be combined into the connotations of economic growth quality [5]. In 2015, the United Nations put forward 17 sustainable development goals to thoroughly solve the development problems regarding social, economic, and environmental development in an integrated way, and to shift to the path of sustainable development [6,7]. In the same year, China proposed a new development philosophy including innovative, coordinated, green, open, and shared development based on sustainable development, and then put forth high-quality economic development (HQED), which embodies the concrete implementation of the five new development concepts [8]. To achieve a more efficient, equitable, and sustainable form of development, China declared that China's economy had transformed from high-speed development to high-quality development in October 2017. In 2020, the Chinese government reiterated that the theme of economic and social development during the 14th Five-Year Plan (2021–2025) should facilitate HQED. This emphasis is aimed at promoting sustainability of the ecological environment, meeting the aspirations for a better life, and enhancing the economy's overall competitiveness. Compared with the sustainable development goals proposed by the United Nations, the HQED advocated by the Chinese government represents a specific implementation of these development goals tailored to address the challenges faced by China and many other developing countries. The promotion of HQED is not only an essential necessity for maintaining healthy economic development, but also a crucial guarantee for achieving sustainable development.

Against this background, it is particularly important to understand what HQED is and how to promote HQED and, then, promote sustainable development. A number of scholars have conducted research on HQED [9–11], investigating both the comprehensive factors [12] that influence it and the individual influencing factors [13–15]. Notably, some found that transportation is one of the significant influencing factors [16]. As HQED has been gradually implemented, China's high-speed railway (HSR) is also undergoing rapid development. Since 2012, China has added over 3000 kms of HSR lines annually. By the end of 2022, the operating mileage had reached 42,000 kms, significantly impacting the country's socioeconomic development. However, existing research on the impact of HSR on HQED is still insufficient. At this point, further questions are yet to be explored: whether and how HSR improves HQED, what are the detailed impacts, what are the impact mechanisms of the HSR on HQED, and further, if the HSR can drive HQED, is there any heterogeneity in the impact of the HSR on HQED in a specific region? This study aims to explore these questions and empirically test the impact of the HSR on HQED and the regional differences based on 41 cities in the Yangtze River Delta (YRD) region in China, which is one of the three most economically developed regions in China and a pioneer in practising HQED. A deep examination of these questions would be beneficial for amplifying the effect of HSR, enhancing regional HQED, and achieving sustainable development.

The remaining sections in this study are organised as follows. Section 2 presents the literature review. Section 3 defines the connotations of HQED, and builds and illustrates the theoretical mechanism and hypotheses. Section 4 describes the study area, econometric model, variable design, and data sources, as well as the distribution characteristics and spatiotemporal evolution characteristics of the HQED in the YRD. Section 5 presents the empirical testing and the analytical results. Section 6 contains the discussion and policy implications and in Section 7 we make the main conclusions.

2. Literature Review

2.1. The Research on High-Quality Economic Development

After the quality of economic development became a concern in the 1990s, some scholars began to focus on it [3,17,18]. Xu argued that an increase in production efficiency is equivalent to a degree of economic quality; that is, the ability of a given input to generate more output [3]. Thomas et al. considered economic growth quality to be a supplement to the speed of the development and that the quality of the economic development should also encompass the distribution of welfare, the ecological environment, risk resistance, and governance [17]. Barro believed that the quality of economic growth should cover the life expectancy, environmental conditions, social welfare, political institutions, and religious beliefs [18].

As Chinese economic development shifted to a new norm (*xinchangtai*) in 2012, there has been increased focus on the quality of economic development. After HQED was introduced in 2015, many scholars have explored its connotations, the measurement methods, and influencing factors. Most researchers consider the connotations of HQED based on the report from the 19th National Congress of the Communist Party of China, which states that development must adhere to quality first and prioritise efficiency; implement structural reforms on the supply side as the main aim; promote quality, efficiency, and power change in the economic development; and improve total factor productivity (TFP) [14]. The connotations include not only economic factors, but also social, environmental, and some other detailed factors, such as governance, quality of life, livelihood, employment status, education level, and national life expectancy [11,15,19,20]. As the new development concept, namely, innovative, coordinated, green, openness, and sharing are put into practice gradually, the five aspects are regarded as the fundamental pillars of China's HQED [21].

Regarding the measurement of HQED, earlier studies primarily focused on the quality of economic development. Many scholars adopted total factor productivity (TFP) as a measurement method [22]. With the clarification of the HQED concept, some scholars have added green development to TFP [23]. Given the extensive connotations of HQED, it is unscientific to measure it from one aspect. Therefore, more scholars have adopted a multi-dimensional approach to measure high-quality development or HQED. Yang et al. evaluated HQED from the economic structure, economic efficiency, and ecological environment aspects [24]. Pan et al. constructed a high-quality development index from five aspects: economic development, innovation efficiency, environmental impact, ecological services, and people's livelihoods [25]. Liu et al. built a comprehensive indicator system based on the five development concepts to measure the HQED level [26]. Kong et al. built an economic growth quality index system with the dimensions of efficiency, stability, and sustainability [27]. Li et al. set up a multiple-index system to calculate HQED, encompassing five dimensions and 24 detailed indexes, completely corresponding to the five development concepts [15]. Guo et al. constructed an evaluation index system for HQED at the city level from five dimensions, including the industrial structure, inclusive TFP, technology innovation, the ecological environment, and residents' living standards [28]. Li, F. and Li, M. built an evaluation system, including economic operation, social development, and ecological sustainability, to calculate the HQED in 41 cities in the YRD and found that there is a prominent divergence in cities [20].

In addition, there are some studies on the HQED pattern and influencing factors. Researchers have found that the level of HQED is influenced by multiple factors, including transportation, the suitability of the natural environment, the urbanisation level and population agglomeration, the investment pattern, and the digital economy [12,14,15,29].

2.2. The Research on the Effect of High-Speed Railway on Economic Development

The ever-improving HSR network has made a huge impact on social and economic development, and has sparked widespread debate among government, industry, and academia [30,31]. It is widely recognised that the opening of HSR reduces transportation costs and travel time [32], improves accessibility, accelerates the flow of economic factors, stimulates urban economic vitality, and promotes urban economic growth [33–35]. Sasaki et al. [36,37] found that HSR have positive effects on services and manufacturing: it accelerates the growth of tertiary industries and also improves the production efficiency of enterprises or industries, thereby promoting economic growth and development. Sahu and Verma found that productivity was increased by the introduction of HSR at an institutional or industrial level [38]. Moreover, HSR can indirectly promote economic growth by impacting on labour, industrial agglomeration, the stimulation of new consumption and employment, and other aspects [39,40].

Existing literature on the effect of HSR on regional economic development is relatively rich and, among them, some empirical studies have examined the impact of transportation infrastructure improvement on the HQED in cities [41]. Li and Wang [42] investigated the impact of the introduction of HSR on the growth quality of the Yangtze River Economic Belt. Hu et al. [43] discovered that the introduction of HSR or the improvement of access facilities would enhance the mutual effect of urbanisation and socioeconomic growth. Kong et al. [44] studied the influence of HSR on the quality of urban economic growth based on Chinese cities.

However, the combined impact of HSR on HQED, particularly in the YRD region, and its regional differences have been insufficiently studied. In addition, the existing studies have not yet reached a consistent conclusion; some indicate that HSR have no positive effects on the economy and may even have a negative effect [45]. The most noticeable effect is the 'metropolis effect' brought in by the introduction of HSR [46], which promotes business expansion in mega cities, creating a polarising effect, and reducing economic growth rates in the periphery, hence increasing externalisation [47]. An et al. also found that HSR enormously improves the overall connectivity of urban networks in the YRD, while it also aggravates regional economic disparities [34].

The YRD region has the densest HSR network and was among the first to launch HSR. At the same time, the region is one of the most economically dynamic regions in China. It has a pivotal strategic position in the country's modernisation, and it is regarded as a suitable experimental area for comprehensive research on geographical and sustainability science at the trans-regional scale [48].

This study focuses on the impact of HSR on the HQED in the YRD in China, and its contributions are as follows. First, it constructs a comprehensive impact mechanism of HSR on HQED from five dimensions, namely innovation, coordination, green, openness, and sharing. This can enrich the research literature on HQED in general, provide a new perspective for promoting HQED with a new economic norm, and also enlarge the research perspective on the HSR effect.

Second, based on the rich connotations of regional HQED and the present literature, we built an evaluation system which investigates both the input and output sides, and includes six aspects in the second level and 19 detailed indicators in the third level. Based on the evaluation system, we focus on specific cities within the YRD in China, instead of just eastern, central, and western China, as the research scope for a larger scale. We first calculate the HQED index for 41 cities in the YRD from 2011 to 2019, then explore the spatiotemporal evolution characteristics, and find that the HQED improves over time and takes on a distance attenuation law centred on Shanghai and Suzhou. These innovate the existing measurement methods for HQED and also optimise its applicability in practice.

Third, regarding HSR opening as an exogenous shock, we examine the impact of HSR on HQED in the YRD, and carry out heterogeneity tests in the YRD by using the difference-in-differences (DID) model, which effectively strips out the impact of the HSR on HQED. We find that HSR have an obviously different impact on HQED in cities with different locations and sizes. This enlarges the breadth of the present studies on the effects of HSR on HQED. Given that the HQED in the YRD has a benchmarking and leading role in China, it can provide experience for other regions in China, and even other countries, on aspects for amplifying the impact of HSR and promoting HQED.

3. Theoretical Framework and Research Hypotheses

Since the introduction of HQED, its connotations have gradually been refined by a number of scholars from different perspectives and emphases, according to different research motivations [11,14,15,21,23,23,28], yet a unified expression remains elusive. This study also tends to join the debate according to the five new development concepts of "innovation, coordination, green, openness, and sharing" [15]. Among them, innovative development focuses on solving the problem of development momentum and motivation, coordinated development emphasises solving the problem of unbalanced development, green development addresses solving the problem of harmony between human beings and nature, open development means solving the problem of internal and external linkage in the development, while shared development focuses on solving the problem of social justice [49]. In this study, HQED is regarded as a comprehensive development encompassing high economic efficiency, innovation driven, coordination among regions (including urban and rural, developed and underdeveloped areas), harmony between human beings and nature, positive interaction between internal and external linkage, social justice, and sharing (all people have access to education, health care and other rights, sharing the fruits of social and economic development).

Based on previous studies, this study builds the impact mechanism of HSR on HQED from the five new development concepts and the nature and role of HSR. Compared with other transportation, the HSR has many advantages, including higher speed, larger transportation capacity, better safety, more comfort and convenience, lower energy consumption, and better economic benefits. These advantages reduce the travel time and perceived psychological distance between cities, boost the accessibility of cities and, consequently, hasten the flow of labour, information, knowledge, technology, and capital [50,51]. All economic entities are able to allocate the factors of production on a larger scale, thereby increasing the allocation efficiency of resources. Through the effects of agglomeration and knowledge spillovers, the compression effect of time and space [1] caused by the improvement in accessibility will have a positive impact on regional HQED. The rapid development of HSR leads the factors to flow to fit the market, eliminating geographical constraints and accelerating industry renewal. Simultaneously, the development of HSR will induce changes in the productive efficiency of production organisation [38]. It will also accelerate the transformation and updating of industrial structures and the enhancement of research and development expenditures (R&D) and human capital. This will enhance the quality of economic development in multiple ways. Figure 1 presents the detailed impact mechanism of HSR on the five aspects of HQED.

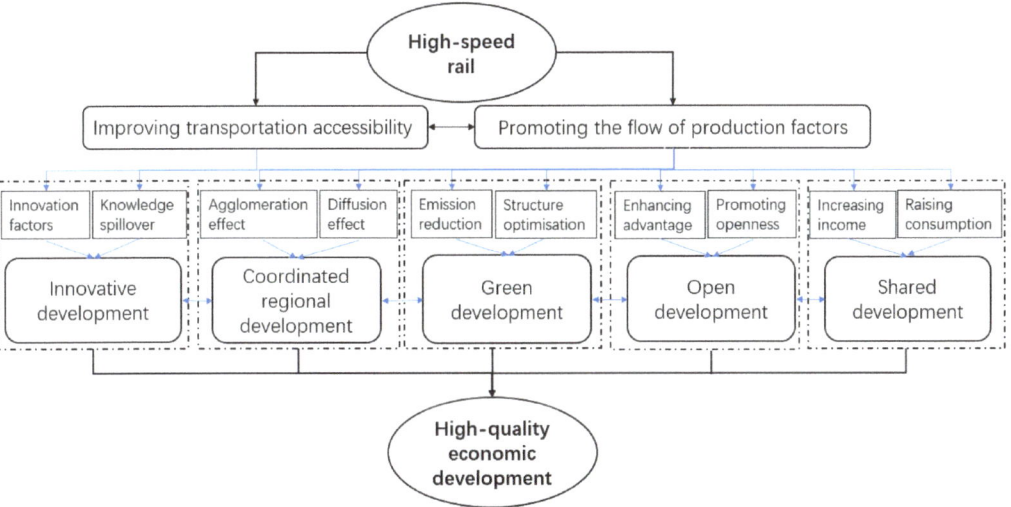

Figure 1. Impact mechanism framework for HSR on regional HQED.

3.1. The Impact of HSR on Innovative Development

The opening of HSR improves regional accessibility and boosts innovative development [52] in two ways. First, it improves the flow efficiency of innovation factors. Second, it promotes knowledge spillovers and enhances knowledge externalities, ultimately enhancing innovative regional development.

HSR will accelerate the concentration of innovation factors, comprised of labour and capital [40], to cities connected by the HSR. This is an important path to promote regional innovation development. HSR attracts more highly educated and skilled talent, along with high-tech industry employees, to flow into enterprises in cities along the HSR. When combined with the improvement in the information environment, these factors promote the innovation in enterprises and promote the cities' innovation [52]. The flow of innovation factors triggered by the HSR may also show a 'syphon effect', which can suppress innovation in peripheral cities. The opening of the HSR saves transportation costs, accelerates the dissemination of elements and technologies, and breaks down regional borders. This process shortens distances between regions and fosters cross-regional collaborative innovation [53], thereby enhancing face-to-face communication and promoting the tacit knowledge spillover among regions, cities, and enterprises along the HSR. Consequently, this increases the patent output and innovation ability of enterprises, further promotes the innovation performance of enterprises and industries, and ultimately regional innovation levels [54].

In general, HSR can effectively improve the spillover of technological innovation and accelerate the high-quality innovation and development of cities. This is realised by promoting agglomeration economy spillovers, learning activities among enterprises, and the flow of human resources and capital, as well as trade and cooperation [55]. Additionally, Hanley et al. also found that HSR increased innovation collaboration between enterprises at the city level [56].

3.2. The Impact of HSR on Regional Coordinated Development

Faber found that HSR links lead to a drop in GDP growth in outlying counties because industrial output dramatically falls in these areas in China. Faber found that transportation connected the central and surrounding cities and reduced trade costs by strengthening communication and exchanges between regions [45], which contributes to coordinated development. HSR significantly improves cities' accessibility, thereby reducing the time and economic costs for transregional economic activities. It enables various production factors to flow more conveniently among regions. The factors gather or diffuse in different cities and regions, trigger the redistribution of economic activities, and affect the coordinated development in the regions. The economic redistribution caused by HSR is manifested in two opposite effects: agglomeration and diffusion. A larger agglomeration effect enlarges the regional development gap, and large cities continue to 'syphon' resources, as well as development opportunities, from small cities. In contrast, a larger diffusion effect narrows the gap between regions and promotes the process of coordinated regional development.

Researchers have made different discoveries on the actual impact of HSR on regional coordinated development. Some found that HSR increases the aggregation effect because HSR induces resources to accumulate in central cities [57]; thus, HSR gradually enlarges regional differences. Other scholars believe that HSR enhances the location advantages of surrounding small cities, causing spillover effects from large cities to small cities, thus narrowing regional differences. Yao et al. found that HSR have consistently reduced interregional and urban–rural inequalities by fostering spillovers from larger urban areas to smaller peripheral areas [58]. Jiang and Kim found that HSR contribute to realise the convergence objectives in China and Korea [59], and gradually promote coordinated regional development.

The agglomeration effect and diffusion effect induced by HSR exist simultaneously. Under different temporal and spatial horizons, the two effects show the tendency for one to be stronger and the other weaker. But, in the long run, the spatiotemporal compression effect provides more opportunities for less developed regions and improves coordinated regional development. Just as Jin and Wang [60] found that during the formation period for the main lines, the agglomeration effect is the main force and strengthens the imbalance in the regional development, while during the branch line construction period, the outflow of factors from developed cities becomes the main trend, which benefits the balance of the regional development. Zhang et al. also proved that HSR improve the inequity at the national and also help to promote provincial economic equity in China [61].

3.3. Impact of HSR on Green Development

The impact of HSR on green development is mainly achieved through carbon reduction and industrial structure optimisation, because HSR have the characteristics of relatively low energy consumption, high cleanliness, green environmental protection, high efficiency, and comfort. Compared to other traditional ground transportation, HSR obtain its driving force from electricity, which produces less pollution than traditional transportation, thereby reducing energy consumption and vehicle exhaust emissions, air pollutants, and industrial waste [62]. For example, the concentration of PM2.5 in air could decrease about 2.8% after the county connects to the HSR network [63]. Zhang et al. found that HSR are more beneficial to green development and greatly facilitate GTFP (green total factor productivity) by diminishing energy consumption, environmental pollution, and improving technological innovation [64]. For short- and medium-distance travel activities, HSR travel has greatly replaced the use of private cars, coaches, and traditional trains.

The introduction of HSR is conducive of the continuous optimisation of the industrial structure [65], which is not only one of the important ways to reduce environmental pollution and ecological damage, but also one of the cores of HQED. HSR are sensitive to the degree of convenience for passenger transportation in the service industry [66]. HSR have a critical role in accelerating employment and the number of consumers within higher service industries and can also greatly enhance the agglomeration of higher service industries in central cities, and can effectively optimise and upgrade the industrial structure [67]. Since the construction and opening of a large scale HSR networks in China, the employment rate in the service industry has significantly increased and the structure of the industry has shifted from labour intensive to capital intensive.

Of course, the construction of HSR requires intensive consumption of pollutant-intensive resources. In exploring the impact of the opening of HSR on the five dimensions of HQED, Liu et al. found that the coefficient of the impact of HSR on green development is statistically insignificant in both the long and short term [41].

3.4. Impact of HSR on Open Development

The improvement to transportation will decrease the cost of cargo transportation and the cost of time and inventory, allowing enterprises to obtain the price advantage from export products [68]. This can also promote and improve the structure and quality of regional export products. Although HSR are mainly used for passengers, these conclusions cannot simply be transplanted to the analysis on HSR, while HSR can have a positive impact on trade and exports through at least two channels. First, by improving regional accessibility, it facilitates communication between people and corporates, and reduces enterprises' costs for searching for information, communication, outsourcing, and other marginal spending. This enables enterprises to strengthen their ties with outside areas by enhancing matching efficiency with suppliers, thereby improving the performance of enterprises along the route [69,70]. Consequently, it enhances the international competitive advantage of products. Zhou et al. also found that HSR can decrease information barriers and costs to enter the international market and promote the export growth of agriculture-related companies around HSR stations [71]. Charnoz et al. also found that HSR improve commuting efficiency between regions and the exchange of information in French between corporate headquarters and branches [32]. Cosar and Demir discovered that the improvement to domestic transportation has reduced access barriers to international markets, which can help a country take part in the international supply chain, boost international competitive advantages and, hence, contribute to the country's international trade [72].

Second, the substitution effect on traditional transportation releases transportation resources for freight transportation. It increases the efficiency of freight transportation and the availability of cargo space, thereby reducing the cost of cargo transportation and enhancing the competitiveness of products for enterprises. HSR also make it easier for manufacturers to find more suitable suppliers to maximise their profits, which, in turn, significantly promotes exports [73]. Additionally, HSR will significantly promote the trade opening of Chinese cities through the tariff transmission mechanism [74]. Overall, the opening of HSR can promote competitive advantages, increase the scale and quality of exports [71,75], and further boost open development.

3.5. Impact of HSR on Shared Development

HSR meet the needs of the public for fast travel because it reduces travel time and increases the public's opportunities to travel. The introduction of HSR also affects two detailed aspects of shared development, income, and consumption. According to Gil-Pareja et al. [76], HSR have both direct and indirect impacts on the income of urban residents along the railway lines. Direct impacts include increasing investment and business income, creating more employment opportunities, and reducing the time cost of tourist travel. Indirect impacts are mainly reflected in employment and income aspects through resource reallocation, which is caused by changes in high-speed transportation. Researchers have found that HSR can reduce income inequality in China [77] and Italy [78].

From the perspective of general consumption, HSR promote consumption by reducing circulation costs and commodity prices. The opening of HSR generally contributes to the growth in consumption expenditure and the upgrading of consumption structures in cities along the railway line [44,79]. HSR also push consumption-related industries to decrease their prices for market competition [80]. Of course, the promotional impact of HSR on consumption varies depending on the station location and the scale of the city [81]. At the same time, by compressing the spatiotemporal distance, some residents' working, living, and consumption needs across cities are met. In total, HSR enlarge the overall consumption by expanding the scope of consumption [82].

It is worth noting that the above five dimensions are not absolutely isolated, rather, they are interrelated and reinforced in some ways. From a horizontal perspective, among the five dimensions, innovative development is the driving force of HQED [49]. It plays an important role in the development of the other four dimensions and, conversely, the other dimensions also rely on the driving force of innovation. Coordinated development focuses on the coordinated development of urban/rural areas and developed/developing regions from a spatial perspective, and shared development aims to include people in sharing the fruits of development and, ultimately, realising common prosperity [8]. These two dimensions are highly complementary to each other, and are also the ultimate goals of HQED, leading the development of the other three dimensions. Green development is reflected in all aspects of economic growth and social development [8], which is both the inheritance of sustainable development [6] and the concrete embodiment of sustainable development in China. This dimension runs through the other four dimensions and is also an important symbol of China's economic development, shifting from the pursuit of speed to the pursuit of quality [23]. Open development aims to foster extensive engagement in the global industrial division of labour and cooperation, while upholding a diverse and stable international economic pattern and economic and trade relations. It serves as a key guarantee for achieving innovative, coordinated, green, and shared development. Taking the interactive relationship between innovative development and green development as an example, the opening of HSR accelerates the flow of talent and knowledge within wider regions, and then improves the level of regional innovation and human capital. This further promotes the improvement of pollution control technology and the progress of green and environmental protection production technology, and it realises the suppression effect of HSR on air pollution. Meanwhile, the innovative development caused by HSR promotes the transformation and upgrading of industries, which, in turn, contributes to energy conservation and pollution emissions reduction and further accelerates green development and sustainable development. Similar interactions among other dimensions exist. According to this theoretical analysis on the mechanism, the study puts forward two research hypotheses, as follows:

Hypothesis 1. *The opening of HSR has a significant positive impact on regional high-quality economic development.*

Hypothesis 2. *There are significant regional differences in the impact of HSR on regional high-quality economic development.*

4. Research Design

4.1. Study Area

The study area, the Yangtze River Delta (YRD), is located in the lower reaches of the Yangtze River in China, including four provincial-level administrative units, those of the Shanghai municipality, Jiangsu province, Zhejiang province, and Anhui province, containing forty-one prefectural-level cities, covering an area of 358,000 square km^2. The YRD is one of the most economically developed and innovative regions in China. By the end of 2022, this area had a permanent population of 231 million and a total GDP of CNY 29.03 trillion. With an area less than 4% of the whole country, this region produces nearly one-fourth of the total economic output and one-third of the total imports and exports in China. The YRD is the only super-large urban agglomeration in China that ranks among the top six in the world. It plays a pioneering and exemplary role in HQED in China. At the same time, the YRD has the highest density of HSR networks in China with over 13,749.7km of railway, of which more than 6700 km and nearly 30 routes (Figure 2) are HSR [2]. Thus, the YRD has become a model region for exploring the impact of HSR on HQED.

4.2. Methodology

4.2.1. Model Construction

The method of difference-in-differences (DID) can validly control the endogenous problems in policy variables and acquire an unbiased net effect for policy evaluation [67], and it has been used to examine the effect of HSR between an HSR city and a non-HSR city [83]. This study identifies cities that have implemented HSR as the treatment group and those that have not implemented HSR as the control group, based on panel data collected from 41 prefecture-level cities in the YRD during the period of 2011–2019. The grouping dummy variable D_{it} is used as a proxy: $D_{it} = 1$ means that city i launched HSR in year t, and $D_{it} = 0$ means that city i did not launch HSR in year t. The staging dummy variable T_{it} is also set, $T_{it} = 1$ denotes the period subsequent to the commencement of the HSR, whereas $T_{it} = 0$ denotes the period preceding the commencement of the HSR. The focus of the study is on the interaction term $D_{it} \times T_{it}$, which indicates the actual impact of the HSR opening on the level of the quality of the economic development in the city. Since the time of opening of the HSR varies across the prefectures, this paper replaces the interaction term $D_{it} \times T_{it}$ with the variable $treat_{it}$, which indicates whether city i was affected by the shock of opening the HSR in year t. If city i opened the HSR in year t, its corresponding $treat_{it}$ value is 1, otherwise $treat_{it} = 0$. Based on the theoretical analysis and Liu et al. [39], we constructed the following multi-period DID model.

$$Hq_{it} = \alpha + \beta treat_{it} + \gamma X_{it} + \mu_i + \tau_t + \varepsilon_{it} \tag{1}$$

In Equation (1), the explanatory variable Hq_{it} denotes the level of HQED in city i in year t, $treat_{it}$ is the product of $D_{it} \times T_{it}$. X_{it} represents several other control variables that have an impact on the HQED in the city, μ_i is an individual fixed effect, τ_t is the time fixed effect, and ε_{it} is the random disturbance term. The coefficient β in Equation (1) is the focus in this study, which indicates the cumulative impact of the inauguration of the HSR on the HQED in the city. β is greater than 0 if the opening of the HSR boosts the HQED in the city, otherwise it is less than 0. In order to estimate the degree of impact of the opening of the HSR on the HQED, this study adds five control variables X, namely the level of economic development, openness, financial level, industrial advancement, living standard, and innovation; where the level of economic development (lrgdp) is represented as the gross domestic product per capita in logarithmic form [24], the openness (lforeign) is characterised by taking the (log) actual amount of foreign capital spent in the year from Xu and Pan [84], the expression of the financial level (lfin) is derived from the (log) year-end deposit balance of the financial institutions based on Li and Wang [42], and the industry advancement (ind) is measured by the Theil index from [28], the living standard (lwage) is represented by the (log) average employee salary from [34], and the innovation level (linn) is expressed as the logarithm of the number of patent applications from [14].

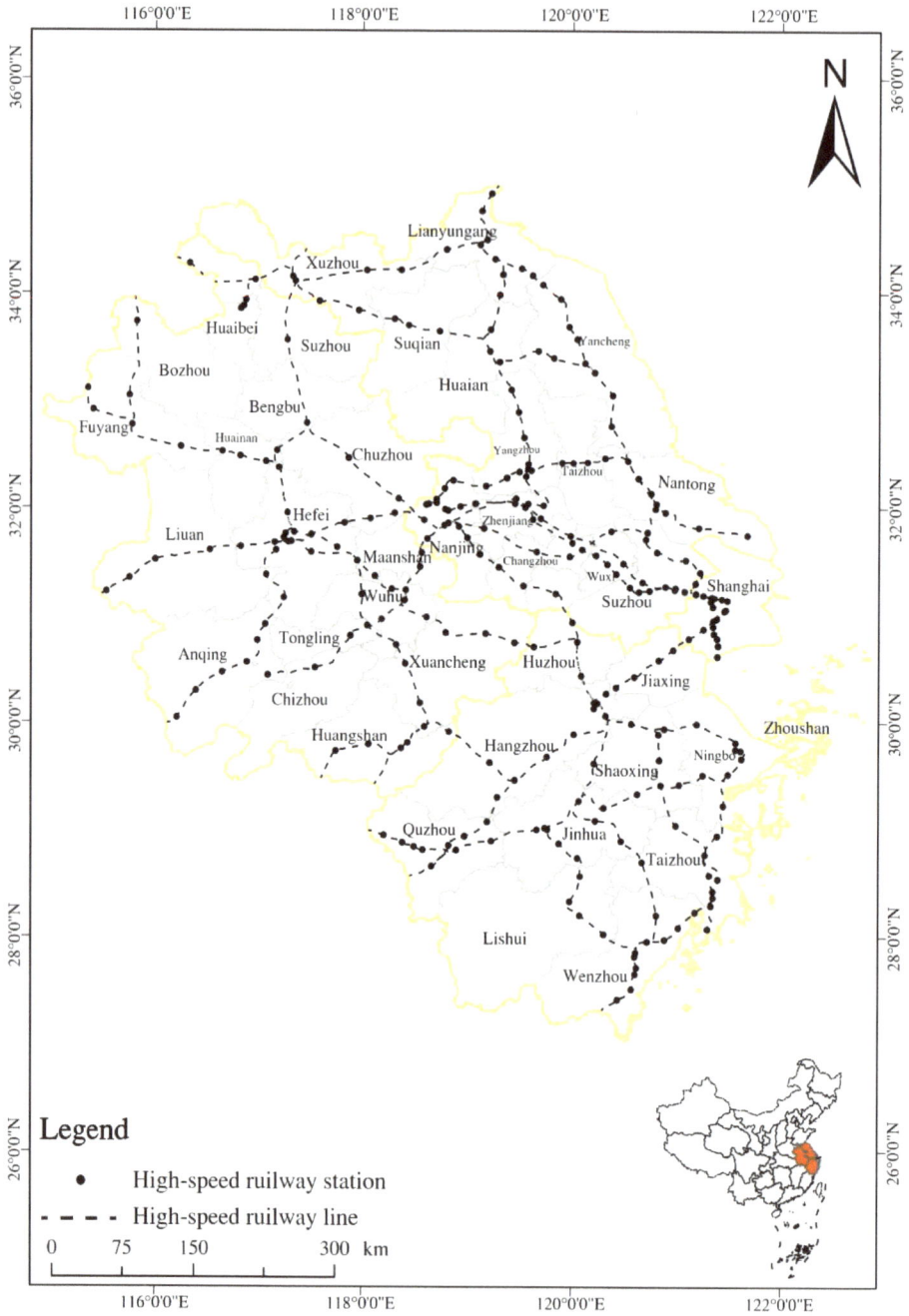

Figure 2. The distribution of HSR in the Yangtze River Delta in 2022. Source: The website of the National Railway Administration and the Special Issue on Railway Passenger and Cargo Transport.

The reasons for selecting these variables are: (1) the level of economic development is one of the most important macroeconomic indicators that objectively reflects the dynamics of economic and social development, and "GDP per capita" measures the main material basis for the per capita income and living standard of a country's residents. (2) Improving the level of openness and constructing a new open economic system at a higher level can promote HQED in the YRD, and because the YRD's export trade volume and outward investment rank among the top in the country, it is necessary to include openness in the control variables. (3) The financial level reflects the marginal productivity of capital, which is an important manifestation of the optimal allocation of capital. Increasing financial support in key areas can consolidate and strengthen the foundation of the real economy. (4) Industrial advancement has a "structural dividend" on economic growth, which helps to improve the green total factor productivity growth rate and is, overall, conducive to promoting high-quality development in the YRD. The high-quality development of the economy needs to satisfy the growing needs of the people. (5) HQED is supposed to meet the people's growing needs for a better life, and it is necessary to include the living standard in the control variables. The "average worker's wage" is a suitable indicator of the living standard. (6) The HQED of the YRD economy needs to realise the shift from the primary factors of production to scientific and technological innovation. As the first driving force for development, innovation is the engine of HQED.

4.2.2. Data Sources and Data Processing

As mentioned in the literature review, HQED has rich connotations, and researchers believe that more attention should be paid to improving the economic structure, including the industrial structure, consumption structure, import and export structure, and openness [85]. The measurement index should cover various aspects, and many scholars have built index systems including three, five, or six aspects [15,24,27,28,65]. For example, Kong et al. constructed an index system including three dimensions of efficiency, stability, and sustainability [27]. Li et al. built an index system including the aspects of the economic growth structure and economic efficiency, the ecological environment, and social coordination [20]. Guo and Sun set up an evaluation index including 6 primary indexes, namely economic growth, innovation, coordination, green, openness, and sharing [21].

Based on the previous literature, this study further expands the index system for HQED, mainly originating from the five dimensions of HQED, also considering factor input efficiency. Besides, we also try to cover the sustainable development goals [6] proposed by the United Nations that are closely related to the reality in China. Of course, when determining specific indicators, we fully consider the availability of the data. Table 1 shows the specific evaluation system, which includes three levels, with 2 dimensions in level 1, namely the input index and output index, 6 indexes in level 2, namely the factor input and innovative development, coordinate development, green development, open development, and shared development, and 19 indicators in level 3 (see Table 1).

According to Li and Li [20], Guo et al. [28], Liu et al. [41], and Kong et al. [44], the entropy value method is applicable to comprehensively evaluate the HQED in the YRD in this study. After standardising the data, calculating the weight of the samples for each indicator, the entropy value and coefficient of variation for each indicator are derived. After calculating the weights of each indicator in Table 1, the comprehensive score for the HQED is further estimated according to the weights of each indicator and standardised statistics.

The original data are from the China City Statistical Yearbook, the China Statistical Yearbook, the yearbooks of each province, and the website of the National Bureau of Statistics. The opening information on HSR (Figure 2) comes from the State Railway Administration and the Special Issue on Railway Passenger and Freight Transportation (available on request).

Table 1. Index system for regional HQED.

Level 1 Index	Level 2 Index	Level 3 Index		Attribute
		Specific Measurement Index	Index Interpretation and Index Unit	
Input index	Factor input efficiency	Labor input	Total number of employees in the whole society/GDP	Positive
		Capital input	Total fixed assets investment/GDP	Positive
		Government investment	Government expenditure/GDP	Positive
Output index	Innovative development	R&D investment intensity	R&D expenditure/GDP (%)	Positive
		Number of invention patents owned by 10,000 people	Authorised number of invention patents/resident population at the end of the year (pieces/10,000 people)	Positive
		Science and technology investment intensity	Fiscal expenditure on science/total fiscal expenditure (%)	Positive
	Coordinated development	Urban–rural income ratio	Urban per capita disposable income/rural per capita disposable income (%)	Negative
		Coordinated pressure index	Unemployment rate + CPI (%)	Negative
	Green development	Green coverage rate of built-up area	Green coverage rate of built-up area (%)	Positive
		Harmless treatment rate of domestic garbage	Harmless treatment rate of domestic garbage (%)	Positive
		Sewage treatment rate	Sewage treatment quantity/total sewage quantity (%)	Positive
		Energy consumption per unit, GDP	Total energy consumption/GDP (tonne/CNY 10,000)	Negative
		Industrial wastewater discharge per unit, GDP	Industrial wastewater discharge/GDP (tonne/CNY 10,000)	Negative
	Open development	Openness to foreign capital	Actual utilisation of foreign investment/GDP	Positive
		Foreign trade dependence degree	Total import and export/GDP	Positive
	Shared development	Per capita disposable income	Per capita disposable income (CNY 10,000)	Positive
		Per capita consumption expenditure	Per capita consumption expenditure (CNY 10,000)	Positive
		Per capita expenditure on education and culture	Per capita expenditure on education and culture (CNY 10,000)	Positive
		Intensity of medical and health investment	Medical and health expenditure of local finance/general finance budget expenditure (%)	Positive

Notes: Among the 17 goals on sustainable development, number 8 and number 16 relate to decent work and economic growth, and institutional justice, respectively, part of 8 and the government function corresponding to the factor input efficiency. Number 9 relates to industrial technology innovation, part of 8 and 9 corresponding to innovative development. Number 10 relates to reducing inequality, so part of 8 and 10 corresponding to coordinated development. Number 6 refers to clean drinking water, 7 refers to clean energy, 11 refers to sustainable communities, 12 refers to sustainable supply, ensuring responsible production and consumption, 13 refers to climate action, 15 refers to the protection of ecosystems, all these corresponding to green development. Number 17 relates to the partnership for implementation and revitalisation of global sustainable development, corresponding to open development. Number 3 is about health and well-being, 4 is about quality education, these two goals and part of 10 correspond to shared development.

Based on the above methods, we calculated all the HQED indexes for the 41 prefecture-level cities in the YRD region from 2011 to 2019 and selected 2011, 2015, and 2019 as the years to analyse the spatiotemporal evolution characteristics (Figures 3–5).

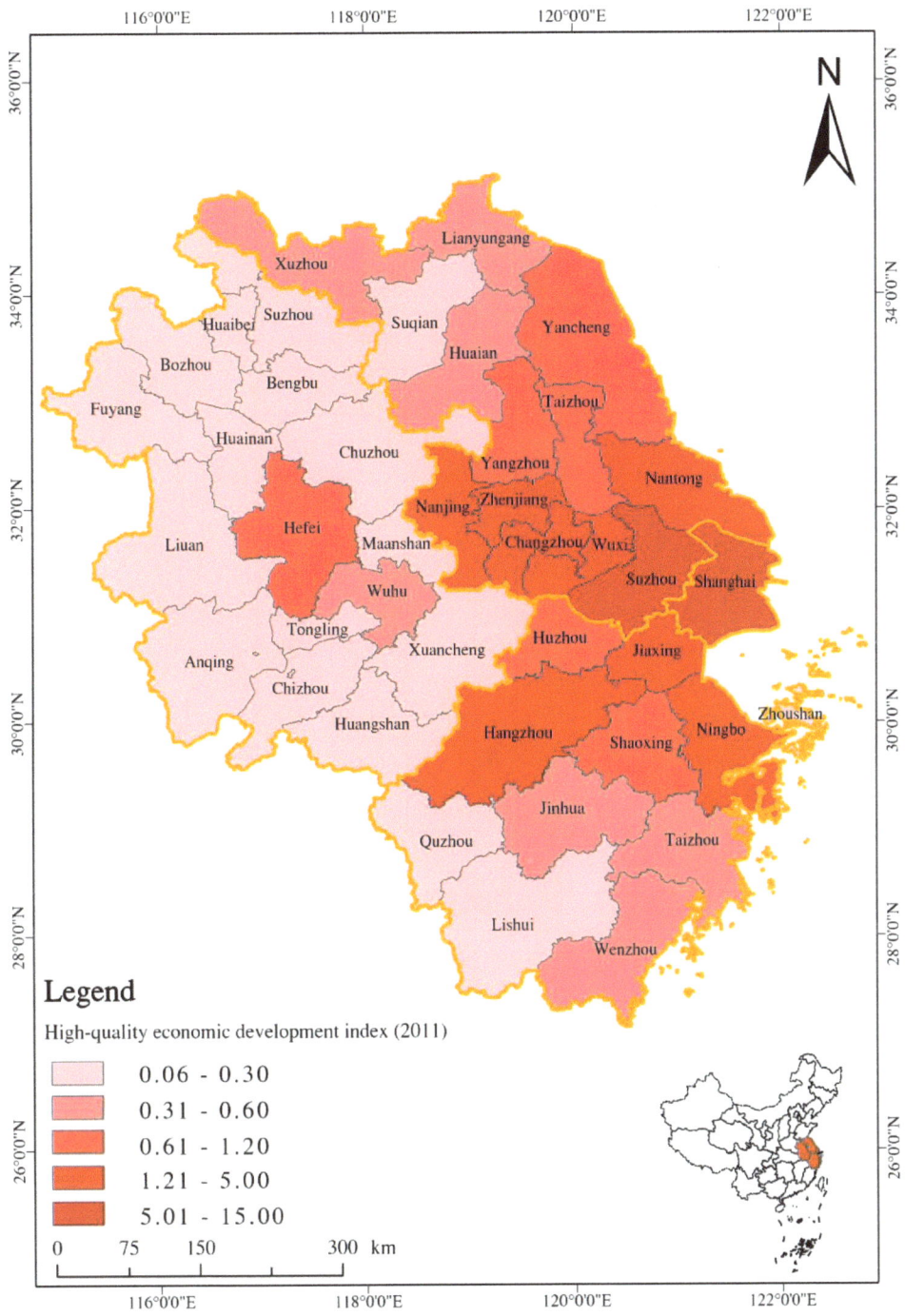

Figure 3. Distribution of the HQED index in the YRD (2011).

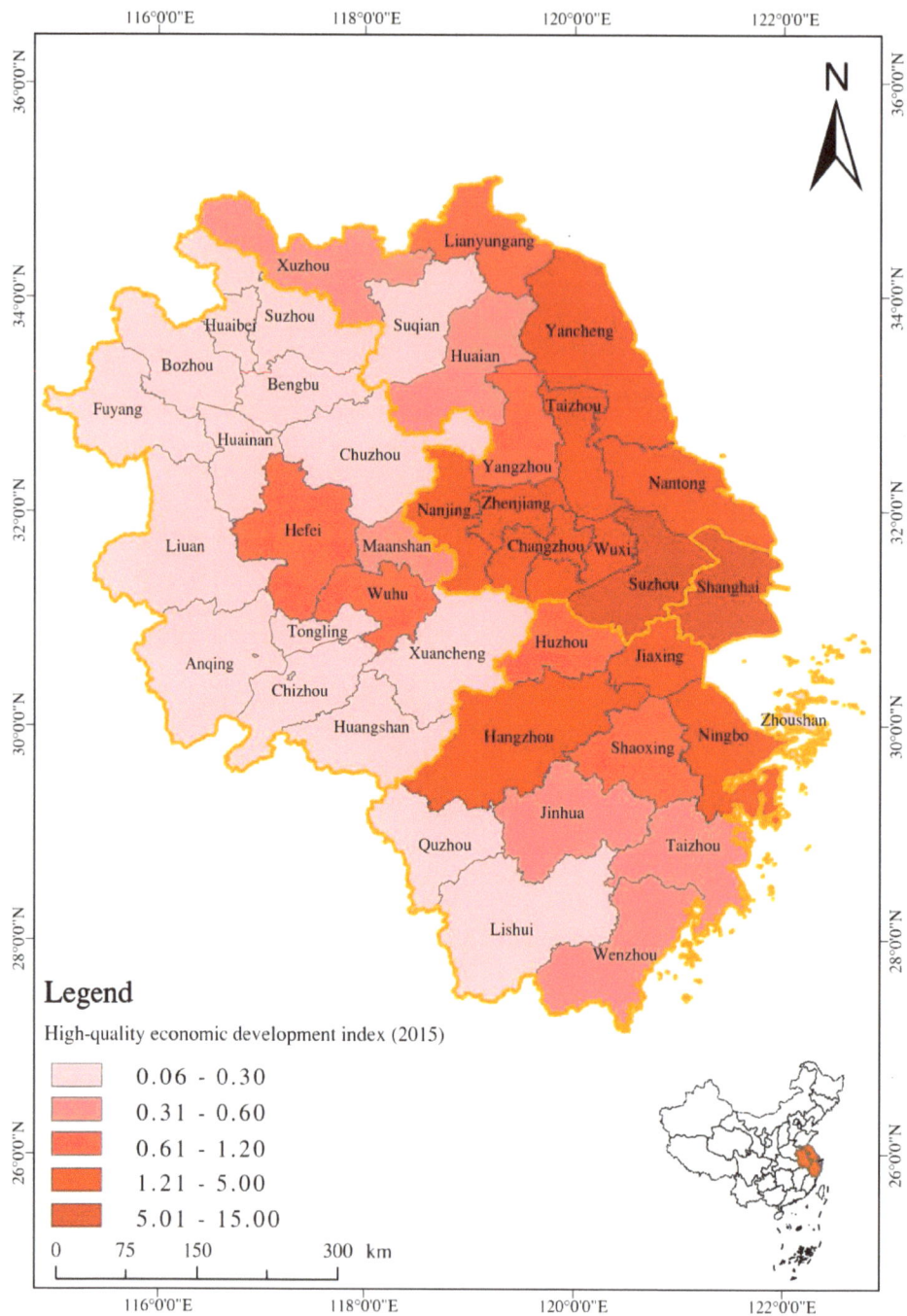

Figure 4. Distribution of the HQED index in the YRD (2015).

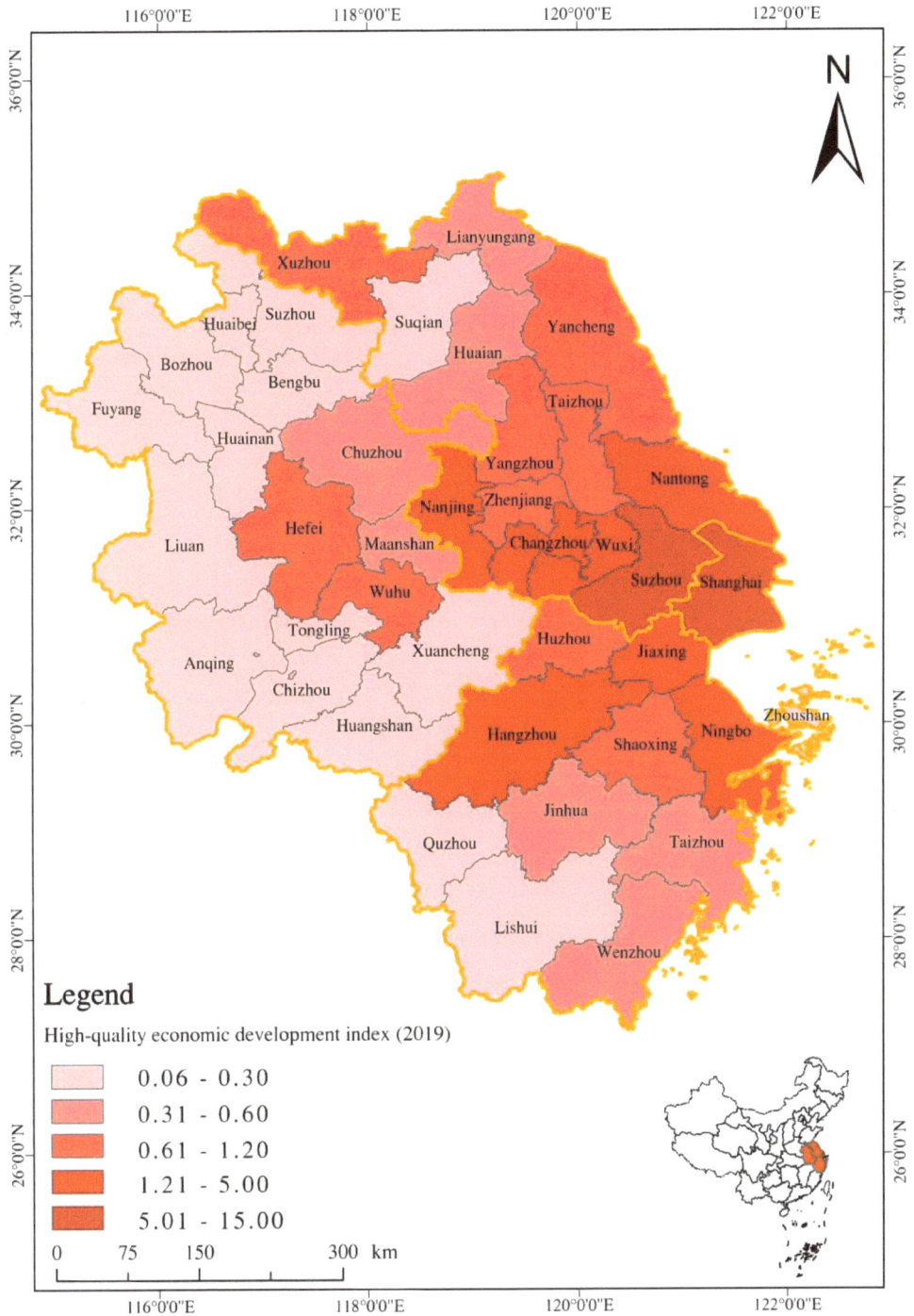

Figure 5. Distribution of the HQED index in the YRD (2019).

As shown in Figures 3–5, the HQED index for the cities in the YRD are much higher than that of the central and western regions, but the regional differences within the YRD are obvious. From 2011 to 2019, there is a significant trend of radiating down to the north, west, and south, with Shanghai and Suzhou at the centre. The two cities are always the first tier, and the second and third tiers are mainly concentrated in the areas close to Shanghai and Suzhou, in southern Jiangsu province and in eastern and northern Zhejiang province and Hefei in Anhui province. Except for Xuzhou, the cities in the peripheral areas of the YRD are in the fourth and fifth tiers. From 2011 to 2019, the HQED index for cities in Anhui province significantly improved because these cities are close to Nanjing in Jiangsu province. They are Wuhu, jumping from the fourth to the third tier, and Chuzhou and Maanshan, jumping from the fifth to the fourth tier. The rate of quality improvement in these cities is significantly higher than some similar cities in Jiangsu and Zhejiang province that are closer to Shanghai. All the other cities in the YRD basically maintained their original positions. This indicates that Shanghai has successfully driven quality improvement in the surrounding regions, which, in turn, has further driven quality improvement in the surrounding areas, showing a clear principle of distance decay. This distance attenuation law also presents a wave-like feature.

The other control variables in this study are also derived from the aforementioned databases, and the descriptive statistics for all the data are shown in Table 2.

Table 2. Descriptive statistics for the variables.

Variable	Observation Number	Mean	Standard Deviation	Minimum	Maximum
Hq	369	1.407	2.759	0.034	13.901
treat	369	0.629	0.484	0	1
lrgdp	369	1.763	0.604	0.009	2.991
lfin	369	1.31	1.111	−0.831	4.815
lforeign	369	2.165	1.226	−0.814	5.25
ind	369	6.668	0.311	5.863	7.488
lwage	369	1.815	0.297	1.011	2.774
linn	369	3.759	1.272	0.572	6.541

5. Empirical Results

5.1. Regression Results

Table 3 shows the regression results for Equation (1). Column (1) presents the regression result without adding the control variables; the coefficient of variable *treat* is 0.104 and it is significant. Columns (2) to (7) are the regression results adding in order the six control variables, namely the economic development level, openness, financial level, industrial advancement, living standard, and innovation.

The results show that the coefficients of the variable *treat* are statistically significant, with coefficient values between 0.092 and 0.123. The results indicate that the opening of a HSR has a significant positive contribution to HQED in cities in the YRD, which has led to an increase in the HQED in cities with a HSR station from 0.092 to 0.123. Comparing the regression results in columns (2)–(7), it is found that the coefficient of influence tends to be first upwards and then downwards, which indicates that it is necessary to include the control variables in Equation (1). The coefficients of the control variables are significantly positive except for *lwage and linn*, which means that the level of economic development, openness, the financial level, and industrial advancement are also conducive of HQED; this result matches with our expectations and the mechanism explained earlier. Thus, Hypothesis 1 is confirmed. It is consistent with the mainstream view [13,44] that the opening of HSR stimulates economic vitality and promotes the urban economy.

Table 3. Empirical results on the impact of HSR opening on HQED in the YRD.

	(1) FE1	(2) FE2	(3) FE3	(4) FE4	(5) FE5	(6) FE6	(7) FE7
treat	0.104 **	0.121 ***	0.120 ***	0.123 ***	0.100 **	0.100 **	0.092 **
	(4.552)	(4.475)	(4.451)	(4.395)	(4.505)	(4.488)	(4.535)
lrgdp		0.414 ***	0.416 ***	0.339 ***	0.243 **	0.280 **	0.267 **
		(10.67)	(10.61)	(10.77)	(11.64)	(11.76)	(11.79)
lforeign			0.023 **	0.020 *	0.021 **	0.025 **	0.024 **
			(1.080)	(1.070)	(1.065)	(1.079)	(1.080)
lfin				0.482 ***	0.393 **	0.373 **	0.327 **
				(15.73)	(16.20)	(16.17)	(16.57)
ind					0.451 **	0.485 **	0.458 **
					(21.40)	(21.39)	(21.48)
lwage						−0.418 *	−0.442 *
						(22.57)	(22.63)
linn							0.0629
							(5.000)
_cons	1.358 ***	0.761 ***	0.709 ***	0.420 **	−2.299 *	−1.957	−1.869
	(3.923)	(15.86)	(15.97)	(18.38)	(130.2)	(131.0)	(131.0)
N	369	369	369	369	369	369	369
R^2	0.083	0.124	0.137	0.161	0.173	0.182	0.186

Notes: *, **, *** indicate statistical significance at the 10%, 5%, and 1% levels; the values in parentheses are robust standard deviations.

The HSR opening drives the concentration of human and material capital in cities, promotes the optimal allocation of production factors, and indirectly promotes the HQED in cities by leading consumption and creating jobs. The level of innovation is expected to have a positive effect on high-quality development, but the results are not significant. This may be due to the relatively singular way of measuring the innovation level or the existence of differences in the innovation levels across cities, which may smooth out the influence of innovation on the HQED in the regression. Considering that innovation is indeed an indispensable factor affecting HQED, it is retained in the model.

Following this, we further explored the impact of the opening of HSR on the five dimensions of HQED in the YRD individually. Taking innovative, coordinated, openness, green, and shared development as dependent variables, we conducted regression analysis using Equation (1). The results are shown in Table 4. It was found that the opening of HSR have a significant positive effect on innovation and openness development in the YRD, with impact coefficients of 0.022 and 0.057, respectively. The impact coefficients on coordinated, green, and shared development are relatively small, while still positive. The results indicate that the opening of HSR greatly accelerates the level of R&D innovation and external openness in the YRD, and improves the coordinated, green, and shared development of cities to some extent, too.

5.2. Robustness Tests

5.2.1. Parallel Trend Test

Assessing the effects of policy implementation by using DID is subject to the basic assumption of parallel trends. The DID estimation results are considered credible if the experimental and control groups exhibit a similar trend prior to the commencement of the HSR and exhibit no significant differences, that is, the regression coefficient of *treat* before the opening time of the HSR is not significant, while the coefficient of *treat* after the opening time of the HSR is significantly different from zero. Therefore, the study defined dummy variables for 9 years to re-estimate the coefficients, namely *Pre-4*, *Pre-3*, *Pre-2*, *Current*, *Post-1*, *Post-2*, *Post-3*, *Post-4*, and *Post-5*, denoting the period spanning from four years prior to five years subsequent to the opening of the HSR.

Table 4. Empirical results on the impact of HSR opening on five dimensions of HQED in the YRD.

	(1) Innovative Development	(2) Coordinated Development	(3) Green Development	(4) Open Development	(5) Shared Development
treat	0.022 **	0.005 **	0.001 **	0.057 **	0.006 **
	(0.011)	(0.002)	(0.001)	(0.028)	(0.003)
lrgdp	0.065 **	0.013 **	0.004 **	0.167 **	0.018 **
	(0.029)	(0.006)	(0.002)	(0.074)	(0.008)
lforeign	0.006 **	0.001 **	0.0003 **	0.015 **	0.002 **
	(0.003)	(0.001)	(0.0001)	(0.007)	(0.001)
lfin	0.079 **	0.016 **	0.004 **	0.205 **	0.022 **
	(0.040)	(0.008)	(0.002)	(0.104)	(0.011)
ind	0.111 **	0.023 **	0.006 **	0.287 **	0.031 **
	(0.052)	(0.011)	(0.002)	(0.134)	(0.015)
lwage	−0.107 *	−0.022 *	−0.006 *	−0.277 *	−0.030 *
	(0.055)	(0.011)	(0.003)	(0.142)	(0.016)
linn	0.015	0.003	0.001	0.039	0.004
	(0.012)	(0.002)	(0.001)	(0.031)	(0.003)
_cons	−0.453	−0.093	−0.025	−1.171	−0.128
	(0.318)	(0.065)	(0.017)	(0.821)	(0.090)
N	369	369	369	369	369
R^2	0.186	0.186	0.186	0.186	0.186

Notes: *, ** indicate statistical significance at the 10%, 5% levels.

As seen in Figure 6, the coefficients of the dummy variables in the four periods before the policy shock are not significantly different from zero, indicating that the control and treatment groups have similar development trends; the coefficients are significantly higher than zero after the policy shock occurs, and they perform particularly obviously in the first to fourth periods. The test results confirm that the model in this study can reasonably reflect the policy effects.

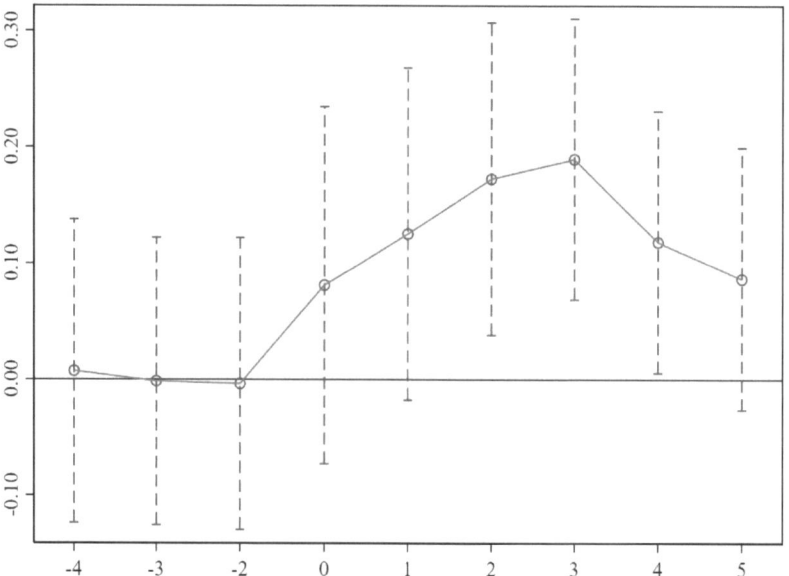

Figure 6. Parallel trend test.

5.2.2. Placebo Test

If the HQED in the city is caused by factors other than the HSR, then assume that the city does not have a HSR in the year of its opening, and the opening of the HSR will also bring about significant results. Otherwise, it can be assumed that the improvement in HQED comes from the effect of HSR shocks. For this reason, this study used a counterfactual approach to conduct placebo tests for both the region and time, that is, a random sample period was selected for each sample subject as its policy time. In the multi-period DID, each of the 41 cities in the YRD was randomly selected from 2011 to 2019 as the time when the policy was implemented, which was used as the basis for the grouping and staging of the 'experiment' of HSR opening. The evaluation of the robustness of the benchmark regression was judged by the significance and distribution of the double difference term estimates.

The red curve in Figure 7 is the kernel density distribution of the estimated coefficients, and the orange dots are the p-values corresponding to the estimated coefficients. It is evident from Figure 7 that the coefficients of the randomised DID terms are concentrated around zero, and most of the p-values are higher than 0.1. Randomly advancing the opening time of the HSR in each city leads to a significant weakening of the driving effect of the HSR opening on the HQED in terms of the significance and strength of the effect. This also confirms from the counterfactual perspective that the HSR does improve the HQED in the city, and Hypothesis 1 is further confirmed.

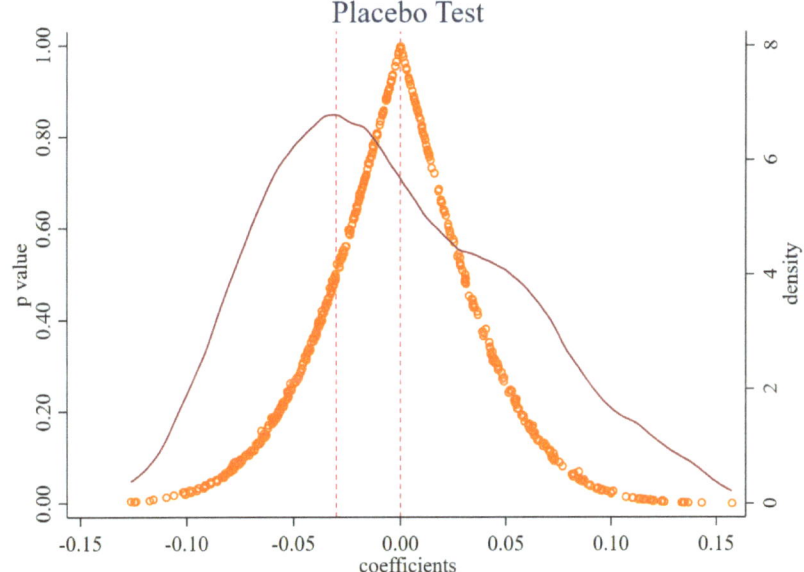

Figure 7. Placebo test.

5.2.3. Robust Test

In order to check whether the conclusion still has stable explanatory power when the conditions change, this paper conducts a robustness test by means of the Winsorise treatment, adjusting the sample period, dynamic regression, and IV, respectively. Table 5 reports the results. The impact of the HSR opening on the HQED is significantly positive in all the columns. Therefore, it is verified that the basic hypothesis, that the opening of a HSR can promote HQED in the YRD, is stable.

Table 5. Robust test results.

	(1) Winsorise	(2) 2012–2018	(3) Dynamic	(4) IV: Slope
treat	0.086 *	0.083 *	0.453 *	0.446 *
	(4.417)	(5.002)	(24.088)	(1.836)
lrgdp	0.226 **	0.257 **	0.180	0.460 **
	(11.49)	(12.74)	(16.116)	(2.277)
lforeign	0.024 **	0.023 *	0.019	0.028 ***
	(1.052)	(1.199)	(4.458)	(2.724)
lfin	0.338 **	0.327 *	0.760	0.118
	(16.14)	(19.78)	(84.671)	(0.950)
ind	0.467**	0.586**	1.278	−0.152
	(20.92)	(22.88)	(96.403)	(−0.513)
lwage	−0.423 *	−0.587 **	0.507	−1.031 ***
	(22.04)	(26.35)	(130.825)	(−3.666)
linn	0.066	0.042	−0.010	0.066
	(4.870)	(6.067)	(15.150)	(1.356)
L.hq			0.006 ***	1.625
			(0.112)	(0.867)
_cons	−1.915	−2.408 *		1.624
	(127.6)	(140.7)		(0.867)
N	369	328	287	369
R^2	0.187	0.189		0.994

Notes: *, **, *** indicate statistical significance at the 10%, 5%, and 1% levels.

Specifically, considering the large gap between the HQED in some cities and the existence of extreme value cases, this study first tries to test the extreme value treatment at the 1% and 99% quantile of *hq*. The results show that the coefficient of the impact of the opening of a HSR on HQED is 0.086, which is reduced compared with the coefficient estimate of the benchmark regression. Second, since some cities have opened a HSR before 2011, the sample period is adjusted to 2012–2019, and the results in column (2) indicate that the effect of a HSR opening on HQED after 2012 is still positive. Third, the city's development in the previous year may influence the current period's HQED index, this study lags the explanatory variables by one period to establish a dynamic panel data model, and the results are shown in column (3). The regression results for *treat* and L.hq are both significantly positive, and the coefficient of *treat* is higher when compared to the static regression results.

In order to overcome the possible endogeneity problem, this study refers to Ji and Yang [86] and selects an interaction term for the slope and year dummy variables as the instrumental variable, since the slope can affect the difficulty of HSR construction, and the slope is an exogenous geographic variable, which does not have a direct effect on the HQED index. The regression results in column (4) are still positive after correcting for potential endogeneity issues (Table 5).

5.3. Endogeneity Test

The selection of high-speed rail stations is usually considered to be non-random and biased towards cities with better economic development. There may be endogeneity issues in the model. In view of the close relationship between the opening of the post station and urban traffic, the opening of the post station is not related to the current situation. Therefore, the dummy variable (*ifyz*) for the opening of the Ming Dynasty post station can be used as an instrumental variable for the opening of the HSR (*treat*), referring to Li et al. [87].

After using the treatment effect model and MLE method, it can be seen that *ifyz* is significantly positive in the first step of the regression, which shows that the choice of an instrumental variable is effective. In the second step of the regression, the core explanatory variable treat is significantly positive, which is shown as *1.treat* in column (2), indicating that the sample with the treat value of 1 has a higher level of HQED by 0.266 units. This is about three times the benchmark regression result of 0.092, which shows that the influence of *treat* on *hq* is underestimated without considering the self-selection bias (Table 6). After MLE processing of the models with self-selection bias, the estimation bias was alleviated. The results prove that the opening of a HSR still has an important role in promoting the HQED in the YRD.

Table 6. Endogeneity test results.

	(1) Treat	(2) Hq
lrgdp	−0.715	0.363 ***
	(0.359)	(0.064)
lforeign	−0.107 **	0.026 ***
	(0.042)	(0.010)
lfin	1.036	0.053
	(0.232)	(0.106)
ind	0.755 ***	0.192 ***
	(0.036)	(0.018)
lwage	1.627	−0.420
	(0.231)	(0.231)
linn	−0.435 ***	0.098 **
	(0.075)	(0.043)
ifyz	0.374 *	
	(0.213)	
1.treat		0.266 ***
		(0.048)
_cons	−5.971	−1.182
	(4.286)	(1.776)

Notes: *, **, *** indicate statistical significance at the 10%, 5%, and 1% levels.

5.4. Heterogeneity Analysis

The previous parts of this study mainly examined the overall effect of HSR opening on HQED. Since the initial endowment of the transportation facilities and economic characteristics of cities vary greatly in the YRD region, it is helpful to further examine the heterogeneous performance of diverse types of cities after the opening of a HSR in order to test in-depth the differential impact of HSR on HQED. Specifically, this study groups the 41 cities according to geographic distance, referred to by Gao et al. [88], and city size. Geographic distance is measured by the distance between each city and the capital city of the province in which it is located. Within a boundary of 150,000 m, it divides the sample (excluding Shanghai, Nanjing, Hangzhou, and Hefei) into 18 cities close to the provincial capital city and 19 cities far from the provincial capital city. In addition, in terms of city size, according to the national urban scale classification standards, 41 cities in the YRD are classified into three categories based on the permanent population size at the end of 2019: 1–5 million people, 5–10 million people, and more than 10 million people. The results from the sub-sample regression are shown in Table 7.

As for the impact in cities in different locations, the result shows that the opening of HSR have a significantly stronger promotional effect for the areas farther away from the provincial capital city, while the promotional effect for the areas closer to the provincial capital city is smaller and not significant. This result may be due to the siphon effect. The provincial capital cities are the political, economic, and cultural centres of the provinces in the YRD, and they are still in the stage of agglomeration development and have comparative advantages in regard to resource allocation. Due to the drainage effect caused by the opening of a HSR, provincial capital cities have an overtaking effect on neighbouring cities, which relatively reduces the HQED level of nearby cities. But in cities 150,000 m away, this effect basically disappears, and those cities are more likely to receive the combined impact of the HSR and further improve their HQED level.

Table 7. Heterogeneity analysis results.

Dependent	Distance		Population		
Variable	(1) close	(2) far	(3) 1–5 million	(4) 5–10 million	(5) 10 million +
treat	0.016	0.164 ***	0.023	0.110 **	0.092 **
	(6.042)	(5.921)	(0.084)	(0.049)	(0.045)
lrgdp	0.312 ***	0.312 ***	0.272 **	0.272 **	0.267 **
	(11.99)	(11.99)	(0.118)	(0.118)	(0.118)
lforeign	0.025 **	0.025 **	0.025 **	0.025 **	0.024 **
	(1.077)	(1.077)	(0.011)	(0.011)	(0.011)
lfin	0.358 **	0.358 **	0.304 *	0.304 *	0.327 **
	(16.58)	(16.58)	(0.167)	(0.167)	(0.166)
ind	0.466 **	0.466 **	0.471 **	0.471 **	0.458 **
	(21.39)	(21.39)	(0.215)	(0.215)	(0.215)
lwage	−0.394 *	−0.394 *	−0.439 *	−0.439 *	−0.442 *
	(22.68)	(22.68)	(0.226)	(0.226)	(0.226)
linn	0.053	0.053	0.066	0.066	0.063
	(5.006)	(5.006)	(0.050)	(0.050)	(0.050)
_cons	−2.046	−2.060	−1.936	−1.943	−1.869
	(130.8)	(130.9)	(1.312)	(1.313)	(1.310)
N	369	369	369	369	369
R^2	0.195	0.195	0.189	0.189	0.186

Notes: *, **, *** indicate statistical significance at the 10%, 5%, and 1% levels; the values in parentheses are robust standard deviations. "10 million +" in column (5) represents a resident population of more than 10 million people.

As for the impact of different sized cities, the *treat* coefficient for cities with a population size exceeding 10 million is equal to the estimated coefficient of 0.092 in the basic regression, and the *treat* coefficient for cities with a population size of 5–10 million is greater than 0.092. This indicates that the HSR enhances the hub position of cities with a population over 5 million as growth poles in regional development, and the convenient HSR network brings more capital, human resources, and information to these mega cities, making positive contributions to the HQED in these cities. Meanwhile, the coefficient of treat in column (3) is 0.023, but fails to pass the significance test, indicating that the opening of the HSR has not yet brought about a significant positive impact effect on the HQED in cities with a population of 1–5 million in the YRD, which is possibly because provincial capitals and other mega cities themselves are more attractive than surrounding cities, and human and physical capital flow to big cities because of the opening of a HSR. Thus, Hypothesis 2 is verified.

The coefficients of the control variables in the heterogeneity tests are significantly positive except for *linn*, which means that the level of economic development, openness, the financial level, industrial advancement, and living standard are also conducive of HQED in the YRD region.

6. Discussion

6.1. Theoretical Value and Practical Value

Based on the five dimensions of the new development concepts, sustainable development goals, and the characteristics of HSR, this study defines the connotations of HQED and builds a comprehensive theoretical framework that includes five interacting aspects of the influence mechanism. It then sets up an evaluation system for HQED, including both input and output indexes in the first level, six dimensions in the second level, and 19 detailed indicators in third level. Compared to the present evaluation system and mechanism analysis, the mechanism framework and the evaluation system are more holistic, more comprehensive, and more practical. Subsequent empirical study confirms that HSR have a significant impact on HQED in the YRD and the robustness tests confirm the results.

The results support previous studies [13,20,30,44,84] in that the introduction of HSR have improved the HQED in the cities along the railway lines.

The heterogeneity test based on city location and size indicates that the impact of a HSR on HQED in the YRD has obvious regional differences. HSR have a stronger impact on cities far away from the provincial capital city and positively improves HQED in cities with a population more than 5 million. However, for those smaller cities, especially for those cities with population less than 5 million and close to the provincial capital city, the impact of a HSR is not significant. The heterogeneity results partly confirm the present studies, while also present inconsistent result. Some researchers find that major cities benefit more from HSR than smaller cities, and eastern cities benefit more than central and western cities [44,47]. Research on HSR in Europe indicates that the impact of a HSR on a region is complex, bringing about positive impacts on the core areas of central cities, especially those around the stations, but possibly posing a threat to the status of peripheral cities, with an impact on the renewal of edge cities as well [46]. Our study indicates that although HSR can theoretically improve the local transportation situation and bring more development opportunities to cities, the cities with a smaller population may lose their resources and production factors to large cities with higher economic agglomeration and advanced industrial structures. The loss of factors, to some extent, offsets the positive impact. But for large cities far away from the provincial capital city, such as Xuzhou city in northeast of Jiangsu province, and Ningbo city in southeast Zhejiang province, HQED benefits from the HSR.

In addition, we also found that the control variables, including the level of economic development, openness, financial level, and industry advancement contribute to the HQED in the YRD. The study not only extends the research on the economic effect of HSR and the factors affecting regional HQED, especially on the impact of HSR on HQED, but also introduces the case of the Yangtze River Delta, which is also valuable for sustainable and regional HQED in other regions and countries.

6.2. Policy Enlightenment

As the fastest and most technologically advanced ground transportation infrastructure, HSR set up inter-provincial and inter-city links and play an important role in economic development, regional integration, and urban cluster construction [89]. Against the background of the completion of the "eight vertical and eight horizontal" network, and in the context of the new economic norm that emphasises the needs of the people and solving the contradictions and problems of unbalanced and insufficient development, this study offers three policy enlightenments based on the empirical findings.

Firstly, HSR play a crucial role in facilitating regional HQED, which fits the needs of the contemporary era and the objectives of national strategic development; thus, the planning, construction, and investment in HSR networks should be further improved so as to amplify the effect of the HSR and promote HQED and sustainable development in cities along the routes, making the HSR network a more important part of the new economic circulation system. Except for HSR, a variety of factors contribute to HQED, therefore, when promoting regional HQED, other factors should be considered jointly to promote HQED in the future.

Secondly, the heterogeneity test results indicate that flexible supporting policies should be made and make full use of the dividends brought by the HSR according to local context. The influence of a HSR opening on HQED in cities with different locations and sizes varies significantly. Therefore, different countermeasures should be conducted, taking specific problems, various comparative advantages, various composition structures, various degrees of development, and accurate positioning into consideration. As the HSR comes into operation, large cities should still make full use of the high-quality resources and industrial agglomeration effect, reasonably formulate active industrial planning, and cultivate the development opportunities of high value-added industries. Meanwhile, cities with a population of less than 5 million should make full use of the dividends of HSR, and vigorously develop secondary and tertiary industries, such as tourism and transportation logistics. At the same time, they could actively utilise the spillover effects of large cities according to their own advantages, and continuously and effectively improve HQED.

Thirdly, there is still much room to enhance the function of HSR operation in promoting innovation, openness, green, coordinated, and sharing development. The supporting infrastructure around HSR stations should be synchronously constructed to build a green transportation system with 'zero-distance interchange and seamless connection' and fully release the effects of HSR. In addition, the systemic constraints and fetters should be further broken down to achieve the potential of HSR in resource sharing. The flow and sharing of intellectual resources should be accelerated to furnish human resources and technical support for HQED.

7. Conclusions
7.1. Main Conclusions

In this study, we established the influence framework of HSR on regional HQED, built the system of indicators for measuring HQED, and then measured and analysed the temporal and spatial variation in the HQED index in the YRD. A DID model and robustness tests were then constructed to analyse whether the opening of a HSR promotes regional HQED, based on panel data from 41 cities in the YRD from 2011 to 2019. Mechanism analysis shows that HSR can improve the level of economic quality development by improving inter-city accessibility, accelerating factor flow, reducing environmental pollution, promoting industrial agglomeration, optimising income distribution, and promoting consumption. Meanwhile, the impact of HSR on the five dimensions of HQED is mutually reinforcing and interrelated.

The empirical study finds that a HSR opening made significant contributions to the HQED in the YRD cities and had significant heterogeneity in view of the location and cite size. The findings remain reliable after parallel trend tests, multiple robustness tests, placebo tests, and endogeneity tests. In addition, the level of economic development, openness, the financial level, and industry advancement contribute together to the HQED in the YRD. Secondly, there are notable differences in the contribution of HSR on the HQED in cities of different scales and locations in the YRD. The impact is significant in cities with a population greater than 5 million, while the contribution to cities with a population lower than 5 million is not obvious. At the same time, the opening of a HSR has a significant stronger promotional effect on the cities farther away from the provincial capital city.

7.2. Limitations and Future Directions

This study has the following limitations. First, due to data limitations, this empirical study is conducted on the prefecture level and only adopts the opening of a HSR as the introduction of the HSR instead of service frequency and total travel time of the HSR. If county-level data and detailed data on HSR can be obtained, the research conclusions will be more targeted and will comprehensively reflect regional differences on multiple levels. Second, although this study finds that the spatial distribution of HQED presents obvious spatial differences, and the impact of HSR on HQED is very divergent within different cities in the YRD, it has not yet explored the spatial spillover effects. In the future, the spatial spillover effects from HSR on HQED can be explored in further studies.

Author Contributions: Supervision, C.G.; Conceptualization and visualization, C.G. and D.L.; data curation, methodology and test analysis, L.C. and C.G.; Funding acquisition, C.G.; writing—original draft preparation, C.G. and L.C.; writing—review and editing, C.G., L.C. and D.L. All authors have read and agreed to the published version of the manuscript.

Funding: The research was funded by the National Natural Science Fund of China, grant number 41971165.

Data Availability Statement: The publicly available sources for the data used in this study have been described in 4.4.2 of the article. The original data are from the China City Statistical Yearbook, the China Statistical Yearbook, the yearbooks of each province, and the website of the National Bureau of Statistics. The opening information on HSR comes from the State Railway Administration and the Special Issue on Railway Passenger and Freight Transportation. The slope used in 5.2.3 is from https://mp.weixin.qq.com/s/9qZMImMyMbQIgwI7HB7EUg. The Ming Dynasty post station used in 5.3 is from https://mp.weixin.qq.com/s/CqCEexVbYrz6V6JIFEiSvg. The data used in this paper are available on request from the corresponding author.

Acknowledgments: Thanks to anonymous reviewers for their valuable suggestions and the support of National Natural Science Fund of China.

Conflicts of Interest: The authors declare no conflict of interest.

Notes

1. The concept of 'space–time compression' was first proposed by American sociologist R.D. McKenzic in his book Urban Community in 1933. It is a theory on temporal and spatial changes in human interaction caused by advances in transportation and communication technology. According to the theory, the time and distance required for human interaction within a given geographical area have been reduced by improvements in transportation and communication technology.
2. Data resource: Wang Li. The Yangtze River Delta plans to invest more than 90 billion yuan in railway construction this year and add 2500 km of HSR in the next three years [N]. Jiefang Daily, 12th, Feb, 2023.

References

1. Hansmann, R.; Mieg, H.A.; Frischknecht, P. Principal sustainability components: Empirical analysis of synergies between the three pillars of sustainability. *Int. J. Sustain. Dev. World Ecol.* **2012**, *19*, 451–459. [CrossRef]
2. Mebratu, D. Sustainability and Sustainable Development: Historical and Conceptual review. *Environ. Impact Assess. Rev.* **1998**, *18*, 493–520. [CrossRef]
3. Xu, X.M. Quality is the Key to Economic Development. *Res. Financ. Econ. Issues* **1998**, *12*, 10–12. (In Chinese). Available online: https://kns.cnki.net/kcms2/article/abstract?v=HbIh-_fAmwQf3D_5MmXoC4bbhD4eN0MY4duLx0SXB4UlNyLhiqPt5E8a4DfyxfgRzqFAMTRTqTBKiD1YJb8AzlJFQpw1ZCEp-FJmPhcOF-MD1hzqQS3aCQ==&uniplatform=NZKPT&language=CHS (accessed on 30 December 1998).
4. Thomas, V. The Quality of Growth. In Proceedings of the IMF Conference on Second Generation Reforms, Washington, DC, USA, 8–9 November 1999.
5. Boyle, D.; Simms, A. *The New Economics: A Bigger Picture*; Earthscan: London, UK, 2009.
6. United Nations. Millennium Development Goals UNDP. 2015. Available online: https://www.undp.org/content/undp/en/home/sdgoverview/mdg_goals.html (accessed on 2 November 2020).
7. Schmidt-Traub, G.; Kroll, C.; Teksoz, K.; Durand-Delacre, D.; Sachs, J.D. National Baselines for the Sustainable Development Goals Assessed in the SDG Index and Dashboards. *Nature Geosci.* **2017**, *10*, 547–555. [CrossRef]
8. He, L.F. Comprehensively implementing new development concepts and promoting China's economy towards high-quality development. *Macroecon. Manag.* **2018**, *4*, 14. (In Chinese)
9. Sun, J.W.; Jiang, Z. The path to high-quality development in China's coastal areas. *Acta Geogr. Sin.* **2021**, *76*, 277–294. (In Chinese)
10. Li, L.R.; Huang, X.J.; Liu, P.F. High-Quality Development of Science and Technology Finance and Logistics Industry in the Yangtze River Economic Belt: Coupling Analysis Based on Deep Learning Model. *Comput. Intell. Neurosci.* **2022**, *2022*, 5185190. [CrossRef]
11. Zhang, J.K.; Hou, Y.Z.; Liu, L.P.; Zhuo, J.; He, X. The goals and Strategy Path of High-quality Development. *Manag. World* **2019**, *35*, 1–7. (In Chinese) [CrossRef]
12. Li, Y.R.; Pan, W.; Wang, J.; Liu, Y.S. Spatial Pattern and Influencing Factors of High-Quality Development of China at the Prefecture Level. *Acta Ecol. Sin.* **2022**, *42*, 2306–2320. (In Chinese). Available online: https://kns.cnki.net/kcms2/article/abstract?v=ECJfaSgxqGddO7iWrVP9XoukOF9_mh9_u6jjf1ygLm-XZj7_4xRAHVRdtg9WDLOZJIX9RlXljyQa91TLhVHR_UoCWhvpw56tjcA5GjBkCpNPABwtwWCUGoIkHNpR-p-S&uniplatform=NZKPT&language=CHS (accessed on 23 March 2022).
13. Wu, Y.; Li, H.; Luo, R.; Yu, Y. How digital transformation helps enterprises achieve high-quality development? Empirical evidence from Chinese listed companies. *Eur. J. Innov. Manag.* **2023**, *ahead-of-print*. [CrossRef]
14. Ma, T.; Cao, X.X.; Zhao, H. Development zone policy and high-quality economic growth: Quasi-natural experimental evidence from China. *Reg. Stud.* **2023**, *57*, 590–605. [CrossRef]
15. Li, X.; Lu, Y.; Huang, R. Whether foreign direct investment can promote high-quality economic development under environmental regulation: Evidence from the Yangtze River Economic Belt, China. *Environ. Sci. Pollut. Res.* **2021**, *28*, 21674–21683. [CrossRef] [PubMed]
16. Pokharel, R.; Bertolini, L.; te Brömmelstroet, M.; Acharya, S.R. Spatio-temporal evolution of cities and regional economic development in Nepal: Does transport infrastructure matter? *Transp. Geogr.* **2021**, *90*, 102904. [CrossRef]

17. Thomas, V.; Dailami, M.; Mansoor, D.; Dhareshear, A.; Kaufmann, D.; Kishor, N.; Lopez, R.; Wang, Y. *The Quality of Growth*; Oxford University Press: Oxford, UK, 2000.
18. Barro, R.J. Quantity and Quality of Economic Growth. *Chil. Econ.* **2002**, *5*, 17–36.
19. Alexandra, S.S.L. An empirical approach of social impact of debt on economic growth: Evidence from the European union. *Ann.–Econ. Ser.* **2016**, *5*, 189–198.
20. Li, F.L.; Li, M.D. Research on Comprehensive Evaluation of High-Quality Development of Yangtze River Delta Urban Agglomeration. *West Forum Econ. Manag.* **2021**, *32*, 36–48. (In Chinese). Available online: https://kns.cnki.net/kcms2/article/abstract?v=ECJfaSgxqGeMVq1pdsuwadXF4_rGYAZHNNkO_E8-VrQPLwFLg-IE3LnmCY9jSv-28m0jcmfYjXkK-5IjEjJ-go5yE_Ka52bNDykvuTpUoBn-Z1FQ8PweBndlXLhxxCse&uniplatform=NZKPT&language=CHS. (accessed on 22 July 2021).
21. Guo, J.; Sun, Z. How does manufacturing agglomeration affect high-quality economic development in China? *Econ. Anal. Policy* **2023**, *78*, 673–691. [CrossRef]
22. Chow, G.C.; Li, K.W. China's economic growth: 1952–2010. *Econ. Dev. Cult. Chang.* **2002**, *51*, 247–256. [CrossRef]
23. Yu, Y.Z.; Yang, X.Z.; Zhang, S.H. Research on the time-space Transformation Characteristics of China's Economy from high-speed growth to high-quality development. *Res. Quant. Econ. Technol. Econ.* **2019**, *6*, 3–21.
24. Yang, Y.X.; Su, X.; Yao, S.L. Nexus between green finance, fintech, and high-quality economic development: Empirical evidence from China. *Resour. Policy* **2021**, *74*, 102445. [CrossRef]
25. Pan, W.; Wan, J.; Lu, Z.; Liu, Y.S.; Li, Y.R. High-quality development in China: Measurement system, spatial pattern, and improvement paths. *Habitat Int.* **2021**, *118*, 102458. [CrossRef]
26. Liu, Y.X.; Tian, C.S.; Cheng, L.Y. Measurement and Comparison of the High Quality Development Level of the World Economy. *Economist* **2020**, *5*, 69–78. (In Chinese)
27. Kong, Q.; Peng, D.; Ni, Y.; Jiang, X.; Wang, Z. Trade openness and economic growth quality of China: Empirical analysis using ARDL model. *Finance Res. Lett.* **2021**, *38*, 101488. [CrossRef]
28. Guo, B.N.; Wang, Y.; Zhang, H.; Liang, C.Y.; Feng, Y.; Hu, F. Impact of the digital economy on high-quality urban economic development: Evidence from Chinese cities. *Econ. Model.* **2023**, *120*, 106194. [CrossRef]
29. Xu, G.; Yang, H.; Liu, W.; Shi, F. Itinerary choice and advance ticket booking for high-speed railway network services. *Transp. Res. Part C Emerg. Technol.* **2018**, *95*, 82–104. [CrossRef]
30. Vickerman, R. Can high-speed rail have a transformative effect on the economy? *Transp. Policy* **2018**, *62*, 31–37. [CrossRef]
31. Willigers, J.; Wee, B. High-speed rail and office location choices. A stated choice experiment for the Netherlands. *J. Transp. Geogr.* **2011**, *19*, 745–754.
32. Charnoz, P.; Lelarge, C.; Trevien, C. Communication costs and the internal organisation of multi-plant businesses: Evidence from the French high-speed rail. *Econ. J.* **2018**, *128*, 949–994. [CrossRef]
33. Ahlfeldt, G.M.; Feddersen, A. From Periphery to Core: Measuring Agglomeration Effects Using High-Speed Rail. *J. Econ. Geogr.* **2018**, *18*, 355–390. [CrossRef]
34. An, Y.Y.; Dennis Wei, Y.H.; Yuan, F.; Chen, W. Impacts of high-speed rails on urban networks and regional development: A study of the Yangtze River Delta, China. *Int. J. Sustain. Transp.* **2021**, *16*, 483–495. [CrossRef]
35. Wenner, F.; Thierstein, A. High speed rail as urban generator? An analysis of land use change around European stations. *Eur. Plan. Stud.* **2022**, *30*, 227–250. [CrossRef]
36. Sasaki, K.; Ohashi, T.; Ando, A. High-speed Rail Transit Impact on Regional Systems: Does the Shinkansen Contribute to Dispersion? *Ann. Reg. Sci.* **1997**, *31*, 77–98. [CrossRef]
37. Matteo, D.D.; Mariotti, I.; Rossi, F. Transport Infrastructure and Economic Performance: An Evaluation of the Milan–Bologna High-speed Rail Corridor. *Socio-Econ. Plan. Sci.* **2022**, *85*, 101304. [CrossRef]
38. Sahu, S.; Verma, A. Quantifying Wider Economic Impacts of High-speed Connectivity and Accessibility: The Case of the Karnataka High-speed Rail. *Transp. Res. Part A Policy Pract.* **2022**, *158*, 141–155. [CrossRef]
39. Banerjee, A.; Duflo, E.; Qian, N. On the Road: Access to Transportation Infrastructure and Economic Growth in China. *J. Dev. Econ.* **2020**, *145*, 102442. [CrossRef]
40. Guirao, B.; Lara-Galera, A.; Luis Campa, J. High Speed Rail Commuting Impacts on Labour Migration: The Case of the Concentration of Metropolis in the Madrid Functional Area. *Land Use Policy* **2017**, *66*, 131–140. [CrossRef]
41. Liu, J.; Huang, X.F.; Chen, J. High Speed Rail and High-Quality Development of Urban Economy: An Empirical Study Based on the Data of Prefecture-Level Cities. *Contemp. Financ. Econ.* **2021**, *1*, 14–26. [CrossRef]
42. Li, Q.; Wang, Y.C. Opening of High-Speed Rail and High-Quality Development of the Yangtze River Economic Belt. *J. Nanjing Univ. Financ. Econ.* **2021**, *3*, 25–35. (In Chinese). Available online: https://d.wanfangdata.com.cn/periodical/njjjxyxb202103004 (accessed on 20 June 2021).
43. Hu, J.; Ma, G.; Shen, C.; Zhou, X. Impact of Urbanization through High-Speed Rail on Regional Development with the Interaction of Socioeconomic Factors: A View of Regional Industrial Structure. *Land* **2022**, *11*, 1790. [CrossRef]
44. Kong, Q.X.; Li, R.R.; Jiang, X. Has Transportation Infrastructure Development Improved the Quality of Economic Growth in China's cities? A quasi-natural Experiment Based on the Introduction of High-Speed Rail. *Res. Int. Bus. Financ.* **2022**, *62*, 101726. [CrossRef]
45. Faber, B. Trade Integration, Market Size, and Industrialization: Evidence from China's National Trunk Highway System. *Rev. Econ. Stud.* **2014**, *81*, 1046–1070. [CrossRef]

46. Hall, P. Magic Carpets and Seamless Webs: Opportunities and Constraints for High-Speed Trains in Europe. *Built Environ.* **2009**, *35*, 59–69. [CrossRef]
47. Bruzzone, F.; Cavallaro, F.; Nocera, S. The Effects of High-Speed Rail on Accessibility and Equity: Evidence from the Turin–Lyon Case-Study. *Socio-Econ. Plan. Sci.* **2023**, *85*, 101379. [CrossRef]
48. Wu, J.; Sun, W. Regional Integration and Sustainable Development in the Yangtze River Delta, China: Towards a Conceptual Framework and Research Agenda. *Land* **2023**, *12*, 470. [CrossRef]
49. Xi, J. Grasp the New Development Stage, Implement the New Development Concept, and Build a New Development Pattern. *Chin. Civ. Aff.* **2021**, 4–11. (In Chinese) [CrossRef]
50. Liu, L.; Zhang, M. High-speed rail impacts on travel times, accessibility, and economic productivity: A benchmarking analysis in city-cluster regions of China. *J. Transp. Geogr.* **2018**, *73*, 25–40. [CrossRef]
51. Cascetta, E.; Cartenì, A.; Henke, I.; Pagliara, F. Economic growth, transport accessibility and regional equity impacts of high-speed railways in Italy: Ten years expost evaluation and future perspectives. *Transp. Res. Part A Pol. Pract.* **2020**, *139*, 412–428. [CrossRef]
52. Bian, Y.C.; Wu, L.H.; Bai, J.H. Does High-Speed Rail Improve Regional Innovation in China? *J. Financ. Res.* **2019**, *468*, 132–149. (In Chinese). Available online: https://kns.cnki.net/kcms2/article/abstract?v=ECJfaSgxqGc7vUREPLY2FfuDkQTqonkiuTQ10EqvRnSB4pChyx5sz9E5uogJYzWHNLHEaN41HMBhXWbY8QCetSo3q7_GKgV3XaeqlGuZkHLCQ-fAnlS4kKJoctMqhN8X&uniplatform=NZKPT&language=CHS (accessed on 25 June 2019).
53. Yang, Y.; Ma, G.C. How can HSR promote inter-city collaborative innovation across regional borders? *Cities* **2023**, *138*, 104367, ISSN 0264-2751. [CrossRef]
54. Yang, X.H.; Zhang, H.R.; Lin, S.L.; Zhang, J.P.; Zeng, J.L. Does high-speed railway promote regional innovation growth or innovation convergence? *Technol. Soc.* **2021**, *64*, 101472. [CrossRef]
55. Yan, Y.G.; Ni, P.F.; Liu, X.L. High-Speed Train, Immobility Factors and the Development of Periphery Regions. *China Ind. Econ.* **2020**, *389*, 118–136. (In Chinese) [CrossRef]
56. Hanley, D.; Li, J.C.; Wu, M.Q. High-speed railways and collaborative innovation. *Reg. Sci. Urban Econ.* **2022**, *93*, 103717. [CrossRef]
57. He, L.Y.; Tao, D.J. Measurement of the Impact of High-Speed Rail Opening on Knowledge Spillover and Urban Innovation Level. *J. Quant. Tech. Econ.* **2020**, *37*, 125–142. (In Chinese) [CrossRef]
58. Yao, S.; Zhang, F.; Wang, F.; Ou, J. Regional economic growth and the role of highspeed rail in China. *Appl. Econ.* **2019**, *51*, 3465–3479. [CrossRef]
59. Jiang, M.; Kim, E. Impact of high-speed railroad on regional income inequalities in China and Korea. *Int. J. Urban Sci.* **2016**, *20*, 393–406. [CrossRef]
60. Jin, Y.; Wang, S.B. Spatial Effects of High-Speed Railway and Its Hot Issues. *Urban Plan. Int.* **2020**, *35*, 27–33. (In Chinese) [CrossRef]
61. Zhang, F.N.; Yang, Z.W.; Jiao, J.J.; Liu, W.; Wu, W.J. The effects of high-speed rail development on regional equity in China. *Transp. Res. Part A Policy Pract.* **2020**, *141*, 180–202. [CrossRef]
62. Chang, Y.; Lei, S.H.; Teng, J.J.; Zhang, J.X.; Zhang, L.X.; Xu, X. The Energy Use and Environmental Emissions of High-Speed Rail Transportation in China: A bottom-up modelling. *Energy* **2019**, *182*, 1193–1201. [CrossRef]
63. Chang, Z.; Deng, C.; Long, F.; Zheng, L. High-speed rail, firm agglomeration, and PM2.5: Evidence from China. *Transp. Res. Part D Transp. Environ.* **2021**, *96*, 102886. [CrossRef]
64. Zhang, F.; Yao, S.; Wang, F. The role of high-speed rail on green total factor productivity: Evidence from Chinese cities. *Environ. Sci. Pollut. Res.* **2023**, *30*, 15044–15058. [CrossRef]
65. Chen, L.M.; Ye, W.Z.; Huo, C.J.; James, K. Environmental Regulations, the Industrial Structure, and High-Quality Regional Economic Development: Evidence from China. *Land* **2020**, *9*, 517. [CrossRef]
66. Qin, Y. 'No County Left Behind?' The Distributional Impact of High-Speed Rail Upgrades in China. *J. Econ. Geogr.* **2017**, *17*, 489–520. [CrossRef]
67. Xuan, Y.; Lu, J.; Yu, Y.Z. The Impact of High-Speed Rail Opening on the Spatial Agglomeration of High-End Service Industry. *Financ. Trade Econ.* **2019**, *40*, 117–131. (In Chinese) [CrossRef]
68. Bougheas, S.; Demetriades, P.O.; Morgenroth, E.L.W. Infrastructure, transport costs and trade. *J. Int. Econ.* **1999**, *47*, 169–189. [CrossRef]
69. Bernard, A.B.; Moxnes, A.; Saito, Y.U. Production Networks, Geography, and Firm Performance. *J. Political Econ.* **2019**, *127*, 639–688. [CrossRef]
70. Francois, J.; Manchin, M. Institutions, infrastructure, and trade. *World Dev.* **2013**, *46*, 165–175. [CrossRef]
71. Zhou, J.J.; Fan, X.Y.; Yu, A.Z. The Impact of high-speed railway accession on agricultural exports: Evidence from Chinese agriculture-related enterprises. *Complexity* **2021**, *2021*, 4225671. [CrossRef]
72. Cosar, A.K.; Demir, B. Domestic road infrastructure and international trade: Evidence from Turkey. *J. Dev. Econ.* **2016**, *81*, 232–244. [CrossRef]
73. Xu, M.Z. *Riding on the New Silk Road: Quantifying the Welfare Gains from High-Speed Railways*; University of Virginia: Charlottesville, VA, USA, 2017.
74. Sun, P.Y.; Zhang, T.T.; Yao, S.J. Tariff Transmission, Domestic Transport Costs and Retail Prices. *Econ. Res. J.* **2019**, *54*, 135–149. (In Chinese). Available online: http://lib.cqvip.com/Qikan/Article/Detail?id=7001545429 (accessed on 25 March 2019).

75. Tang, Y.H.; Yu, F.; Lin, F.Q. Research on high-speed rail, trade cost and Enterprise export in China. *Econ. Res. J.* **2019**, *7*, 158–173. (In Chinese)
76. Gil-Pareja, S.; Llorca-Vivero, R.; Paniagua, J. Does High-Speed Passenger Railway Increase Foreign Trade? An Empirical Analysis. *Int. J. Transp. Econ.* **2015**, *42*, 357–376. Available online: https://econpapers.repec.org/RePEc:jte:journl:2015:3:42:4 (accessed on 3 July 2023).
77. Chen, F.L.; Xu, K.N.; Wang, M.C. High Speed Railway Development and Urban–Rural Income Inequality: Evidence from Chinese Cities. *Econ. Rev.* **2018**, *210*, 59–73. (In Chinese) [CrossRef]
78. Matteo, D.D.; Cardinale, B. Impact of high-speed rail on income inequalities in Italy. *J. Transp. Geogr.* **2023**, *111*, 103652. [CrossRef]
79. Elhorst, J.P.; Oosterhaven, J. Integral Cost–Benefit Analysis of Maglev Rail Projects Under Market Imperfections. *J. Transp. Land Use* **2008**, *1*, 65–87. [CrossRef]
80. Masson, S.; Petiot, R. Can the high speed rail reinforce tourism attractiveness? The case of the high speed rail between Perpignan (France) and Barcelona (Spain). *Technovation* **2009**, *29*, 611–617. [CrossRef]
81. Zhang, Y.Y.; Ma, W.L.; Yang, H.J.; Wang, Q. Impact of high-speed rail on urban residents' consumption in China—From a spatial perspective. *Transp. Policy* **2021**, *106*, 1–10. [CrossRef]
82. Blum, U.; Haynes, K.E.; Karlsson, C. The regional and urban effects of high-speed trains. *Ann. Reg. Sci.* **1997**, *31*, 1–20. [CrossRef]
83. Lu, X.; Tang, Y.; Ke, S. Does the Construction and Operation of High-Speed Rail Improve Urban Land Use Efficiency? Evidence from China. *Land* **2021**, *10*, 303. [CrossRef]
84. Xu, J.; Pan, J. Analysis of Time and Space Differences and Mechanisms of Transportation Infrastructure Promoting Economic Growth—Research Based on Two-way Fixed Effect Model. *Inq. Econ. Issues* **2019**, *449*, 29–42. (In Chinese). Available online: https://kns.cnki.net/kcms/detail/detail.aspx?FileName=JJWS201912003&DbName=CJFQ2019 (accessed on 1 December 2019).
85. Hartig, T.; Kahn, P.H. Meta-principles for developing smart, sustainable, and healthy cities. *Science* **2016**, *352*, 938–940. [CrossRef]
86. Ji, Y.; Yang, Q. Can the High-speed Rail Service Promote Enterprise Innovation? A Study Based on Quasi-natural Experiments. *J. World Econ.* **2020**, *43*, 147–166. (In Chinese)
87. Li, L.B.; Yan, L.; Huang, J.L. Transportation Infrastructure Connectivity and Manufacturing Industries in Peripheral Cities in China: Markup, Productivity and Allocation Efficiency. *Econ. Res. J.* **2019**, *54*, 182–197.
88. Gao, Y.; Song, S.; Sun, J.; Zang, L. Does high-speed rail really promote economic growth? evidence from China's Yangtze River delta region. *SSRN Electron. J.* **2018**, *24*, 316–338.
89. He, Y.; Sherbinin, A.D.; Shi, G.; Xia, H. The Economic Spatial Structure Evolution of Urban Agglomeration under the Impact of High-Speed Rail Construction: Is There a Difference between Developed and Developing Regions? *Land* **2022**, *11*, 1551. [CrossRef]

Disclaimer/Publisher's Note: The statements, opinions and data contained in all publications are solely those of the individual author(s) and contributor(s) and not of MDPI and/or the editor(s). MDPI and/or the editor(s) disclaim responsibility for any injury to people or property resulting from any ideas, methods, instructions or products referred to in the content.

Article

The Impact of Urbanization on Food Security: A Case Study of Jiangsu Province

Jiayu Kang [1,2,3], Xuejun Duan [1,2,3],* and Ruxian Yun [4]

1. Nanjing Institute of Geography and Limnology, Chinese Academy of Sciences, Nanjing 210008, China; jykang@niglas.ac.cn
2. Key Laboratory of Watershed Geographic Sciences, Nanjing Institute of Geography and Limnology, Chinese Academy of Sciences, Nanjing 210008, China
3. University of Chinese Academy of Sciences, Beijing 100049, China
4. School of Education, Nanjing University, Nanjing 210046, China; jssyrx@163.com
* Correspondence: xjduan@niglas.ac.cn

Citation: Kang, J.; Duan, X.; Yun, R. The Impact of Urbanization on Food Security: A Case Study of Jiangsu Province. *Land* **2023**, *12*, 1681. https://doi.org/10.3390/land12091681

Academic Editor: Hossein Azadi

Received: 12 July 2023
Revised: 14 August 2023
Accepted: 23 August 2023
Published: 28 August 2023

Copyright: © 2023 by the authors. Licensee MDPI, Basel, Switzerland. This article is an open access article distributed under the terms and conditions of the Creative Commons Attribution (CC BY) license (https://creativecommons.org/licenses/by/4.0/).

Abstract: Food security has received extensive academic attention in recent years. However, research results analyzing cultivated land pressure from the perspective of urbanization are relatively few. This study used Jiangsu Province as the study area and analyzed the spatial pattern evolution of cultivated land pressure from 2005 to 2019 by constructing a formula for a cultivated land pressure index. The study used a spatial econometric approach to analyze the spatial relationship between urbanization and cultivated land pressure. Based on the spatial Durbin model, the impact of urbanization on the pressure on cultivated land is analyzed. According to the results, Jiangsu Province showed an obvious north–south divergence in the spatial distribution of the cultivated land pressure index, with the low-value areas of the cultivated land pressure index mainly distributed in northern and central Jiangsu, and the high-value areas mainly distributed in southern Jiangsu. The urbanization level and cultivated land pressure level in Jiangsu Province showed obvious spatial clustering characteristics, and there was a certain overlap between the high- and low-value clustering areas of the two, with significant positive spatial correlation features. The total urbanization had no significant effect on the cultivated land pressure. Population urbanization and industry urbanization showed a significant negative effect on cultivated land pressure in Jiangsu Province, while land urbanization showed a positive effect. Both population and land urbanization had a significant negative spatial spillover effect on cultivated land pressure. Plausible explanations of these results were provided and policy implications were drawn.

Keywords: urbanization; cultivated land pressure; spatial distribution; spatial Durbin model; Jiangsu Province

1. Introduction

Cultivated land resources are the material basis for sustaining human social development; therefore, changes in the quantity and quality of cultivated land resources significantly affect regional food production and food security [1–3]. According to research results, by 2030, 3.7% of the world's cultivated land will be encroached upon due to urbanization [4,5]. The rapid development of urbanization and industrialization in China has profoundly affected the use of cultivated land and food production, especially under the situation of global climate change and the currently tightening international trade environment, and the seriousness of the food security issue has begun to be highlighted. Urbanization profoundly affects cultivated land use and food production, directly impacting Chinese food security. In the 21st century, the Chinese urbanization level has increased rapidly, with the urbanization rate exceeding 50% in 2011 and 60% in 2020, and the urban resident population already outnumbers the rural resident population [6,7]. Since 2004, Chinese grain production has increased continuously, but with the development of the economy and the continuous increase in food demand, the problem of food security in

China has not been fundamentally alleviated [8]. China's urban sprawl has resulted in massive encroachment on prime cultivated land [4]. China lost more than 14.5 million hectares of cultivated land between 1979 and 1995 and about 8.32 million hectares between 1996 and 2008 due to urbanization after the reform and opening up, which seriously threatens China's food security [9,10]. Based on the obvious reduction in cultivated land, the Chinese government has introduced a series of relevant policies, such as setting a red line of 1.8 billion mu (120 million hectares) of cultivated land and achieving a balance of cultivated land through land preparation [11]. However, the issue of food security has not been fully resolved, most newly reclaimed arable land is of low quality, and the cultivated land reserve is still insufficient [12,13].

With the rapid development of urbanization in China, the impact of urbanization on food security may not only be negative. On the one hand, farmers' income from farming in the countryside is significantly lower compared to the income from entering urban work [14,15]. With the progress of urbanization, the reduction in cultivated land and the loss of rural labor will make food production decrease. In the context of increasing food demand, this directly worsens the problem of food security in China. On the other hand, with the rapid increase in China's urbanization, urban areas and industries already have the ability to feed rural areas and agriculture [16,17]. Higher urbanization may imply higher technology levels, stronger agricultural investment levels, and more cultivated land per capita, promoting the mechanization and modernization of food production, which in turn boosts the China's food production. Thus, the relationship between urbanization and food security may be open to debate.

In China, the current situation of food security is serious; the ecological footprint of cultivated land in most of the major grain-producing provinces exceeds the ecological carrying capacity, resulting in an ecological deficit and showing an expansion [18]. Among these provinces, Jiangsu Province, as a major grain-producing and economically developed province in China, has a representative food security problem. Jiangsu Province has the most serious ecological deficit in the cultivated land system among the major grain-producing regions [18], and the province's grain supply and demand are already in a tight balance [19]. Jiangsu Province is also a province with rapid urbanization and urban ecological well-being performance, where urban ecological well-being continues to rise and consumption of resources continues to decline [20,21]. In addition, due to the long-term uneven development of the economy in Jiangsu Province, there is an imbalance in regional food production and consumption: central and northern Jiangsu are relatively less economically developed as the major food production regions, while southern Jiangsu is more economically developed as the main food marketing region. In this context, it is typical to investigate the relationship between urbanization and food security in Jiangsu Province, which is an economically developed province and a major grain-producing area. Simultaneously, counties in Jiangsu Province are economically developed, and as the smallest functional area of the study unit, they are more reflective of the relationship between regional urbanization and food security compared to municipal and provincial areas. This study took Jiangsu Province as the case area and counties as the study unit, focusing on analyzing the spatial pattern evolution of urbanization and food security in counties of Jiangsu Province, exploring the spatial correlation between urbanization and food security, using the spatial Durbin model to measure and calculate the influence of urbanization and food security. Based on a spatial perspective, this study utilized a spatial approach to more effectively reflect the impact of urbanization on food security. We expect to benefit Jiangsu Province and China's future urbanization development strategy.

Scholars generally use cultivated land pressure as a measure of food security. In the measurement of cultivated land pressure, non-Chinese scholars measure simple indicators and various perspectives, and they rarely explore the spatial and temporal evolution of cultivated land pressure [11]. One scholar measured the cultivated land carrying capacity under ten dietary structures through simulations using a biophysical model [22]. The pressure on the food supply for cultivated land was examined from the perspective of increasing food demand due to continuous population growth [23]. One scholar measured the risk posed by urban expansion to regional cultivated land resources mainly from the perspective of changes in the amount of cultivated land [24]. Through continuous innovation of index models reflecting cultivated land pressure, Chinese scholars summarized a modified cultivated land pressure index model based on the interaction of region, population, cultivated land, and food, which was accepted by academia and widely used in regional food security evaluation [25]. At the scale of study, the main focus is on provincial and municipal study units.

In terms of the impact of urbanization on food security, scholars have mainly measured the role of urbanization on food production and thus analyzed its impact on food security [26]. Scholars have conducted numerous studies on the relationship between urbanization and food production, mainly forming two views. First, there is a negative relationship between urbanization and food production; as urban areas continue to expand, the area of cultivated land gradually decreases, which leads to a decrease in the area of food sown, weakening the food supply capacity [27,28]. Simultaneously, the concentration of the rural population in cities has led to the abandonment of large areas of cultivated land, resulting in a decrease in regional food supply capacity [29]. In addition, urbanization can also cause soil environmental pollution of cultivated land, crowd out water resources, and constrain food production [30,31]. Second, there is a positive or equilibrium relationship between urbanization and food production, and some scholars found that population urbanization has a significant positive effect on food production at the Chinese national scale [32]. In addition, rural labor migration and reduction in cultivated land resources do not significantly threaten China's food security [33]. Moreover, urbanization can promote food production by upgrading food production technology [34]. From the case area of Jiangsu Province, both industry and population urbanization have a significant contribution to food production [35].

2. Materials and Methods

2.1. Study Area

For this paper, 65 county study units in Jiangsu Province were studied. They included Xuzhou City, Lianyungang City, Suqian City, Huai'an City, Yancheng City, Yangzhou City, Taizhou City, Nantong City, Nanjing City, Zhenjiang City, Changzhou City, Wuxi City, and Suzhou City. The northern Jiangsu region includes Xuzhou City, Lianyungang City, and Suqian City; the central Jiangsu region includes Huai'an City, Yancheng City, Yangzhou City, Taizhou City, and Nantong City; and the southern Jiangsu region includes Nanjing City, Zhenjiang City, Changzhou City, Wuxi City, and Suzhou City (Figure 1).

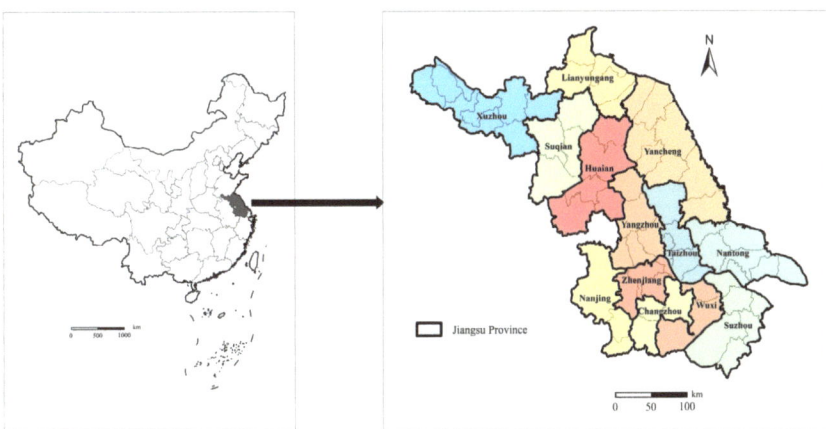

Figure 1. Location of Jiangsu Province.

2.2. Date Source

The data in this paper include two aspects: socio-economic data and land urbanization data. The socio-economic data were obtained from the Jiangsu Statistical Yearbook from 2006 to 2020. The land urbanization data were mainly obtained from the China Earth System Science Data Sharing Platform, and the ENVI5.3 software was used to decode and classify the land-use remote sensing data, extract and vectorize the spatial extent of construction land development in each region, finally use it as the basis to calculate the area of regional construction land.

2.3. Research Methods

2.3.1. Research Idea

The coordination of the population, industry, and land urbanization in a region is an important criterion for measuring the urbanization quality [36]. In terms of the relationship between the three, population urbanization promotes the flow of labor among different sectors and industries, which promotes industrial adjustment and structural upgrading and increases the development demand for urban land, and in turn reacts to the population and industrial development, forming a circular process [37]. Factor mobility between urban and rural areas is an important entry point for analyzing the relationship between urbanization and food security [32]. Based on this perspective, urbanization (population urbanization, industry urbanization, and land urbanization) affects regional food security by influencing agricultural labor, agricultural technology and capital, and regional cultivated land quantity and quality, respectively (Figure 2): ① The transformation of the rural population status causes changes in food production. As the process of population urbanization advances, capital, technology, population, and information continue to gather in the central towns, and the rural population flows from the countryside to the cities. This may lead to a reduction in food production due to insufficient rural labor inputs, or an increase in food production due to the promotion of large-scale cultivated land management as per capita cultivated land increases. ② Industrial investment preferences affect food production. In the process of industry urbanization, the inertia of urban priority development leads to the one-way concentration of capital in favor of secondary and tertiary industries in cities [38]. This leads to slow rural development and agricultural technology updating, due to the existence of the industrial–agricultural scissors difference, and a decrease in the food production, which in turn causes a decrease in the regional food security. However, if higher industrial urbanization is developed, the marginal effect of capital's returns in the cities decreases and promotes capital investment in rural areas, which also possibly promotes higher food production, thus ensuring regional food security. ③ Changes in the quantity and quality of cultivable land affect food production. As land urbanization continues, the area of cultivated land continues to shrink as construction land in regional cities expands. In recent years, although China has put forward the requirement of "balanced occupation" of cultivated land in the urbanization process, the status of "occupying the best and compensating the worst" is common. The area of high-quality cultivated land has been reduced, which has directly led to a decline in food production, thus threatening regional food security.

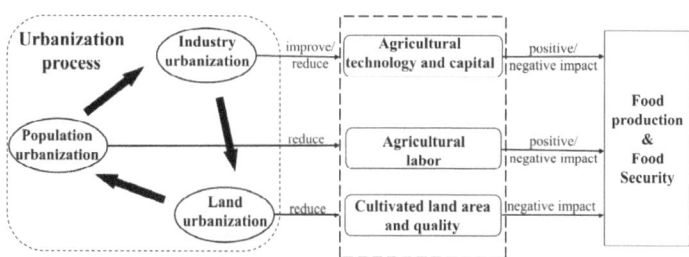

Figure 2. Mechanisms of the impact of urbanization on food security.

China's grain output increased continuously for 16 years from 2005 to 2020, while the years 2005–2020 were China's 11th, 12th, and 13th Five-Year Plan periods, the periods of rapid development of China's urbanization, and considering the impact of the COVID-19 in 2020, 2005–2019 was chosen as the time period for this study.

2.3.2. Indicator Construction

Cultivated Land Pressure Index

In terms of measuring food security at the country–regional level, there are mainly measures such as Prevalence of Undernourishment, Global Hunger Index, and Global Food Security Index [39]. Prevalence of Undernourishment, tools for measuring food availability, such as food balance sheets, have traditionally drawn from nationally aggregated data on food supply (i.e., total amount of food produced and imported) and utilization (i.e., the quantity of food exported, fed to livestock, used for seed, processed for food and non-food uses, and lost during storage and transportation) [40]. These indicators are used to estimate food shortages and surpluses, to develop projections of future food demand, and to set agricultural production targets. The Global Hunger Index (GHI), developed by IFPRI, aims to measure "hunger" using 3 equally weighted indicators: (1) undernourishment (i.e., the proportion of undernourished people as a percentage of the population); (2) child underweight (i.e., the proportion of children younger than 5 y who have a low weight for their age); and (3) child mortality (i.e., the mortality rate for children younger than age 5 y) [41]. This indicator reflects the extent of acute food insecurity in the region. The Global Food Security Index (GFSI) is another multi-dimensional tool for assessing country-level trends in food security. It was designed by the Economist Intelligence Unit (one of several companies of a publicly traded multinational, the Economist Group) and sponsored by DuPont. The index uses a total of 30 indicators within 3 domains of food security, affordability (6 indicators), availability (10 indicators), and quality and safety (14 indicators), to provide a standard against which country-level food security can be measured [42].

China's rapid urbanization, growing wealth, and emphasis on health are placing higher demands on food quality and safety [43]. China has a high rate of food self-sufficiency and is not dependent on imports; food security has been ensured [44,45]. However, in the long run, food security is most urgently faced with a shortage of cultivated land resources [46]. Therefore, in China, it is most appropriate to analyze food security from the perspective of cultivated land resources.

The cultivated land pressure index can be used to judge the cultivated land resource tension in a certain population area by the balance level of supply and demand of cultivated land, reflecting regional food security [47]. Specifically, the cultivated land demand can be expressed in terms of the minimum cultivated area per capita to guarantee food security; the cultivated land supply can be obtained from the actual cultivated area per capita [11]. In general, different geographical environments lead to differences in the efficiency and quality of cultivated land utilization, that is, the heterogeneity of cultivated land, which is the basic property of cultivated land [11]. Therefore, the cultivated land pressure index is corrected by introducing a cultivated land quality factor, and the corrected cultivated land pressure index is a combined pressure threshold that integrates the quantity and quality of cultivated land; the result is more effective [48]. This index can be calculated using the following equation:

$$K_i = \frac{AD_i}{AS_i} = \frac{\rho \cdot (F/W_i) \cdot Q_i \cdot R_i}{H_i} \tag{1}$$

$$G_i = \frac{K_i}{\sigma} = \frac{K_i}{(C_i \cdot R_i)/(C_n \cdot R_n)} \tag{2}$$

where K_i is the cultivated land pressure index of the ith county, AD_i reflects the food demand of the ith county and is the minimum per capita cultivated area that can guarantee food security, AS_i is the actual per capita cultivated area of the ith county, ρ is the food self-sufficiency rate (%), F is the per capita food demand (kg/person), W_i is the unit sown area of the ith county grain yield (kg/hm^2), Q_i is the ratio of the sown area of grain crops in the ith county to the total sown area of crops (%), R_i is the replanting index of the ith county (%), G_i is the modified cultivated land pressure index of the ith county, σ is the correction coefficient of cultivated land quality, C_i is the grain yield per unit sown area in city i (kg/hm^2), R_i is the replanting index of the ith county (%), C_n is the grain yield per unit sown area in Jiangsu Province (kg/hm^2), and R_n is the replanting index of Jiangsu Province (%). According to related studies and the China's Grain Security Program for Medium and Long-Term (2008–2020) [49], it is proposed that the self-sufficiency rate of grain should have been kept above 95% by 2020. Thus, ρ was set at 95% in this study. The State Food and Nutrition Consultant Committee (https://sfncc.caas.cn/ accessed on 15 April 2023) proposed that the objective per capita grain demand for a well-off society in an all-round way should be 437 kg/person; therefore, F was set as 437 kg/person.

With reference to the evaluation criteria for the cultivated land pressure index and the findings of related studies, the cultivated land pressure can be graded at five levels [11]: grain security zone (K ≤ 0.9), alarm-pressure zone (0.9 < K ≤ 1), low-pressure zone (1 < K ≤ 1.5), medium-pressure zone (1.5 < K ≤ 2) and high-pressure zone (K > 2).

Population–Industry–Land Urbanization Coupling Coordination Degree

Population urbanization, industry urbanization, and land urbanization referred to in this paper are all measured using single-dimension indicators, which are calculated using the following equations:

$$P_i = \frac{PU_i}{PT_i} \tag{3}$$

$$L_i = \frac{LU_i}{LT_i} \tag{4}$$

$$I_i = \frac{IS_i + IT_i}{GDP_i} \tag{5}$$

where P_i, I_i, and L_i are the population urbanization, industry urbanization, and land urbanization levels of administrative unit i, respectively; PU_i and PT_i are the urban population and total population of administrative unit i, respectively; LU_i and LT_i are the built-up area and the total land area of administrative unit i, respectively; and IS_i, IT_i, and GDP_i are the output value of the secondary and tertiary industries and GDP of administrative unit i, respectively.

The population–industry–land urbanization coupling coordination degree refers to the normalization of population urbanization, industry urbanization, and land urbanization, and the coupled coordination degree formula is applied to derive the result.

$$C_i = 3 \times \frac{(P_i \times L_i \times I_i)^{1/3}}{(P_i + L_i + I_i)} \tag{6}$$

$$D_i = \sqrt{C_i \times T_i}, \ T_i = \alpha P_i + \beta L_i + \gamma I_i \tag{7}$$

where C_i represents the population–industry–land urbanization coupling degree of administrative unit i; D_i represents the coupling coordination degree of administrative unit i; P_i, I_i, and L_i are population urbanization, industry urbanization, and land urbanization levels of administrative unit i, respectively; and α, β, and γ are all taken as 1/3.

Based on the research demand, the coupling coordination degree was divided into five intervals according to the value domain, i.e., low-level coupling coordination degree (0–0.2), lower-level coupling coordination degree (0.2–0.4), medium-level coupling degree (0.4–0.6), higher-level coupling degree (0.6–0.8), and high-level coupling degree (0.8–1.0).

2.3.3. Impact Factor Selection

The factors influencing food security are multifaceted. The natural background is one of the essential factors of grain production and a possible condition influencing the magnitude of cultivated land pressure [50]. Socio-economic factors, according to the type of action, can be divided into agricultural production conditions and urbanization [11]. Agricultural production conditions directly influence cultivated land pressure by changing the production performance of cultivated land [11]. In contrast, urbanization acts on cultivated land pressure either directly or indirectly through encroaching on cultivated land resources, stimulating the transfer of agricultural population, etc. [11].

(1) Natural background. Jiangsu Province is located in a plain area with small slope variations and little difference in arable land fertility, and water resources are an important factor affecting grain yield. Therefore, we chose the water resource level to characterize the natural background.

(2) Agricultural production conditions. The economic level, fertilizer application level, agricultural mechanization level, and labor input level are taken as influencing factors. That is, the regional economic level implies the strength of the local farmers' input to food production, which affects cultivated land pressure. The fertilizer application level affects cultivated land pressure by influencing cultivated land production performance. The agricultural mechanization level implies the input of agricultural machinery, which affects cultivated land pressure. The labor input level implies the number of people involved in food production and reflects the efficiency of food production.

(3) Urbanization. Urbanization plays an important role in food production and is the focus of this paper. A healthy and efficient urbanization process is inevitably the coordinated development of urbanization in the three dimensions of population urbanization, industry urbanization, and land urbanization [36]. The population–industry–land urbanization coupling coordination degree can reflect the new urbanization development degree [36]. Therefore, we chose the population–industry–land urbanization coupling coordination degree, population urbanization, industry urbanization, and land urbanization as the core explanatory variables.

In order to clarify the impact of urbanization on food security, based on the regression model, the impact of overall urbanization on food security was first calculated with the population–industry–land urbanization coupling coordination degree as the core explanatory variable, and then the impact of the three different urbanization dimensions on food security was calculated with the degree of population urbanization, industry urbanization, and land urbanization as the core explanatory variables. Specific indicators are shown in Table 1.

Table 1. Definitions and descriptions of variables.

Types		Indicator Selection	Measurement Method	Symbol
Core explanatory variables	Urbanization	Population–industry–land urbanization coupling coordination degree	Calculation of the coupling coordination degree of population urbanization, industry urbanization, and land urbanization	X1
		Population urbanization	Ratio of urban population to resident population	X2
		Industry urbanization	Secondary and tertiary sectors as a proportion of GDP	X3
		Land urbanization	Ratio of the built-up area and total land area	X4
Control variables	Agricultural production conditions	Economic level	Per capita GDP	X5
		Fertilizer application level	Fertilizer application per unit area of cultivated land	X6
		Agriculture modernization level	Total mechanical power per unit area of cultivated land	X7
		Labor input level	Percentage of agricultural workers	X8
	Natural background	Water resource level	Water resources per capita	X9

2.3.4. Spatial Measurement Methods

Global Spatial Autocorrelation

Global spatial autocorrelation expresses the spatial dependence of units within their spatial extent in terms of their total spatial extent. The use of univariate global spatial autocorrelation (univariate Moran's I) allows for the exploration of the spatial correlation characteristics of a given variable. The spatial association characteristics of the independent and dependent variables can be explored using bivariate global spatial autocorrelation (bivariate Moran's I). Moran's I is often used as an index to characterize the spatial association of units, and the formula is as follows:

$$Moran's\ I = \frac{\sum_{i=1}^{n}\sum_{j=1}^{n}(Y_i - \overline{Y})(Y_j - \overline{Y})}{S^2 \sum_{i=1}^{n}\sum_{j=1}^{n} W_{ij}} \quad (8)$$

where $S^2 = \frac{1}{n}\sum_{i=1}^{n}(Y_i - \overline{Y})$, $\overline{Y} = \frac{1}{n}\sum_{i=1}^{n} Y_i$, Y_i and Y_j denote the observed values of the ith and jth regions, n is the total number of each research unit, and W_{ij} is the spatial weight matrix. The range of Moran's I index values is $[-1, 1]$; the closer the value is to 1, the stronger the positive spatial correlation of the research units, and the closer the value is to 0, the stronger the negative spatial correlation; if the value of Moran's index is 0, it means that the spatial correlation of each research unit in the spatial range is 0.

Local Spatial Autocorrelation

Global spatial autocorrelation is a measure of the overall spatial correlation of the variables and ignores local instabilities and local patterns in space, so bivariate local spatial autocorrelation is used to measure the spatial relationship between the independent variable in region i and the dependent variable in region j.

$$I_i = \frac{(Y_i - \overline{Y})}{S^2} \sum_{j=1}^{n} W_{ij}(Y_j - \overline{Y}) \quad (9)$$

where I_i represents the localized Moran's index for the ith region, $S^2 = \frac{1}{n}\sum_{i=1}^{n}(Y_i - \overline{Y})$, $\overline{Y} = \frac{1}{n}\sum_{i=1}^{n} Y_i$, Y_i and Y_j denote the observed values of the ith and jth regions, n is the total number of each research unit, and W_{ij} is the spatial weight matrix. Combined with the significance of I_i, local spatial autocorrelation maps can be drawn to identify cold hotspots where geographic elements are spatially clustered. There are four main types of spatial clustering: high–high (H-H type), low–high (L-H type), low–low (L-L type), and high–low (H-L type).

Spatial Durbin Model

The spatial Durbin model can examine the endogenous correlation of dependent variables and detect the direct and interactive effects of external factors, enabling a more accurate estimation of the spatial correlation of housing affordability in the Yangtze River Economic Belt and the degree of influence of the population–industry–land urbanization coupling degree, population urbanization, industry urbanization, land urbanization, and other impact factors [51].

$$y_{it} = \rho \sum_{j=1}^{n} w_{ij} y_{it} + \beta x_{it} + \theta \sum_{j=1}^{n} w_{ij} x_{it} + \mu_i + \varphi_t + \varepsilon_{it} \quad (10)$$

where y_{it} is the observed value of the dependent variable; x_{it} is the observed value of the independent variable; ρ reflects the spatial lag of the explanatory variable; θ is the spatial regression coefficient of the explanatory variable; β is the regression coefficient of the independent variable; φ_t and μ_i denote time fixed effects and spatial fixed effects, respectively; ε_{it} is a random error term obeying independent identical distribution, denoting other factors not included in the econometric model; and w_{ij} is the spatial weight matrix.

3. Results

3.1. Spatial Distribution and Evolution Characteristics

As shown in Figures 3 and 4, the spatial differentiation pattern and evolution characteristics of the cultivated land pressure index and urbanization level indicated that the spatial differentiation effect of the two is obvious.

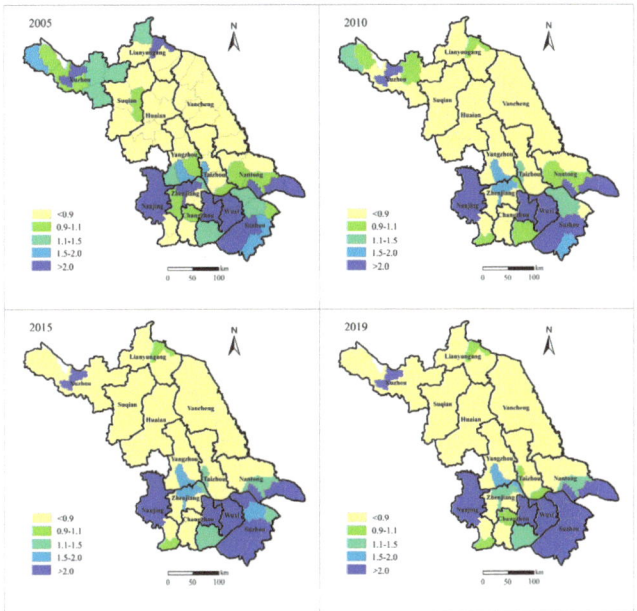

Figure 3. Spatial distribution of cultivated land pressure index.

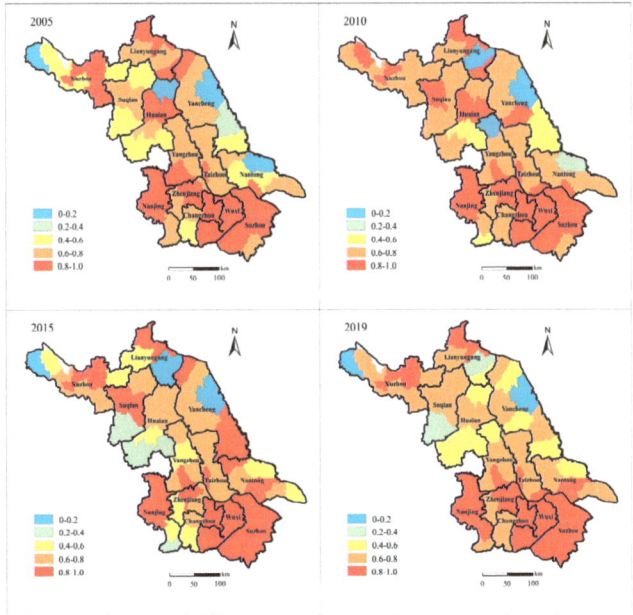

Figure 4. Spatial distribution of population–industry–land urbanization coupling coordination degree.

From 2005 to 2019, the spatial distribution of the cultivated land pressure index in Jiangsu Province has obvious spatial differentiation between the north and the south, with little change in spatial evolution. The areas with larger cultivated land pressure index values were mainly distributed in the municipal districts of the north-central Jiangsu Province region and in the southern Jiangsu Province region; while the areas with smaller index were mainly distributed in the north and center of Jiangsu Province as well as the main grain-producing areas in the south of Jiangsu Province.

From 2005 to 2019, there are certain characteristics in the spatial distribution of the population–industry–land urbanization coupling coordination degree in Jiangsu Province. The areas with high coupling coordination degrees were mainly distributed in the southern Jiangsu Province region, the municipal districts were also distributed in the central and northern Jiangsu Province region and clustered in the southern Jiangsu Province region. Study units with low and medium coupling coordination degrees were mainly distributed in central and northern Jiangsu Province, and the number of such units is gradually increasing.

As a result of the comparison, it is found that the cultivated land pressure index and the population–industry–land coupling coordination degree of high- and low-value agglomerations in the counties of Jiangsu Province showed certain overlapping characteristics. Study units with a high coupling coordination degree tend to have a higher cultivated land pressure index.

3.2. Exploratory Spatial Correlation Analysis of Urbanization and Cultivated Land Pressure Indices

In order to deeply analyze the spatial relationship between urbanization and cultivated land pressure in Jiangsu Province, based on the exploratory spatial analysis method, the univariate and bivariate Moran's I of urbanization and the cultivated land pressure index in Jiangsu Province were measured. Simultaneously, the local spatial autocorrelation of the urbanization and cultivated land pressure index was visualized and characterized. The results are shown in Table 2 and Figure 5. From 2005 to 2019, Jiangsu Province's cultivated land pressure index and population–industry–land urbanization coupling coordination degree had strong positive spatial accumulation, with the former showing an upward trend and the latter showing a downward trend. The global spatial autocorrelation between the population–industry–land urbanization coupling coordination degree and the cultivated land pressure index had an obvious positive spatial correlation and showed an increasing trend. It reflected that the spatial distribution of urbanization in Jiangsu Province showed a certain positive influence on the spatial distribution of cultivated land pressure.

Table 2. Moran's I value of cultivated land pressure index and population–industry–land urbanization coupling coordination degree.

Moran's I	Variable	2005	2010	2015	2019
Univariate	cultivated land pressure index	0.359 ($p = 0.001$)	0.235 ($p = 0.001$)	0.255 ($p = 0.001$)	0.343 ($p = 0.001$)
	population–industry–land urbanization coupling coordination degree	0.264 ($p = 0.001$)	0.22 ($p = 0.001$)	0.295 ($p = 0.001$)	0.322 ($p = 0.001$)
Bivariate	cultivated land pressure index–coupling coordination degree	0.235 ($p = 0.001$)	0.218 ($p = 0.001$)	0.251 ($p = 0.001$)	0.281 ($p = 0.001$)

From the results of the local spatial autocorrelation of the bivariate spatial autocorrelation of the population–industry–land urbanization coupling coordination degree and the cultivated pressure index in Jiangsu Province, the spatial distribution of the urbanization positively affects the cultivated land pressure index in adjacent study units in an absolute majority; the significant H-H zones (high urbanization–high cultivated pressure) are mainly located in Yancheng City, Huai'an City, Suqian City, and other areas in north and central Jiangsu Province, while the significant L-L zones (low urbanization-low cultivated pressure) are concentrated in Suzhou City, Wuxi City, Changzhou City, and other cities in southern Jiangsu Province. From 2005 to 2019, the significant L-L zone showed an increasing trend, which indicated that the positive spatial spillover effect of the overall urbanization level on the cultivated land pressure in the northern and central regions of Jiangsu was enhanced.

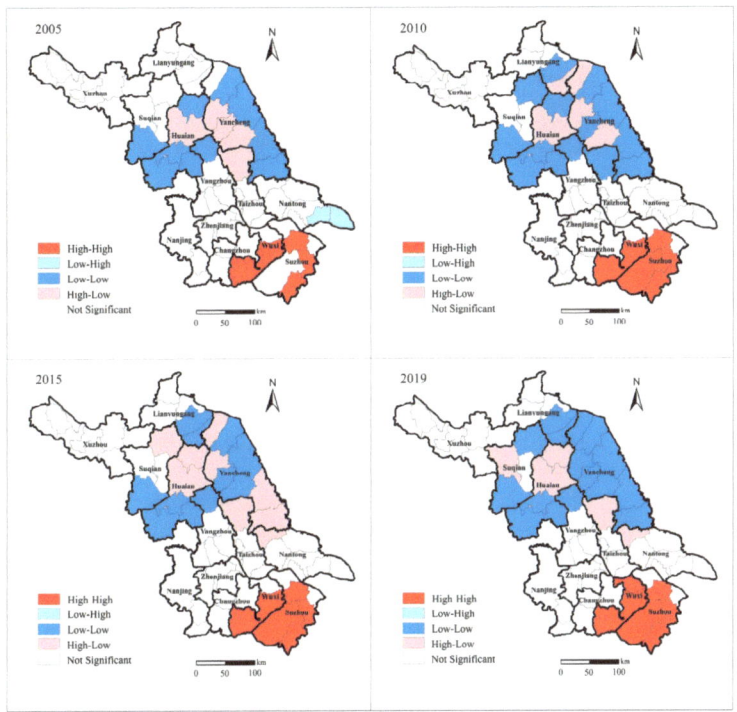

Figure 5. Spatial correlation pattern of population–industry–land urbanization coupling coordination degree and cultivated land pressure from 2005 to 2019.

3.3. Spatial Regression Analysis of Urbanization and Cultivated Land Pressure

3.3.1. Spatial Regression Analysis of the Impact of Total Urbanization on the Cultivated Land Pressure

Based on the panel data of 65 study units in Jiangsu Province from 2005 to 2019, the spatial Durbin model (SDM) was used to analyze the impact of total urbanization on the cultivated land pressure by taking the population–industry–land urbanization coupling coordination degree as the core explanatory variable and the cultivated land pressure index as the dependent variable. The number of observations was 975.

Firstly, the traditional mixed panel data model without spatial interaction effects was applied to the panel data of Jiangsu Province for estimation and residual testing to determine whether the spatial error model (SEM) and spatial lag model (SLM) were superior to the non-spatial model; then, the likelihood ratio (LR) test was conducted to determine whether SDM can be degraded to SEM or SLM; and finally, the Hausman test was conducted to determine whether to choose the fixed effects or random effects. After testing, the results showed that the SDM with fixed effects was the most appropriate. Generally, the fixed-effects estimates are given in the SDM, but due to the incorporation of the spatially lagged dependent and independent variables in the model, its marginal effects cannot be directly reflected, and it is difficult for its estimates to accurately measure the effect of the independent variables on the dependent variables [52]. Therefore, partial differential equations need to be used to calculate the direct and indirect effects of the respective variables [53]. The results of SDM parameter estimation are shown in Table 3.

Table 3. Regression results of SDM with the population–industry–land urbanization coupling coordination degree as the core explanatory variable.

Variable	Direct Effect	Indirect Effect
Population–industry–land urbanization coupling coordination degree	−0.018 ($p = 0.491$)	0.008 ($p = 0.923$)
Economic level	−0.000007 ($p < 0.001$)	0.000006 ($p = 0.869$)
Fertilizer application level	−1.234 ($p < 0.001$)	0.513 ($p = 0.558$)
Agriculture modernization level	0.021 ($p = 0.444$)	0.142 ($p = 0.031$)
Labor input level	−0.001 ($p = 0.909$)	0.01 ($p = 0.488$)
Water resource level	0.002 ($p < 0.001$)	0.008 ($p = 0.923$)

Note: We set the impact of influences to be significant when the p-value is less than 0.1; number of observations: 975.

It was found that the total urbanization level does not have a significant effect on cultivated land pressure. This indicated that the impacts of population urbanization, industry urbanization, and land urbanization on cultivated land pressure in Jiangsu Province are differentiated. Therefore, in the next step, population urbanization, industry urbanization, and land urbanization should be taken as the core explanatory variables to conduct regression analysis on the cultivated land pressure index again.

3.3.2. Spatial Regression Analysis of the Impact of Three Urbanization Dimensions on Cultivated Land Pressure

Taking population urbanization, industry urbanization, and land urbanization as the core explanatory variables, regression analysis was conducted using the SDM based on panel data of 65 research units in Jiangsu Province from 2005 to 2019. The number of observations was 975. After testing, the results showed that the SDM with fixed effects is the most suitable, and the parameter estimation results are shown in Table 4.

Table 4. Regression results of the SDM with population urbanization, industry urbanization, and land urbanization as the core explanatory variables.

Variable	Direct Effect	Indirect Effect
Population urbanization	−0.019 ($p = 0.06$)	−0.002 ($p = 0.02$)
Industry urbanization	−0.024 ($p = 0.07$)	0.058 ($p = 0.13$)
Land urbanization	4.06 ($p < 0.001$)	−1.424 ($p = 0.04$)
Economic level	−0.000013 ($p < 0.001$)	−0.00003 ($p = 0.578$)
Fertilizer application level	−1.32 ($p < 0.001$)	0.426 ($p = 0.507$)
Agriculture modernization level	0.067 ($p = 0.02$)	0.198 ($p = 0.002$)
Labor input level	−0.012 ($p = 0.128$)	−0.015 ($p = 0.34$)
Water resource level	0.0017 ($p < 0.001$)	−0.002 ($p < 0.001$)

Note: We set the impact of influences to be significant when the p-value is less than 0.1; number of observations: 975.

It was found that population urbanization had a significant negative impact on cultivated land pressure. This was due to the changes in food production caused by the transformation of the rural population status. The continued transfer of the rural population to the cities and the migration of surplus rural labor had caused a shortage of labor, but it had also helped the transfer of rural land and the realization of large-scale operations, which had increased the efficiency of food production and improved the food production level.

Industry urbanization had a significant negative impact on the cultivated land pressure index. From 2005 to 2019, the process of industrialization in Jiangsu Province continued to advance, with the industrialization rate increasing from 92.4% to 95.7%, so that industry urbanization reached a high level. With the rapid industry urbanization, the marginal effect of capital profit from the secondary and tertiary industries in the city decreased, and the flow of capital began to shift from the city to the countryside and from the secondary and tertiary industries to the primary industry. Urban areas supported the countryside, industry fed agriculture, and other related policies were implemented more and more vigorously; agricultural management was upgraded, agricultural technology was modernized, human habitat was improved, and rural development was changing rapidly. Therefore, industry urbanization played a positive role in promoting food production, thereby reducing the cultivated land pressure.

Land urbanization showed a significant positive effect on the cultivated land pressure index. Changes in the quantity and quality of the cultivated land affect food production. In 2005–2019, which was a period of rapid urban expansion in Jiangsu Province, the construction land area expanded by 2.5 times. As urbanization continued, regional urban construction land continued to expand, and the cultivated land area continued to shrink. In recent years, although China had put forward the requirement of "balanced occupation and compensation" for the occupation of cultivated land in the urbanization process, the status of "occupying the best and compensating for the worst" has been common. The area of high-quality cultivated land had been reduced, and the reduction in the quantity and quality of cultivated land had led to a decline in the level of food production, thereby exacerbating the pressure on arable land. The reduction in the quantity and quality of high-quality cultivated land directly contributes to the decline in food production, which in turn aggravates cultivated land pressure.

Industry urbanization had no significant effect on adjacent study units, which indicated that industry urbanization had no spatial spillover effect on cultivated land pressure in adjacent areas. Population urbanization and land urbanization had significant negative spatial spillover effects on cultivated land pressure in adjacent areas. With the rapid advancement of population urbanization and land urbanization in the southern Jiangsu region and the important cities in the northern and central Jiangsu regions, there was an agglomeration of population, industry, and capital and a large number of urban land-use indicators. This led to the loss of agricultural employees, the slow process of land urbanization, and a continuous increase in per capita arable land area in the surrounding relatively underdeveloped areas. As advanced agricultural technologies and management models were promoted and large-scale cultivation was realized, the food production capacity of these areas increased.

4. Discussion and Conclusions

4.1. Discussion

The spatial distribution of urbanization and cultivated land pressure in the counties of Jiangsu Province had a clustering effect and an obvious positive spatial correlation characteristic. Therefore, it is necessary to consider the spatial effect when analyzing the impact of urbanization on cultivated land pressure. It was found that total urbanization had no significant effect on the cultivated land pressure level in Jiangsu Province, and we further analyzed the effects of three different dimensions of urbanization on cultivated land pressure. According to the results of Xu, population urbanization and industry urbanization played a positive role in the level of the food security threshold at the scale of Jiangsu Province, while land urbanization showed a negative role [35]. This study confirms this result at the county scale of Jiangsu Province. This paper has two theoretical contributions. First, this paper considered spatial and temporal factors in analyzing the impact of urbanization on the cultivated land pressure index, and it explored the spatial spillover effect of the former on the latter based on the SDM. Secondly, this study explored not only the impact of total urbanization on the cultivated land pressure, but also the impact of population urbanization, industry urbanization, and land urbanization on the cultivated land pressure. Therefore, this paper analyzed the research problem by adopting a sound methodology and derived reasonable conclusions, which contribute to the study of rural geography.

This paper also has shortcomings. The impact of urbanization on food security is dynamic, and due to limitations of length, this paper does not measure the impact of urbanization on food security for each year, while this measurement idea could be more intuitive for reflecting the changes in the impact of urbanization on food security within an interval. The urbanization and cultivated land pressure between the north and south of Jiangsu Province showed large differences, and in the future, the authors will conduct regression analyses on the samples from the north and the south, separately, and make comparisons. Moreover, the authors will also conduct a comparative analysis of the impact of urbanization on food security between provinces with different economic development levels.

4.2. Conclusions

The impact of regional urbanization on food security has received extensive attention from academics in recent years. For this paper, a cultivated land pressure index model was constructed to analyze the spatial pattern of food security in Jiangsu Province, and the spatial relationship of urbanization and cultivated land pressure as well as the impact of urbanization on cultivated land pressure were analyzed based on spatial econometrics. The main conclusions are as follows:

An obvious north–south divergence was shown in the spatial distribution of the cultivated land pressure index, with the low-value areas of the cultivated land pressure index mainly distributed in northern and central Jiangsu and the high-value areas mainly distributed in southern Jiangsu. The urbanization level and cultivated land pressure level in Jiangsu Province showed obvious spatial clustering characteristics, and there was a certain overlap between the high- and low-value clustering areas of the two, with significant positive spatial correlation features. The total urbanization had no significant effect on the cultivated land pressure. Population urbanization and industry urbanization showed a significant negative effect on cultivated land pressure in Jiangsu Province, while land urbanization showed a positive effect. Both population and land urbanization had a significant negative spatial spillover effect on cultivated land pressure.

The findings of this paper offer two policy recommendations for Jiangsu Province. First, optimize the urban–rural development paths. In northern and central Jiangsu Province, the government should set out to educate and train the labor force left behind in the countryside to become professional farmers, improve the quality of the labor force, actively renovate the hollow villages, level the land, realize land transfer, and promote the large-scale operation of cultivated land relying on advanced management technology, so as to improve the regional food supply capacity. In southern Jiangsu Province, the government should optimize the land-use mechanism, avoid disorderly urban expansion, conscientiously implement the land-occupancy-supplement balance policy, and implement the most stringent system of cultivated land protection to safeguard the level of food production in counties and cities. Second, build a regional coordination mechanism for food supply and demand. Cultivated land pressure in Jiangsu Province shows a spatially differentiated pattern of high in northern and central Jiangsu and low in southern Jiangsu. The government should promote the coordination of food supply and demand between northern, central, and southern Jiangsu and encourage cooperation between food-producing and food-selling regions. Meanwhile, communication in the agricultural sector between counties should also be strengthened to achieve higher levels of food production through the spatial spillover effects of agricultural and management technologies.

Author Contributions: Conceptualization, J.K. and X.D.; Methodology, J.K. and R.Y.; Formal analysis, R.Y.; Resources, X.D.; Data curation, J.K.; Writing—original draft, J.K.; Writing—review & editing, J.K.; Visualization, J.K. All authors have read and agreed to the published version of the manuscript.

Funding: This research was funded by National Key R&D Program of China, grant number 2018YFD1100100.

Data Availability Statement: The publicly available sources for the data used in this study have been described in Section 2.2 of the article. The data in this paper include two aspects: socio-economic data and land urbanization data. The socio-economic data were obtained from the Jiangsu Statistical Yearbook from 2006 to 2020. The land urbanization data were mainly obtained from the China Earth System Science Data Sharing Platform, and the ENVI5.3 software was used to decode and classify the land-use remote sensing data, extract and vectorize the spatial extent of construction land development in each region, finally use it as the basis to calculate the area of regional construction land.

Conflicts of Interest: The authors declare no conflict of interest.

References

1. Skinner, M.W.; Kuhn, R.G.; Joseph, A.E. Agricultural land protection in China: A case study of local governance in Zhejiang Province. *Land Use Policy* **2001**, *18*, 329–340. [CrossRef]
2. Kastner, T.; Rivas, M.J.I.; Koch, W.; Nonhebel, S. Global changes in diets and the consequences for land requirements for food. *Proc. Natl. Acad. Sci. USA* **2012**, *109*, 6868–6872. [CrossRef] [PubMed]
3. Uddin, G.; Oserei, K. Positioning Nigeria's manufacturing and agricultural sectors for global competitiveness. *Growth Chang.* **2019**, *50*, 1218–1237. [CrossRef]
4. Bren, D.C.; Reitsma, F.; Baiocchi, G.; Barthel, S.; Güneralp, B. Future urban land expansion and implications for global croplands. *Proc. Natl. Acad. Sci. USA* **2016**, *114*, 8939–8944. [CrossRef] [PubMed]
5. Deng, Z.; Zhao, Q.; Bao, H.X.H. The impact of urbanization on farmland productivity: Implications for China's requisition-Compensation balance of farmland policy. *Land* **2020**, *9*, 311. [CrossRef]
6. Bureau of Statistics Website Statistical Bulletin on National Economic and Social Development of the People's Republic of China in 2011[EB/OL]. Available online: http://www.gov.cn/gzdt/2012-02/22/content_2073982.htm (accessed on 29 April 2023).
7. Bureau of Statistics Website Statistical Bulletin on National Economic and Social Development of the People's Republic of China in 2020[EB/OL]. Available online: https://www.gov.cn/xinwen/2021-02/28/content_5589283.htm (accessed on 29 April 2023).
8. Li, F.; Zhang, J.; Yang, X. Research on the Food Security Pattern of Space and Time in China and the Prediction of Security Situation. *Popul. J.* **2016**, *38*, 29–38.
9. Lichtenberg, E.; Ding, C. Assessing farmland protection policy in China. *Land Use Policy* **2008**, *25*, 59–68. [CrossRef]
10. Cheng, L.; Jiang, P.; Chen, W.; Li, M.; Wang, L.; Gong, Y.; Pian, Y.; Xia, N.; Duan, Y.; Huang, Q. Farmland protection policies and rapid urbanization in China: A case study for Changzhou City. *Land Use Policy* **2015**, *48*, 552–566.
11. Li, Y.; Fang, B.; Li, Y.; Feng, W.; Yin, X. Spatiotemporal Pattern of Cultivated Land Pressure and Its Influencing Factors in the Huaihai Economic Zone, China. *Chin. Geogr. Sci.* **2023**, *33*, 287–303. [CrossRef]
12. Yu, D.; Hu, S.; Tong, L.; Xia, C. Spatiotemporal dynamics of cultivated Land and its influences on grain production potential in Hunan Province, China. *Land* **2020**, *9*, 510. [CrossRef]
13. Xin, L.; Li, X. China should not massively reclaim new farmland. *Land Use Policy* **2018**, *72*, 12–15. [CrossRef]
14. Liao, L.; Long, H.; Gao, X.; Ma, E. Effects of land use transitions and rural aging on agricultural production in China's farming area: A perspective from changing labor employing quantity in the planting industry. *Land Use Policy* **2019**, *88*, 104152. [CrossRef]
15. Lu, C. Does household laborer migration promote farmland abandonment in China? *Growth Chang.* **2020**, *51*, 1804–1836. [CrossRef]
16. Liu, Y.; Zhang, Z.; Wang, J. Regional differentiation and comprehensive regionalization scheme of modern agriculture in China. *Acta Geogr. Sin.* **2018**, *73*, 203–218.
17. Long, H.; Tu, S.; Ge, D. The allocation and management of critical resources in rural China under restructuring: Problems and prospects. *J. Rural Stud.* **2016**, *47*, 392–412. [CrossRef]
18. Luo, H.; Zhu, Q.; Luo, Y.; Huang, X. Study on the Ecological Footprint and Ecological Capacity of Cultivated Land—Based on the Panel Data of Main Grain Producing Areas of China: 2007–2016. *East China Econ. Manag.* **2019**, *33*, 68–75.
19. Xu, J.; Ding, Y.; Liu, H. Simulation of Food Security in Jiangsu Province by System Dynamic Model. *J. Cent. Univ. Financ. Econ.* **2014**, *327*, 95–104.
20. Hu, M.; Pang, Y.; Jin, T.; Li, Z. Spatio-Temporal Evolution of EIWB and Influencing Factors: An Empirical Study from the Yangtze River Delta. *Appl. Spat. Anal. Policy* **2021**, *35*, 999–1024. [CrossRef]
21. Hu, M.; Sarwar, S.; Li, Z. Spatio-Temporal Differentiation Mode and Threshold Effect of Yangtze River Delta Urban Ecological Well-Being Performance Based on Network DEA. *Sustainabilty* **2021**, *13*, 4550. [CrossRef]
22. Peters, C.J.; Picardy, J.; Darrouzet-Nardi, A.F.; Wilkins, J.L.; Griffin, T.S.; Fick, G.W. Carrying capacity of US agricultural land: Ten diet scenarios. *Elem.-Sci. Anthr.* **2016**, *4*, 116. [CrossRef]
23. Jayne, T.S.; Chamberlin, J.; Headey, D.D. Land pressures, the evolution of farming systems, and development strategies in Africa: A synthesis. *Food Policy* **2014**, *39*, 1–17. [CrossRef]
24. Kelsee, B.; Eman, G. Modeling Urban Encroachment on the Agricultural Land of the Eastern Nile Delta Using Remote Sensing and a GIS-Based Markov Chain Model. *Land* **2018**, *7*, 114.
25. Luo, X.; Luo, J.; Zhang, L. Farmland pressure and China's urbanisation: An empirical study with the view of geographical differences. *Chin. J. Popul. Sci.* **2015**, *4*, 47–59, 127.
26. Christiansen, F. Food security, urbanization and social stability in China. *J. Agrar. Chang.* **2009**, *9*, 548–575. [CrossRef]
27. Wang, Y.; Bai, J.; Liu, Z.; Qi, J. Theoretical and practical research on the driving force of cultivated land quality improvement. *J. Henan Agric. Univ.* **2020**, *54*, 905–912.
28. Leng, Z.; Fu, C. The influence of the disequilibrium development of urbanization on foods security. *Economist* **2014**, *192*, 58–65.
29. Gong, B.; Wang, L. Study on the Impact of Urbanization on Food Security in Central China: On Empirical Evidence of 14 Prefectures and Municipalities in Hunan Province. *J. Urban Stud.* **2019**, *40*, 88–93.
30. Hanjra, M.A.; Qureshi, M.E. Global Water Crisis and Future Food Security in An Era of Climate Change. *Food Policy* **2010**, *35*, 365–377. [CrossRef]
31. Xu, G.; Wang, Y. A study on food security in China in the urbanization process. *J. Chin. Youth Soc. Sci.* **2013**, *32*, 120–127.

32. Gao, Y.; Zhang, Z.; Wei, S.; Wang, Z. Impact of urbanization on food security: Evidence from provincial panel data in China. *Resour. Sci.* **2019**, *41*, 1462–1474. [CrossRef]
33. Luo, X.; Zeng, J.; Zhu, Y.; Zhang, L. Who will feed China: The role and explanation of China's farmland pressure in food security. *Geogr. Res.* **2016**, *35*, 2216–2226.
34. Zhao, L.; Hou, D.; Wang, Y.; Yin, N. Study on the Impact of Urbanization on the Environmental Technology Efficiency of Grain Production. *China Popul. Resour. Environ.* **2017**, *27*, 106–114.
35. Xu, J.; Zha, T. Provincial Food Security from the Perspective of Urbanization in Jiangsu Province. *Resour. Sci.* **2014**, *36*, 2353–2360.
36. Bian, X.; Chen, H.; Cao, G. Patterns of regional urbanization and its implications: An empirical study of the Yangtze River Delta region. *Geogr. Res.* **2013**, *32*, 2281–2291.
37. Kong, X.; Xie, S.; Zhu, S.; He, Y.; Yin, C. Spatiotemporal Differentiation and Dynamic Coupling of Urbanization of Population, Land and Industry in Hubei Province. *Econ. Geogr.* **2019**, *39*, 93–100.
38. Ni, P.; Dong, Y. Market-Determined New Urbanization: An Analytical Framework. *Reform* **2014**, *246*, 82–93.
39. Jones, A.D.; Ngure, F.M.; Gretel, P.; Young, S.L. What Are We Assessing When We Measure Food Security? A Compendium and Review of Current Metrics. *Adv. Nutr.* **2013**, *4*, 481–505. [CrossRef]
40. FAO. *Food Balance Sheets: A Handbook*; FAO: Rome, Italy, 2001.
41. International Food Policy Research Institute. *Concern Worldwide, Welthungerhilfe. Global Hunger Index 2012. The Challenge of Hunger: Ensuring Sustainable Food Security under Land, Water, and Energy Stress*; International Food Policy Research Institute: Bonn, Germany; Washington, DC, USA; Dublin, Ireland, 2012.
42. Economist Intelligence Unit. *Global food Security Index Examines the Core Issues of Food Affordability, Availability and Quality*; Economist Intelligence Unit: London, UK, 2012.
43. Han, W.; Zhang, X.; Zheng, X. Land use regulation and urban land value: Evidence from China. *Land Use Policy* **2020**, *92*, 104432. [CrossRef]
44. Li, T.; Long, H.; Zhang, Y.; Tu, S.; Ge, D.; Li, Y.; Hu, B. Analysis of the spatial mismatch of grain production and farmland resources in China based on the potential crop rotation system. *Land Use Policy* **2017**, *60*, 26–36. [CrossRef]
45. Zhou, Y.; Li, X.; Liu, Y. Land use change and driving factors in rural China during the period 1995–2015. *Land Use Policy* **2020**, *99*, 105048. [CrossRef]
46. Liu, Y.; Zhou, Y. Reflections on China's food security and land use policy under rapid urbanization. *Land Use Policy* **2021**, *109*, 105699. [CrossRef]
47. Zhang, H.; Wang, Y. Spatial differentiation of cropland pressure and its socio-economic factors in China based on panel data of 342 prefectural-level units. *Geogr. Res.* **2017**, *36*, 731–742.
48. Lu, X.; Liu, R.; Kuang, B. Regional differences and dynamic evolution of cultivated land pressure in Hubei Province. *Trans. Chin. Soc. Agric. Eng.* **2019**, *35*, 266–272.
49. National Development and Reform Commission Website on China's Grain Security Program for Medium and Long-Term (2008–2020)[EB/OL]. Available online: https://www.ndrc.gov.cn/xwdt/xwfb/200811/t20081114_957261.html (accessed on 2 May 2023).
50. Wang, J.; Li, X.; Christakos, G.; Liao, Y.; Zhang, T.; Gu, X.; Zheng, X. Geographical detectors based health risk assessment and its application in the neural tube defects study of the Heshun region, China. *Int. J. Geogr. Inf. Sci.* **2010**, *24*, 107–127. [CrossRef]
51. Zheng, G.; Shimou, Y.; Changyan, W. Spatial-temporal pattern of industrial soot and dust emissions in China and its influencing factors. *Sci. Geogr. Sin.* **2020**, *40*, 1949–1957.
52. Lesage, J.; Pace, R.K. *Introduction to Spatial Econometrics*; CRC Press: New York, NY, USA, 2009.
53. Matlab[EB/OL]. Available online: http://www.regroningen.nl/elhorst/doc/Matlab-paper.pdf (accessed on 2 May 2023).

Disclaimer/Publisher's Note: The statements, opinions and data contained in all publications are solely those of the individual author(s) and contributor(s) and not of MDPI and/or the editor(s). MDPI and/or the editor(s) disclaim responsibility for any injury to people or property resulting from any ideas, methods, instructions or products referred to in the content.

Article

The Regional Effect of Land Transfer on Green Total Factor Productivity in the Yangtze River Delta: A Spatial Econometric Investigation

Wenqin Yan [1,*] and Dongsheng Yan [2]

[1] Center for the Yangtze River Delta's Socioeconomic Development, Nanjing University, Nanjing 210093, China
[2] School of Public Administration, Hohai University, Nanjing 211100, China; yds1223@hhu.edu.cn
* Correspondence: 602022020041@smail.nju.edu.cn

Abstract: This paper investigates the spatial mechanisms and impacts of land transfer on green total factor productivity (GTFP) in the economically dynamic Yangtze River Delta region of China. Using urban-level panel data from 2007 to 2020 and applying spatial econometric models, the study examines the relationship between land transfer and GTFP. The results of the spatial econometric analysis show that land transfer in the overall Yangtze River Delta region contributes positively to the improvement of GTFP. The mediating mechanism of industrial restructuring and upgrading shows statistically significant effects. Further investigation reveals differences in the spatial interdependence of land transfer on the GTFP among cities in different regions. Land transfer in the core area has significant indirect effects on the GTFP of neighboring cities, while the impact of land transfer in peripheral cities on the GTFP of surrounding cities is less discernible. This suggests that there is still a need for further deepening and development of integration in peripheral cities, as factor integration is still insufficient. The findings of this study provide useful insights for local governments in optimizing land transfer practices and promoting industrial transformation, upgrading, and sustainable green development.

Keywords: land transfer; green total factor productivity; Yangtze River Delta; spatial heterogeneity effects

Citation: Yan, W.; Yan, D. The Regional Effect of Land Transfer on Green Total Factor Productivity in the Yangtze River Delta: A Spatial Econometric Investigation. *Land* **2023**, *12*, 1794. https://doi.org/10.3390/land12091794

Academic Editor: Hossein Azadi

Received: 12 August 2023
Revised: 2 September 2023
Accepted: 13 September 2023
Published: 15 September 2023

Copyright: © 2023 by the authors. Licensee MDPI, Basel, Switzerland. This article is an open access article distributed under the terms and conditions of the Creative Commons Attribution (CC BY) license (https://creativecommons.org/licenses/by/4.0/).

1. Introduction

Economic sustainability is a crucial issue for most developing countries, including China. In the years since the launch of China's reform and opening-up policies, its economy has achieved sustained and rapid growth. However, this growth has also brought negative impacts, such as overconsumption of resources and premature depletion of environmental capacity. By promoting green transformation in the economic sector, both the goals of energy conservation and emission reduction and sustainable green development can be achieved [1,2]. In the process of industrial structure adjustment and regional economic development, land use plays a crucial role [3]. As an important component of the land system, land transfer activities can allocate land resources to different industries through land transfer prices and scale and can affect the adjustment and upgrading of industrial structure, thereby affecting macroeconomic sustainable development [4]. Therefore, this study will focus on how land transfer activities affect sustainable green development and provide theoretical support for promoting the dual goals of stable economic development and energy conservation and emission reduction.

Previous research on this topic can be broadly divided into three aspects. The first aspect is land allocation and productivity, which shows that there is a close relationship between land resource allocation and productivity. Optimizing the allocation of agricultural land resources is of great importance in improving agricultural production and productivity [5–8]. In urban industrial production, misallocation of urban construction land resources will significantly reduce urban technological innovation and production efficiency [9]. The second aspect is the study of green total factor productivity. With changes in the stage of economic growth, traditional total factor productivity no longer meets the needs of economic research. It only takes into account the input constraints of factors such as labor and capital while ignoring resource and environmental constraints. This can lead to distortions in the assessment of changes in social welfare and in the evaluation of economic performance, thereby leading to misleading policy proposals [10]. Many scholars have tried to incorporate environmental factors into the efficiency and productivity analysis framework to empirically study the situation of the Chinese economy [11]. With the gradual increase of resource and environmental constraints, scholars have begun to shift their focus to green total factor productivity, which has a "green" connotation [12–14]. The third aspect concerns the impact of land allocation on the sustainable development of the green economy. Optimal land allocation in the industrial sector promotes technological progress and industrial diffusion [2,15,16]. Innovations and improvements in production techniques are conducive to the convergence of human capital in technologically advanced industrial sectors [17]. As a result, a reorganization of the industrial layout can be achieved [18], leading to a reduction in pollutant emissions and the preservation of the ecological environment while promoting sustainable development [19,20].

However, based on a comprehensive review of existing research, there is still a lack of literature exploring the impact of land allocation on GTFP, particularly in terms of analyzing 'spatial spillovers' and delving into the 'heterogeneity effects' of the regions studied. The current literature does not take into account the asymmetric nature of spatial spillovers of economic factors in different regions, nor does it provide an in-depth exploration of the mechanism of the role of land concessions on GTFP on this basis. In response to the above questions, this study uses data from 41 cities at or above the prefecture level in the Yangtze River Delta region of China from 2007 to 2020. First, it explores the mechanisms between GTFP measured by the SBM-DDF model and land transfer behavior, as well as industrial transformation and upgrading. It then empirically tests the spatial effects and heterogeneity of land transfer on GTFP. The aim is to provide policy recommendations for local governments in China on land transfer behavior and improving GTFP.

Compared to related studies, the marginal contributions of this study can be summarized as follows: (1) Integrating land transfer, industrial upgrading, and GTFP into a theoretical framework, exploring the underlying mechanisms of land transfer on GTFP, and conducting mediation effect tests. From the perspective of spatial spillovers, it provides a systematic approach to analyzing the relationships between them and the potential of developing existing theories in land use for green economic growth; (2) Considering the asymmetry of the spatial effects of different regional economic factors in reality, the traditional spatial weight matrices can no longer meet the real economic and social activity connections in the Yangtze River Delta region with the development of synergy. This study adopts a new asymmetric geographic economic weight matrix as a spatial matrix to test the spatial mechanism and impact of land transfer on GTFP. This contributes to the investigation of the mechanism between land transfer and GTFP from a methodological perspective; and (3) In order to comprehensively study the relationship between land transfer activities and GTFP, this paper examines the regional development disparities in the Yangtze River Delta region by dividing it into core and peripheral areas. This allows the regional heterogeneity characteristics of different areas to be examined separately, revealing the insufficient factor flow driven by urban integration in peripheral areas and the need to further deepen synergy development. These studies contribute to improving and enriching the existing literature.

2. Analysis of How Land Transfer Affects GTFP: Theoretical Considerations

Local governments in China have long adopted a land supply strategy of low-price agreements for industrial land and high-price agreements for residential and commercial land to promote rapid local economic development. This model has played a crucial role in increasing local tax revenues and employment opportunities. By attracting industrial enterprises through land supply conditions of zero or even negative land prices, it has driven sustained economic growth and made significant contributions to the stability and development of the local economy [21]. However, the country should not only take into account the increase in production value in the process of economic development but also pay more attention to environmental factors, which is in line with the actual economic production process but also reflects the concept of green development. We are committed to better understanding the economic forces driving these changes [22]. In the long-term "development-oriented land use" model implemented in China, large-scale investment attraction has shown mixed results. If land resources are allocated to production sectors with low value-added and low environmental standards, it leads to inefficiency in land resource allocation [23–25]. The extent of optimization of land allocation factors, which further drives the optimization of other production factor resources and affects regional GTFP [26,27]. Based on these considerations, this study proposes research Hypothesis 1.

Hypothesis 1. *Land transfer activities affect the green total factor productivity of a region.*

The rational allocation of land resources often relies on market-based and competitive pricing, which forces selected land transferees to adjust the factor structure of production inputs according to the principles of comparative advantage. This helps to increase the marginal production value of input factors and maximize cost compensation [28,29]. Overall, this promotes the transformation and upgrading of industrial structures [30–32]. From a specific perspective, on the one hand, the creation of barriers for foreign companies prevents traditional industries with low intensity and low value added from entering the market. On the other hand, this directly increases the production costs of enterprises [33,34], which in turn forces enterprises to upgrade their technology and transform their industries to adapt to higher production costs. Firms that fail to upgrade may choose to relocate from their current locations because they are unable to bear the increased costs, leading to regional shifts and spatial re-planning [16,35].

From the perspective of the overall regional industrial structure, the rational allocation of land transfers plays an important role in the selection of relevant industries within the region, thereby facilitating the increase in the concentration of high-value-added industries in this geographical area. In addition, this process will promote the upgrading and transformation of the regional industrial structure [17,36]. At the same time, the effects of industrial agglomeration, land allocation, and adjustment of industrial structure, among other economic factors, are closely related to the foundation of institutional systems [37]. By optimizing allocation and other methods, these factors influence the output of regional firms and hence GTFP. Based on the above analysis, this study proposes research Hypothesis 2.

Hypothesis 2. *Land transfer activities influence green total factor productivity by promoting the transformation and upgrading of industrial structures.*

The contribution of land and other factors of production to economic development varies across regions and shows spatial concentration phenomena [38]. Due to the existence of communication and interaction, local government actions such as land transfers can affect the development of neighboring cities. On the one hand, there are strong economic linkages between cities, including knowledge and technology diffusion and industrial synergies [39,40]. When a city implements a policy, it not only affects local economic development but also influences the economic efficiency and sustainable development of the surrounding areas through radiation-driven effects and spatial optimization [41,42]. On the other hand, the implementation of specific policies or measures in a region can affect the distribution of spatial resources. Central cities can attract resources from surrounding areas, leading to a so-called 'siphoning effect' on the development of neighboring cities [43]. This could potentially lead to behaviors that are detrimental to specialization and effective competition, such as seeking privileged treatment, thereby preventing neighboring cities from achieving an increase in green total factor productivity [44]. Therefore, based on the above analysis, this study proposes research Hypothesis 3.

Hypothesis 3. *Land transfer activities indirectly affect the GTFP of neighboring areas through spatial spillovers.*

According to the above theory analysis, we advance the following mechanism of land transfer affecting green total factor productivity (Figure 1):

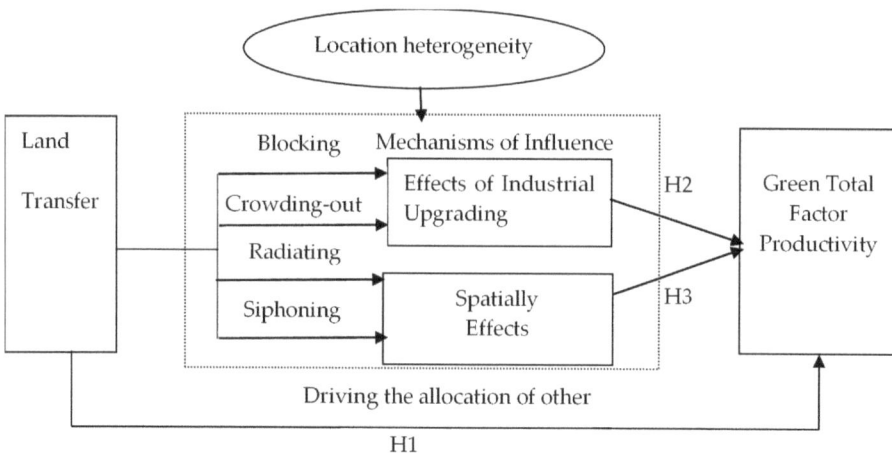

Figure 1. The mechanism of land transfer affecting green total factor productivity.

3. Empirical Analysis

3.1. Spatial Measurement Modeling

In the previous theoretical and mechanistic analyses, it was clear that within the process of land transfer influence on GTFP, there are intricate economic production relationships between regions that inevitably involve spatial correlations. To obtain estimates that are relatively accurate and reflective of reality, it is imperative to use a spatial econometric model, through which this study will analyze the impact effects of land transfer on the evolution of GTFP. The specific spatial econometric model used is as follows:

$$\text{GTFP}_{it} = \alpha + \beta X_{it} + \rho \sum_{j=1, j \neq i}^{N} w_{ij} \text{LAND}_{jt} + \theta \sum_{j=1}^{N} w_{ij} X_{ijt} + \mu_i + \nu_t + \varepsilon_{it}, \quad (1)$$

$$\varepsilon_{it} = \psi \sum_{j=1, j \neq i}^{N} w_{ij} \varepsilon_{jt} + \mu_{it}, \quad (2)$$

where ρ quantifies the indirect effects on GTFP. X represents a collection of explanatory and control variables, including land transfer. β measures the contribution of each explanatory variable to GTFP. θ, on the other hand, quantifies the indirect effects of explanatory variables from neighboring cities on GTFP. It also takes into account temporal and regional effects. Equation (1) represents the most general form of a nested spatial model, known as the GNS model, which includes all types of interaction effects. Since this study focuses primarily on the spatial effects of land transfer on GTFP, four spatial econometric models are used for the analysis: the spatial autoregressive model (SAR), the spatial error model (SEM), the spatial Durbin model (SDM), and the spatial lag model (SLX). The optimal spatial econometric model will be determined through appropriate testing methods.

3.2. Explanation of Variables and Data Sources

This article uses panel data from the Yangtze River Delta region from 2007 to 2020 as the study sample. Given the incompleteness of the data and for the sake of consistency, this paper selects panel data from 41 cities at the prefecture level and above in the Yangtze River Delta region. Smoothing or averaging methods are used to fill in the missing city data. Data are obtained from various sources, including the "Regional Statistical Yearbook" (2008–2021), the "China Urban Statistical Yearbook" the "China Energy Yearbook" the "China Environmental Yearbook" the "National Land and Resources Statistical Yearbook", land market websites, and provincial and municipal statistical yearbooks. Further details on the relevant variables are provided in the following explanations, while basic descriptions of the variables can be found in Table 1.

Table 1. Descriptive statistics of the regression variables.

Variable	Meaning	Number of Observations	Average Value	Standard Deviation	Minimum Value	Maximum Value
Green Total Factor Productivity	GTFP	574	0.9922	0.0463	0.8824	1.4314
Land Transfer	LAND	574	1.9378	1.4200	0.0801	10.5079
Industrial Structure Upgrading	STRU	574	6.6317	0.3234	5.8465	7.5180
Economic Openness	OPEN	574	0.0182	0.0156	0.0001	0.1581
Education	EDU	574	398.2487	291.8996	14.2114	1704.2800
Technological Development	SCI	574	0.0068	0.0135	0.0003	0.1627
Financial Development	FIN	574	1.5931	1.0479	0.4469	6.2709
Infrastructure	INFR	574	13.6756	5.9451	2.4330	37.5701
Per Capita Land Premium	PLAND	574	14,778.1700	13,950.7400	432.6385	86,302.7600

3.2.1. Core Variables

Green total factor productivity (GTFP), the main explanatory variable in this text, is a measure of productivity that incorporates environmental considerations. The traditional directional distance function, which is a radial and directional approach, can overestimate efficiency in the presence of slack variables and cannot simultaneously account for changes in input and output efficiency in a nonproportional manner [45]. Tone (2001) proposed a non-radial, nondirectional SBM model based on slack variables, which effectively addresses the issue of overestimated production efficiency [46]. Fukuyama and Weber (2009) introduced a non-radial, nondirectional SBM-DD on slack measures, combining the SBM model with directional distance functions. This approach allows the simultaneous nonproportional measurement of input and output efficiency factors [47]. Therefore, we use the superefficient SBM-DDF method for calculation purposes.

The indicators for measuring green total factor productivity (GTFP) are as follows:

(1) Input-side indicators: These include labor input, capital input, land input, and energy input. Labor input is measured by the number of employees in each urban jurisdiction. Capital inputs and land inputs are measured by the capital stock and the built-up area of the city jurisdictions, respectively, with the urban capital stock measured using the perpetual inventory method and deflated to the base year 2006 [48].

The energy input is measured by "fitting energy consumption data to global stable nighttime lighting values". The nighttime lighting remote sensing data are now widely used in research work in many fields, such as energy consumption [49]. It is a more reliable international practice to fit energy consumption to the global stable night lighting values [50–52]. The correlation between total energy consumption and total lighting is strong, and the night lighting data can reflect the spatial and temporal dynamics of energy consumption more reliably [53];

(2) Output-side indicators: These include expected output and unexpected output. Expected output is measured by the actual GDP of each city in the Yangtze River Delta region, deflated to the base year of 2006. Unexpected output is measured by the emissions

of three pollutants: industrial wastewater, carbon dioxide emissions, and particulate matter emissions. The number of raw data observations for measuring the GTFP input–output indicator is 4363, and the number of observations used for estimation in this paper is 4592 after filling the gaps using either the smoothing method or the mean method.

The variables of land transfer (LAND) and intermediate variables (STRU) are the main explanatory variables of this paper. Given the availability of data in the Yangtze River Delta region and the need for data consistency, this study examines the impact of land transfer on GTFP. In the economically prosperous Yangtze River Delta region, where the population continues to migrate and the land market is active, land transfer revenue accounts for a relatively high proportion of local comprehensive financial resources. Therefore, the ratio of total land transfer transaction value to general budget revenue is used as an indicator of land transfer. In addition, in a robustness analysis, we use per capita land premiums (PLAND) to investigate the mechanism of the impact of land transfer on GTFP.

The intermediate variable of industrial structure upgrading (STRU) is an important dimension in the process of industrial structural transformation and upgrading. It reflects the dynamic development of industrial structure from a lower level to a higher level in accordance with the historical and logical sequence of economic development. The measurement of industrial structure upgrading can generally be undertaken through indicators such as the coefficient of industrial structure hierarchy, Moore's structural change index, and the proportion of high-tech industries. We can consider the proportion of each industry in GDP as a component of a spatial vector and then combine them into a three-dimensional vector. Then, we can calculate the angles between these three-dimensional vectors and each industry vector separately [54].

$$\theta_j = \arccos\left(\frac{\sum_{i=1}^{3}(x_i, j \times x_i, 0)}{\left(\sum_{i=1}^{3}\left(x_i, j^2\right)\right)^{1/2} \times \left(\sum_{i=1}^{3}(x_i, 0^2)^{1/2}\right)}\right) \quad j = 1, 2, 3, \tag{3}$$

$$STRU = \sum_{k=1}^{3}\sum_{j=1}^{k}\theta_j, \tag{4}$$

in the formula, STRU indicates the upgrading of industrial structure, and its higher value indicates a higher level of advanced industrial structure.

3.2.2. Other Control Variables

The control variables selected for this study are as follows: the degree of openness (OPEN) is measured using the method commonly used in the literature, which consists of expressing the actual amount of foreign investment (converted at the average annual exchange rate) as a ratio of GDP. The level of education (EDU) is measured by the ratio of the number of students enrolled in higher education to the total population at the end of the year. The level of technological development (SCI) is measured by the ratio of scientific expenditure to regional GDP. The level of financial development (FIN) is represented by the ratio of the sum of deposits and loans in financial institutions to GDP. Infrastructure (INFR) is assessed by measuring the area of roads per capita.

3.3. Spatial Distribution Patterns Test

We use ArcGIS11.0 to visually depict the spatial distribution of GTFP and land transfer in the 41 cities at the sub-provincial level or above in the Yangtze River Delta (Figure 2). Representative years are selected for a preliminary examination of their spatial and temporal evolutionary features. Figure 2 shows that the spatial stratification of GTFP and land transfer has become more pronounced over time. The areas with high GTFP in each year are concentrated in Shanghai, Changzhou, Nanjing, Hangzhou, and Ningbo, among others, while the areas with high land transfer are concentrated in Jinhua, Lishui, Shaoxing,

Bozhou, and Suzhou, and the surrounding cities show significant diffusion. This indicates the potential spatial dependence between GTF and land transfer in the Yangtze River Delta region, suggesting the need to further consider the spatial effects of both variables in empirical analysis.

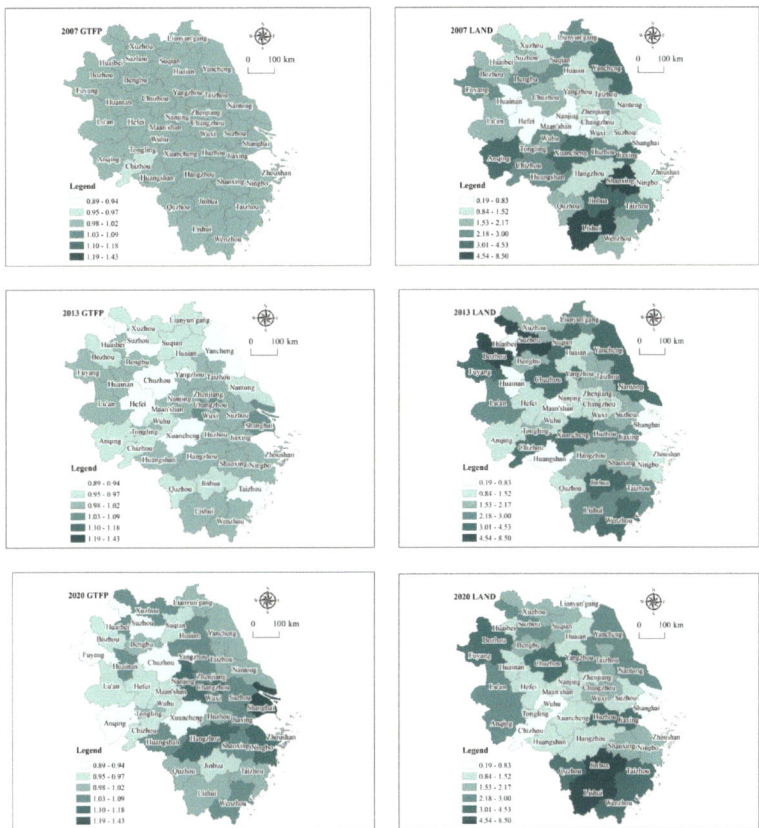

Figure 2. Spatial distribution patterns of GTFP and land transfer in the Yangtze River Delta region in 2007, 2013, and 2020.

3.4. Spatial Weight Matrix and Spatial Correlation Test

The spatial weight matrix represents the degree of connectivity between two cities. This study constructs two types of spatial weight matrices, namely, the traditional symmetric geographic distance weight matrix (W_d) and the novel asymmetric geographic economic weight matrix (W_{d-e}).

First, the geographical distance weight matrix calculates the geographical distances between cities based on the latitude and longitude of their administrative centers and takes the reciprocal of these distances. The form is given as follows: $W_d = 1/d_{ij}, i \neq j$.

Before conducting spatial econometric analysis using panel data, it is necessary to perform a spatial correlation test. In this study, Moran's I index is mainly used to measure spatial correlation, with values between −1 and 1. A Moran's index greater than 0 at a certain level of significance indicates a positive spatial correlation. The formula for calculating Moran's I index is as follows:

$$\text{Moran's I} = \frac{n}{\sum_{i=1}^{n}\sum_{j=1}^{n}W_{ij}} \times \frac{\sum_{i=1}^{n}\sum_{j=1}^{n}W_{ij}(X_i - \overline{X})(X_j - \overline{X})}{\sum_{i=1}^{n}(X_i - \overline{X})^2}. \tag{5}$$

Second, to realistically account for the asymmetry of spatial spillovers of different regional economic factors, we constructed a more realistic asymmetric spatial weight matrix for a more accurate spatial econometric analysis. Following Shao et al. (2020), we developed an asymmetric geographical–economic weight matrix that considers geographical distance and spatial economic linkages [55]. The new asymmetric geographical–economic weight matrix shows that economically prosperous regions have a greater impact on relatively underdeveloped regions, thereby highlighting the spatial weight. The matrix is calculated as follows:

$$W_{d-e} = \frac{1\text{perGDP}_j}{d_{ij}\text{perGDP}_i}, i \neq j. \tag{6}$$

According to Table 2, the panel global Moran's I values for GTFP are positive under both weight matrices, indicating the presence of positive spatial autocorrelation. The estimation results of the panel global Moran's I index in this study are consistent, and the Moran's I index for GTFP passes the 1% significance test with values of 0.1668 and 0.1606. Based on the above analysis, it is necessary to consider the spatial correlation between cities in the Yangtze River Delta when studying GTFP. The impact of land transfers on GTFP should be analyzed using a spatial econometric model.

Table 2. Global Moran's I value for green total factor productivity.

	Geographic Distance Matrix	Economic and Geographic Asymmetric Matrix
Moran's I-value	0.1668	0.1606
Moran's I-statistic	16.0149	14.7915
Significant value	0.0000	0.0000
Standard deviation	0.0106	0.0110

3.5. Model Setting Options

Before applying the appropriate spatial econometric model, it is necessary to carry out the following model selection diagnostics. First, the Lagrange multiplier method (LM) test is used to determine whether the spatial error model (SEM) or the spatial autoregressive model (SAR) should be chosen. Second, the likelihood ratio (LR) test is used to assess the joint significance of time and space effects, allowing the identification of spatial fixed effects, time fixed effects, and spatiotemporal fixed effects. Third, the Hausman test is used to determine the suitability of the spatial Durbin model (SDM). If the null hypothesis is rejected, fixed effects estimation methods should be used; otherwise, random effects estimation methods are appropriate. Fourth, the Wald test is used to determine whether the SDM is superior and more generally applicable than the SAR and SEMs and to further confirm whether the SDM can degenerate into the SAR or SEM. Table 3 presents the diagnostic report of the spatial econometric panel model under the geographical distance weighting matrix.

Based on the results presented in Table 3, first, the results of the Lagrange multiplier (LM) test indicate that both the spatial error model (SEM) and the spatial autoregressive model (SAR) are viable choices, passing the 1% significance threshold. Second, the results of the likelihood ratio (LR) test reject the null hypothesis at the 1% significance level, suggesting the presence of dual fixed effects related to both time and space. Third, the Haus results reject the null hypothesis, indicating the need to implement fixed-effects estimation methods. Finally, both the Wald and LR tests reject the null hypothesis at the 1% significance level, confirming the choice of the spatial Durbin (SDM), which degenerates into the SAR or SEMs. Based on these analyses, this study will use the SDM model with two fixed effects related to time and space for the empirical analysis. Additionally, the empirical results of the SEM, SAR, and SLX models are reported to ensure the robustness of the findings.

Table 3. Spatial econometric model applicability tests under the economic–geography matrix.

Test Content	Test Method	Test Result	
		Statistical Value	Significant Value
SAR and SEM tests	LM-lag test	91.3582	0.0000
	R-LM-lag test	13.3182	0.0000
	LM-err test	193.4393	0.0000
	R-LM-err test	115.3993	0.0000
Fixed-effects test	SFE-LR test (null hypothesis: no spatial-fixed effects)	490.7232	0.0000
	TFE-LR test (null hypothesis: no time-fixed effects)	129.3191	0.0000
	STFE-LR test (null hypothesis: no spatiotemporal-fixed effects)	528.4022	0.0000
Hausman test	Hausman test (null hypothesis: random effects model should be used)	45.9519	0.0000
Simplified test	Wald-lag test (null hypothesis: SDM model can be degraded to SAR model)	23.5968	0.0000
	LR-lag test (null hypothesis: SDM model can be degraded to SEM)	23.8793	0.0000
	Wald-err test (null hypothesis: SDM model can be degraded to SEM)	23.8510	0.0000
	LR-err test (null hypothesis: SDM model can be degraded to SAR model)	23.6513	0.0000

3.6. Results of Model Regression

The spatial weight matrix used in this study adopts an economically and geographically asymmetric matrix. To ensure robustness, the SAR, SEM, SLX, and SDM models were used for econometric estimation. The estimation results are presented in Table 4. From the results listed in Table 4, it can be seen that the spatial autoregressive coefficients of the aforementioned models are significantly greater than zero. Therefore, we can infer the existence of significant spatial interaction effects between the green total factor productivity (GTFP) of different cities in the Yangtze River Delta and their respective explanatory variables. Of particular note is the result derived from the spatial Durbin model (SDM), which shows a significant positive spatial autoregressive coefficient, indicating the presence of both exogenous and endogenous spatial interaction effects. The spatial panel estimation results in Table 4 show that the estimated coefficient of land transfer (Ln LAND) in the Yangtze River Delta region on green total factor productivity (Ln GTFP) is significantly positive at 1%, indicating that land transfer in this region contributes to the improvement of GTFP. In addition, within the spatial Durbin model, the spatial lag term of land transfer (W×Ln LAND) is significantly positive and passes the test of statistical significance. This confirms Hypothesis 3, showing that land transfer in neighboring areas generates spatial effects and consequently influences local GTFP.

However, the marginal impact of land transfer in the Yangtze River Delta region on GTFP cannot be fully and accurately interpreted by this estimation result [56]. It is necessary to further explain the impact of land transfer on the GTFP of a city in the region separately through direct and indirect effects. Therefore, this study analyzes the direct and indirect effects of the spatial Durbin model under an asymmetric economic–geographical distance weight matrix, as detailed in the estimation results in Table 5.

Table 4. Spatial panel estimation results of land transfer and green total factor productivity.

Variable	SAR	SEM	SLX	SDM
LnLAND	0.0046 *	0.0046 *	0.0066 ***	0.0066 ***
	(0.0026)	(0.0026)	(0.0025)	(0.0025)
W×LnLAND			0.0404 **	0.0410 *
			(0.0231)	(0.0228)
log-lik	1338.0017	1338.1157		1349.9414
Obs	574	574	574	574
R^2	0.7116	0.7124	0.7253	0.7251

Note: robustness standard errors are in brackets; * indicates $p < 0.10$, ** indicates $p < 0.05$, and *** indicates $p < 0.01$.

Table 5. Direct and indirect effects of land transfer on green total factor productivity.

Variable	Direct Effect		Indirect Effect		Total Effect	
	Coefficient	Standard Error	Coefficient	Standard Error	Coefficient	Standard Error
LnLAND	0.0067 ***	0.0025	0.0431 *	0.0236	0.0498 **	0.0238
LnOPEN	−0.0040 *	0.0021	−0.0593 **	0.0246	−0.0633 **	0.0249
LnEDU	−0.0144 ***	0.0051	0.1663 ***	0.0619	0.1519 **	0.0638
LnSCI	−0.0098 **	0.0043	−0.0668 *	0.0361	−0.0766 **	0.0362
LnFIN	−0.0400 ***	0.0086	0.0929	0.1007	0.0529	0.1023
LnINFR	−0.0132 **	0.0052	0.0536	0.0558	0.0404	0.0577

Note: * indicates $p < 0.10$, ** indicates $p < 0.05$, and *** indicates $p < 0.01$.

Based on the estimation results in Table 5, it is clear that land transfer in the Yangtze River Delta region not only increases the GTFP of the cities but also contributes to the increase in GTFP in neighboring cities. First, the direct effects are statistically significant at the 1% level and positive in nature. This suggests that land transfer can significantly promote the improvement of GTFP through channels such as technological progress and industrial layout adjustments. Second, the spatial spillovers of land transfer show significant positive effects. This means that land transfer activities in local cities in the Yangtze River Delta region are beneficial for improving the GTFP of neighboring cities. One possible reason for this is that land transfer activities in local cities stimulate further optimization of resource allocation between neighboring cities. In addition, it strengthens economic linkages and promotes healthy competition and effective cooperation between governments and enterprises, leading to positive spatial effects on GTFP.

The analysis of the remaining control variables shows that both the direct and indirect effects of economic openness are significantly negative. This implies that the actual use of foreign investment has a restraining effect on the improvement of GTFP in the Yangtze River Delta region. A possible reason for this is that the relatively high environmental regulatory intensity in the region increases the cost of foreign investment, which to some extent hinders the inflow of foreign capital and the formation of industrial competition. The direct effect of education and technology is negative, while the indirect effect of education on neighboring cities is positive. This reflects the fact that higher education human capital tends to be concentrated in technology- and capital-intensive areas, leading to competition for higher education capital and the outflow of regional human capital and other factors of production. This is detrimental to the growth of the local GTFP. In addition, there are barriers to technological innovation and barriers to technological imitation. In relative terms, actual production capabilities may be more important in influencing regional GTFP than the signaling or screening mechanisms reflected in educational attainment. The Yangtze River Delta region, being relatively developed, relies more on fossil energy for economic development. The economic benefits of technological progress, especially in renewable energy technology, have not yet been fully realized. They may be offset or even aggravated by negative environmental impacts, leading to a decline in GTFP.

The direct impact of finance and infrastructure is negative. This can be attributed to the current problem of resource waste and pollution in infrastructure development, which hinders the improvement of GTFP. The high degree of financial marketization in the Yangtze River Delta region, driven by profit motives, often leads investors to prioritize other options over green innovation projects. Financial institutions do not provide sufficient credit support for such projects, resulting in a misallocation of resources and hindering the development of GTFP. The indirect and total effects were not significant, reflecting the low investment efficiency of financial markets and their weak role in lowering barriers to advanced green technologies and enhancing the spillover of technological progress.

3.7. Robustness Check

In addition to reporting the empirical results of the SEM, SAR, and SLX in the sixth part of this chapter to demonstrate the robustness of the results, we considered replacing the spatial weight matrix and using per capita land premium (PLAND) as a new proxy variable for land transfer to test the robustness of its impact on GTFP.

Moreover, we all know that the year 2020 was affected by the COVID-19 pandemic. Whether the data is still representative? Whether the data still follows the same logic as the earlier years? To prove that, we selected the period ending in 2019 as a robustness check to see whether the findings still hold. The estimated results of the direct and indirect effects of these measures in the spatial Durbin model are presented in Tables 6–8.

Table 6. Robustness test results (I): replaced with geographic distance–spatial weighting matrix.

Variable	Direct Effect		Indirect Effect		Total Effect	
	Coefficient	Standard Error	Coefficient	Standard Error	Coefficient	Standard Error
LnLAND	0.0070 **	0.0027	0.0546 *	0.0331	0.0617 *	0.0337
LnOPEN	−0.0048 **	0.0022	−0.0695 **	0.0280	−0.0743 **	0.0287
LnEDU	−0.0147 ***	0.0052	0.1139 *	0.0667	0.0991	0.0688
LnSCI	−0.0107 **	0.0042	−0.0750 *	0.0384	−0.0857 **	0.0390
LnFIN	−0.0388 ***	0.0087	0.0531	0.1146	0.0143	0.1169
LnINFR	−0.0146 **	0.0059	−0.0312	0.0934	−0.0458	0.0962

Note: * indicates $p < 0.10$, ** indicates $p < 0.05$, and *** indicates $p < 0.01$.

Table 7. Robustness test result (II): replaced with per capita land premium (PLAND).

Variable	Direct Effect		Indirect Effect		Total Effect	
	Coefficient	Standard Error	Coefficient	Standard Error	Coefficient	Standard Error
LnPLAND	0.0037 *	0.0025	0.0524 **	0.0220	0.0562 **	0.0221
LnOPEN	−0.0049 **	0.0021	−0.0660 ***	0.0249	−0.0709 ***	0.0253
LnEDU	−0.0146 ***	0.0053	0.1562 ***	0.0588	0.1416 **	0.0602
LnSCI	−0.0104 **	0.0043	−0.0766 **	0.0368	−0.0870 **	0.0366
LnFIN	−0.0409 ***	0.0086	0.1069	0.0961	0.0660	0.0977
LnINFR	−0.0129 **	0.0052	0.0592	0.0525	0.0464	0.0543

Note: * indicates $p < 0.10$, ** indicates $p < 0.05$, and *** indicates $p < 0.01$.

Table 8. Robustness test result (III): replaced with data ending with 2019.

Variable	Direct Effect		Indirect Effect		Total Effect	
	Coefficient	Standard Error	Coefficient	Standard Error	Coefficient	Standard Error
LnLAND	0.0047 **	0.0023	0.0267	0.0209	0.0314 *	0.0211
LnOPEN	−0.0049 **	0.0020	−0.0242	0.0198	−0.0291	0.0199
LnEDU	−0.0157 ***	0.0044	0.0912 *	0.0510	0.0754	0.0525
LnSCI	−0.0088 **	0.0040	−0.0700 **	0.0312	−0.0787 **	0.0311
LnFIN	−0.0419 ***	0.0084	−0.0329	0.0881	0.0748	0.0895
LnINFR	0.0057	0.0051	0.0688	0.0514	0.0744	0.0529

Note: * indicates $p < 0.10$, ** indicates $p < 0.05$, and *** indicates $p < 0.01$.

The direct, indirect, and total effects of per capita land transfer premiums (PLAND), as a proxy for land transfer, on GTFP can be observed in Table 7. By replacing the spatial weight matrix and selecting the period ending with 2019, the effects of land transfer on GTFP can be observed in Tables 6 and 8 separately.

Overall, these results show the positive impact of land transfer on the improvement of GTFP in the Yangtze River Delta region. In addition, it shows a positive effect on the GTFP of neighboring cities. These results are in line with the estimates presented in Table 5. Moreover, the consistent interpretation of the control variables further strengthens the robustness of this study's investigation in terms of the mechanism of the model's impact on GTFP in the YRD region.

To further address the endogeneity issue arising from the causal relationship between the explanatory and dependent variables, this study adopts the spatially lagged variable model (SLX). Inspired by the approach of Zeng et al. (2019), the lagged one-period Ln LAND(-1) of land transfer, as well as its spatially lagged counterpart W×Ln LAND(-1), are used as instrumental variables in a two-stage least squares (2SLS) estimation [57]. This is an ideal form of testing for endogeneity problems in the spatial panel Durbin model [58]. The estimation results of the SLX model are presented in Table 9.

Table 9. SLX estimation results for land transfer and GTFP.

Variable	LnLAND	W×LnLAND	The Other Control Variables	R^2	F-Test: IV: Ln LAND(-1)	F-Test: IV: W×Ln LAND(-1)	Hausman Test
Estimation result	0.0125 ** (0.0062)	0.0384 * (0.0227)	YES	0.7459	109.66 [0.0000]	44.61 [0.0000]	11.3306 [0.000]

Note: robustness standard errors are in brackets; * indicates $p < 0.10$, ** indicates $p < 0.05$.

The results of the Hausman test in Table 9 indicate the existence of endogenous explanatory variables in the model. The results of the F-test indicate the appropriateness of the selection of instrumental variables, and the selected instrumental variables are highly correlated with the endogenous explanatory variables. The results show that land transfers are conducive to local GTFP enhancement and positive spatial spillovers to GTFP in surrounding areas. The core explanatory variables are consistent with the conclusions of the baseline regression, and the parameter estimates of the other control variables are also basically consistent. Thus, the robustness of the model's estimation results is established.

4. Further Discussion

4.1. Analysis of the Impact Mechanism of Land Transfer on GTFP

According to the hypotheses derived from previous theoretical mechanism research, this paper attempts to further explore the transmission mechanism of land transfer in the Yangtze River Delta on green total factor productivity through the analysis of industrial structural transformation and upgrading effects. Following the methods of Jiang [59], we construct the following model [60] to examine the role of land transfer in promoting industrial structural transformation and upgrading by replacing the dependent variable in the main regression with industrial upgrading (STRU):

$$STRU_{it} = \alpha'' + \beta'' LAND_{it} + \varphi'' z_{it} + \rho'' \sum_{j=1, j\neq i}^{N} w_{ij} STRU_{jt} + \theta'' \sum_{j=1}^{N} w_{ij} LAND_{ijt} + \lambda'' \sum_{j=1}^{N} w_{ij} z_{ijt} + \mu_i'' + \nu_t'' + \varepsilon_{it}'', \quad (7)$$

where z denotes the set of control variables, including the degree of openness to the outside world (OPEN), the level of education (EDU), the level of technological development (SCI), the level of financial development (FIN), and infrastructure (INFR).

From Table 10, it can be seen that the regression in Equation (7) empirically examines the direct and indirect effects of land transfer on the mechanism variable, the transformation and upgrading of industrial structure, and shows a significantly positive impact. This indicates that the improvement of resource allocation through land transfer significantly promotes the process of industrial structural upgrading. The influence of industrial structural transformation and upgrading on green total factor productivity is both direct and obvious. The theoretical analysis section of this paper elucidates this role based on the literature and logical reasoning.

Table 10. Direct and indirect effects of land transfer on industrial structure transformation and upgrading.

Variable	Direct Effect		Indirect Effect		Total Effect	
	Coefficient	Standard Error	Coefficient	Standard Error	Coefficient	Standard Error
LnLAND	0.0048 *	0.0027	0.3529 ***	0.0750	0.3577 ***	0.0770
LnOPEN	−0.0006	0.0013	0.0132	0.0433	0.0126	0.0438
LnEDU	0.0231 ***	0.0025	0.3622 ***	0.0914	0.3853 ***	0.0934
LnSCI	0.0057 ***	0.0019	0.2639 ***	0.0594	0.2696 ***	0.0610
LnFIN	0.0497 ***	0.0036	0.4318 ***	0.0906	0.4815 ***	0.0931
LnINFR	0.0462 ***	0.0044	1.2517 ***	0.1395	1.2979 ***	0.1431

Note: * indicates $p < 0.10$, *** indicates $p < 0.01$.

From the above analysis, it can be inferred that the transfer of land in the Yangtze River Delta region has been validated to improve green total factor productivity through the effects of industrial structural transformation and upgrading. By strategically allocating land resources to industries with high-value-added development needs, enterprises or projects can be attracted, including high-tech enterprises and institutions, facilitating industry–academia cooperation and creating an innovative ecosystem. Optimizing land allocation enhances the service-oriented nature of industries in the Yangtze River Delta while increasing the sophistication of manufacturing, thus driving the evolution of the industrial structure towards higher levels of competitiveness. This development enables industries to move up the value chain, while the application of green production methods and technologies improves efficiency and reduces costs. In addition, the process of industrial transformation and upgrading promotes the development of synergies among industries. Through effective industrial governance, such as the clustering of related industries in industrial parks or technological innovation zones, the provision of appropriate facilities and supporting measures promotes the exchange and sharing of technology, experience, and resources, further promoting sustainable development and the realization of a green economy. Ultimately, this contributes to the growth of GTFP.

4.2. Regional Comparison of Land Transfer and GTFP in the Yangtze River Delta Region

"The National New Urbanization Plan" emphasizes the need to enhance the radiating and driving functions of central cities, the need to promote coordinated integration between different types of cities, and the need to achieve integrated urban development. It has been recognized that there is a significant development gap between cities in the Yangtze River Delta, both in government-led planning and academic research. Therefore, based on current research [61], this paper divides the Yangtze River Delta region into a core area and a peripheral area. The core area comprises 16 cities, including Shanghai, Nanjing, Suzhou, Wuxi, Changzhou, Yangzhou, Zhenjiang, Taizhou, Nantong, Hangzhou, Ningbo, Shaoxing, Jiaxing, Taizhou, Huzhou, and Zhoushan. Historically, the level of development of these core cities, the early initiation of integration processes, and the depth of inter-city linkages have been considered important criteria for classification in macroplanning and related research.

First, we describe the land transfers and GTFP of the core and peripheral cities. From Table 11, it can be seen that the GTFP of the core area cities is higher than that of the peripheral area cities in terms of mean values and maximum–minimum values. This is in line with reality, as the core area cities are more developed and have stronger economic capabilities. They have made significant achievements in economic development, industrial innovation, and urbanization, with considerable efforts in environmental protection and energy conservation, and have also promoted the development of green industries. It is natural that their GTFP is higher than that of peripheral cities. However, when considering the development level and prospects of core area cities, it is necessary to further explore whether the close economic ties and cooperation among cities have facilitated the optimization of resource allocation and the exploitation of complementary advantages, actively promoting low-carbon transformation. Whether there is good interaction with surrounding cities needs to be further discussed before making judgments.

Table 11. Descriptive statistics on urban land transfer and GTFP in core and peripheral areas.

Variable	Meaning	Number of Observations	Average Value	Standard Deviation	Minimum Value	Maximum Value
GTFP in core cities	CG	224	1.0156	0.0566	0.9332	1.4314
GTFP in peripheral cities	PG	350	0.9773	0.0300	0.8824	1.0865
Land transfer in core cities	CL	224	1.6602	1.1190	0.1046	6.6504
Land transfer in peripheral cities	PL	350	2.1155	1.5586	0.0801	10.5079

The level of land transfer in core area cities is lower than that in peripheral area cities, both in terms of mean and maximum values. The minimum value for core area cities is higher than that of peripheral area cities, which may reflect the dependence of peripheral area cities on land fiscal policies compared to the robust land fiscal policies of core area cities. Of course, further judgment and in-depth analysis are needed to determine the specific impact of land transfer on GTFP in both core and peripheral area cities.

Next, looking at the regression results, this paper further measures the relationship between land transfer and GTFP in different regions through spatial analysis, and Table 12 reports the estimation results.

There are significant differences in the relationship between land transfer and GTFP in different regions, and land transfer has a greater impact on the GTFP of core cities. Land transfer in area cities mainly contributes to the improvement of the GTFP in neighboring cities. This suggests that in core cities, regional planning that matches supply and demand benefits coordinated development between core cities and the overall improvement of regional green economic sustainability. The impact of land transfer in peripheral areas on GTFP is mainly manifested in direct effects, while indirect effects are not significant. This suggests that the fiscal efficiency of land transfers in peripheral cities is insufficient. Although land transfer activities may generate some fiscal revenues, the impact of land transfers in cities in the peripheral zone on GTFP in neighboring cities is not yet evident relative to the influence and indirect effects of the core cities and their land markets. Peripheral cities typically have relatively weak economic development and industrial structure and limited competitiveness and spillover potential in land transfer activities. In addition, they often lack advantages in terms of infrastructure, talent, and market scale, resulting in inadequate spatial connectivity with the resources of neighboring cities, which limits the indirect effects of land transfer and hinders the highlighted impact on the sustainable economic development of surrounding cities.

Table 12. Decomposition of effects of spatial Durbin models for different regions.

	Type of Effect	LnLAND	LnOPEN	LnEDU	LnSCI	LnFIN	LnINFR
Core area	Direct effect	0.0044 (0.0054)	−0.0024 (0.0052)	−0.0505 *** (0.0101)	−0.0049 (0.0120)	−0.0796 *** (0.0181)	−0.0055 (0.0093)
	Indirect effect	0.0674 * (0.0397)	−0.0529 (0.0750)	−0.0806 (0.1196)	0.1132 (0.1774)	0.2373 (0.1938)	0.0567 (0.1480)
	Total effect	0.0719 * (0.0389)	−0.0553 (0.0754)	−0.1310 (0.1212)	0.1083 (0.1784)	0.1576 (0.1918)	0.0512 (0.1515)
Peripheral area	Direct effect	0.0033 * (0.0020)	−0.0033 * (0.0017)	0.0030 (0.0040)	−0.0069 ** (0.0030)	−0.0191 *** (0.0065)	−0.0248 *** (0.0047)
	Indirect effect	0.0296 (0.0594)	−0.0684 (0.0609)	0.3529 (0.3117)	−0.1359 (0.1138)	0.2030 (0.1720)	−0.1004 (0.1489)
	Total effect	0.0329 (0.0598)	−0.0717 (0.0614)	0.3559 (0.3126)	−0.1428 (0.1141)	0.1840 (0.1726)	−0.1252 (0.1496)

Note: * indicates $p < 0.10$, ** indicates $p < 0.05$, and *** indicates $p < 0.01$.

5. Conclusions and Policy Implications

5.1. Conclusions of the Study

This article examines the Yangtze River Delta as a research sample. Using panel data from 2007 to 2020, encompassing 41 cities at or above the prefectural level in the Yangtze River Delta region, this study explores the perspectives of industrial upgrading and spatial imbalances. It proposes and verifies that the rational allocation of land transfers promotes the upgrading and transformation of the regional industrial structure, further driving the optimal allocation of other production factors. This facilitates the expansion of high-value-added industries, thus enhancing the local GTFP. Simultaneously, due to the strong mobility of factors and economic interconnections within the Yangtze River Delta region, the radiation effect strengthens knowledge and technological diffusion between regions, optimizes industrial spatial patterns, and subsequently raises the GTFP of neighboring cities.

To examine the heterogeneous effects of regional development, the Yangtze River Delta region is divided into core and peripheral areas for analysis. It is found that land transfer in the core urban areas plays an important role in improving GTFP, while the efficiency of land transfer in the peripheral urban areas needs to be improved and has not significantly improved the green total factor productivity of the surrounding cities. The coordinated development of peripheral urban areas is still inadequate, lacking advantages in market scale and industrial clusters. Therefore, it is necessary to strengthen the free flow of factors and the construction of a unified large market to enhance the competitiveness and premium potential of land transfer activities.

5.2. Policy Implications

The significant spatial spillovers indicate the need to fully consider the formulation of multiple policies and multi-regional coordination, the acceleration of regional spatial factors and technology sharing. In addition, it is essential to enhance the capacity of innovation industry agglomeration to promote green production, promote the optimization and improvement of efficient industrial agglomeration and resource allocation, and ultimately establish a virtuous cycle and development path that enhance green total factor productivity.

The significant mediation mechanism shows the importance of rational allocation of land transfer revenue and the optimization and upgrading of industries. When regulating land transfer, it is necessary not only to stimulate the enthusiasm of local governments, restrict irrational competition for land, and guide the landing of high-quality funds, but also to guide the normal operation of the land market. This will make it possible to allocate land elements to industries with higher efficiency, integrate high-tech industries into the local industrial chain, enhance independent innovation capabilities, and improve the interrelation between industries. This will be more conducive to the improvement of green total factor productivity and ultimately achieve economic modernization and development.

The significant regional heterogeneity indicates that highly adaptable policies should be formulated according to the region's own resource endowments, economic development, and industrial transformation and upgrading needs. The spatial linkage development between cities should be planned based on different degrees of integration to determine the proportion of land transfer. It is necessary to pay more attention to regional coordination, break down administrative barriers that hinder the integration process, deepen regional cooperation, and promote effective interaction between land transfer activities and corresponding urban development. Mobilizing land resources to optimize land allocation, standardizing land transfer activities, attracting high-value-added industries, and stimulating the related effects of advantageous industries are crucial. Strengthening the construction of industrial communities between regions, enhancing the spatial transfer and suitability of industries, strengthening the coordination of industrial chains, and improving the environmental policy system will ultimately drive the improvement of green total factor productivity.

5.3. Shortcomings and Outlook

At present, China's economic development is in a period of major adjustments in green and low-carbon transformation. The report of the Twentieth Congress of the Communist Party of China set out clear requirements for green development, stating that it is necessary to implement a comprehensive environmental protection strategy, develop green and low-carbon industries, coordinate industrial restructuring, pollution control, environmental protection, and response to climate change, and accelerate the green transformation of the development mode. The limitations of this study restrict the completeness of the input–output indicators of GTFP, which are not limited to energy consumption. They can be further improved by adopting comprehensive resource consumption measurements, which will allow for a more scientific and comprehensive calculation of GTFP in the Yangtze River Delta region.

Moreover, this study is based on macro-statistical data and focuses on 41 prefecture-level cities in the Yangtze River Delta region as research units. We attempt to understand the mechanism by which land transfer promotes industrial transformation and upgrading and generates spatial spillovers, thus affecting the regional GTFP at the macro level. However, it remains for future research to investigate whether the same behavioral logic and mechanisms apply at the microlevel of enterprises and to conduct more in-depth and detailed studies in this regard.

Finally, the mechanism analyses in this paper still have certain deficiencies and limitations, and subsequently, we will continue to improve and explore better methods to carry out the study.

Author Contributions: Conceptualization, W.Y. and D.Y.; materials and methods, W.Y.; formal analysis, D.Y.; writing—original draft preparation, W.Y. and D.Y.; writing—review and editing, W.Y. and D.Y.; supervision, D.Y.; project administration, D.Y.; funding acquisition, D.Y. All authors have read and agreed to the published version of the manuscript.

Funding: This paper was funded by the National Natural Science Foundation of China (42101183) and the Major Project of the Research Center for Yangtze River Delta Economic and Social Development, Nanjing University (No. CYD-2020018).

Data Availability Statement: The data used in this study can be found in the relevant publications and have already been cited in the text for illustration. The data presented in this study are available on request from the corresponding author.

Conflicts of Interest: The authors declare no conflict of interest.

References

1. Apergis, N.; Aye, G.C.; Barros, C.P.; Gupta, R.; Wanke, P. Energy efficiency of selected OECD countries: A slacks based model with undesirable outputs. *Energ. Econ.* **2015**, *51*, 45–53. [CrossRef]
2. Chen, W.; Shen, Y.; Wang, Y.A. Does industrial land price lead to industrial diffusion in China? An empirical study from a spatial perspective. *Sustain. Cities Soc.* **2018**, *40*, 307–316. [CrossRef]
3. Gerber, J.F. Conflicts over industrial tree plantations in the South: Who, how and why? *Glob. Environ. Change* **2011**, *21*, 165–176. [CrossRef]
4. Lu, N.C.; Wei, H.J.; Fan, W.G.; Xu, Z.H.; Wang, X.C.; Xing, K.X.; Dong, X.B.; Viglia, S.; Ulgiati, S. Multiple influences of land transfer in the integration of Beijing-Tianjin-Hebei region in China. *Ecol. Indic.* **2018**, *90*, 101–111. [CrossRef]
5. Brandt, L.; Leight, J.; Restuccia, D.; Adamopoulos, T. Misallocation, Selection and Productivity: A Quantitative Analysis with Panel Data from China. In *NBER Working Papers*; No. 23039; NBER: Cambridge, MA, USA, 2017.
6. Duranton, G.; Ghani, S.E.; Goswami, A.G.; Kerr, W.; Kerr, W.R. The Misallocation of Land and Other Factors of Production in India. In *World Bank Policy Research Working Paper*; No. 7221; The World Bank: Washington, DC, USA, 2015.
7. Restuccia, D. Factor misallocation and development. In *The New Palgrave Dictionary of Economics*; Working Paper No. 502; Palgrave Macmillan: London, UK, 2013.
8. Restuccia, D. Misallocation and aggregate productivity across time and space. *Can. J. Econ.* **2019**, *52*, 5–32. [CrossRef]
9. Xie, C.; Hu, H. China's land resource allocation and urban innovation: Mechanism discussion and empirical evidence. *China Indu Econ.* **2020**, *12*, 83–101.
10. Hailu, A.; Veeman, T.S. Environmentally sensitive productivity analysis of the Canadian pulp and paper industry, 1959-1994: An input distance function approach. *J. Environ. Econ. Manag.* **2000**, *40*, 251–274. [CrossRef]
11. Kaneko, S.; Managi, S. Environmental Productivity in China. *Econ. Bull.* **2004**, *17*, 1–10.
12. Jiakui, C.; Abbas, J.; Najam, H.; Liu, J.; Abbas, J. Green technological innovation, green finance, and financial development and their role in green total factor productivity: Empirical insights from China. *J. Clean Prod.* **2023**, *382*, 135131. [CrossRef]
13. Managi, S.; Kaneko, S. Economic growth and the environment in China: An empirical analysis of productivity. *Int. J. Glob. Environ. Issues* **2006**, *6*, 89–133. [CrossRef]
14. Watanabe, M.; Tanaka, K. Efficiency analysis of Chinese industry: A directional distance function approach. *Energ. Policy* **2007**, *35*, 6323–6331. [CrossRef]
15. Yang, Y.T.; Jiang, G.H.; Zheng, Q.Y.; Zhou, D.Y.; Li, Y.L. Does the land use structure change conform to the evolution law of industrial structure? An empirical study of Anhui Province, China. *Land Use Policy* **2019**, *81*, 657–667.
16. Zheng, D.; Shi, M.J. Industrial land policy, firm heterogeneity and firm location choice: Evidence from China. *Land Use Policy* **2018**, *76*, 58–67. [CrossRef]
17. Galor, O.; Tsiddon, D. Technological progress, mobility, and economic growth. *Am. Econ. Rev.* **1997**, *87*, 363–382.
18. Drucker, J. Regional Industrial Structure Concentration in the United States: Trends and Implications. *Econ. Geogr.* **2011**, *87*, 421–452. [CrossRef]
19. Ahmed, A.; Uddin, G.S.; Sohag, K. Biomass energy, technological progress and the environmental Kuznets curve: Evidence from selected European countries. *Biomass Bioenerg.* **2016**, *90*, 202–208. [CrossRef]
20. Jiang, L.; Chen, Y.; Zha, H.; Zhang, B.; Cui, Y.Z. Quantifying the Impact of Urban Sprawl on Green Total Factor Productivity in China: Based on Satellite Observation Data and Spatial Econometric Models. *Land* **2022**, *11*, 2120. [CrossRef]
21. Yang, Q.J.; Yang, Q.; Zhuo, P.; Yang, J. Industrial land grant and bottom-line competition in attracting investment quality—An empirical study based on panel data of Chinese prefecture-level cities from 2007 to 2011. *Manag. World* **2014**, *11*, 24–34.
22. Shapiro, J.S.; Walker, R. Why Is Pollution from US Manufacturing Declining? The Roles of Environmental Regulation, Productivity, and Trade. *Am. Econ. Rev.* **2018**, *108*, 3814–3854. [CrossRef]
23. Brandt, L.; Biesebroeck, J.V.; Zhang, Y. Creative Accounting or Creative Destruction? Firm-level Productivity Growth in Chinese Manufacturing. *J. Dev. Econ.* **2012**, *97*, 339–351. [CrossRef]
24. Restuccia, D.; Rogerson, R. Policy distortions and aggregate productivity with heterogeneous establishments. *Rev. Econ. Dynam.* **2008**, *11*, 707–720. [CrossRef]
25. Yu, B.L.; Fang, D.B.; Pan, Y.L.; Jia, Y.X. Countries' green total-factor productivity towards a low-carbon world: The role of energy trilemma. *Energy* **2023**, *278*, 127894. [CrossRef]
26. Han, Y.; Huang, L.; Wang, X. Does Industrial Structure Upgrading Improve Eco-Efficiency? *J. Quant. Tech. Econ.* **2016**, *33*, 40–598.
27. Li, T.H.; Ma, J.H.; Mo, B. Does the Land Market Have an Impact on Green Total Factor Productivity? A Case Study on China. *Land* **2021**, *10*, 595. [CrossRef]
28. Buera, F.J.; Kaboski, J.P.; Shin, Y. Finance and Development: A Tale of Two Sectors. *Am. Econ. Rev.* **2011**, *101*, 1964–2002. [CrossRef]
29. Lian, H.P.; Li, H.; Ko, K. Market-led transactions and illegal land use: Evidence from China. *Land Use Policy* **2019**, *84*, 12–20. [CrossRef]
30. Friedrich, P.; Nam, C.W. Innovation-oriented Land-use Policy at the Sub-national Level: Case Study from Germany. *Stud. Reg. Sci.* **2013**, *43*, 223–240. [CrossRef]
31. Merikull, J. The Impact of Innovation on Employment Firm- and Industry-Level Evidence from a Catching-Up Economy. *East. Eur. Econ.* **2010**, *48*, 25–38. [CrossRef]

32. Zhao, X.; Nakonieczny, J.; Jabeen, F.; Shahzad, U.; Jia, W.X. Does green innovation induce green total factor productivity? Novel findings from Chinese city level data. *Technol. Forecast. Soc.* **2022**, *185*, 122021. [CrossRef]
33. Albouy, D.; Ehrlich, G. Housing productivity and the social cost of land-use restrictions. *J. Urban Econ.* **2018**, *107*, 101–120. [CrossRef]
34. Kok, N.; Monkkonen, P.; Quigley, J.M. Land use regulations and the value of land and housing: An intra-metropolitan analysis. *J. Urban Econ.* **2014**, *81*, 136–148. [CrossRef]
35. Shu, H.; Xiong, P.P. Reallocation planning of urban industrial land for structure optimization and emission reduction: A practical analysis of urban agglomeration in China's Yangtze River Delta. *Land Use Policy* **2019**, *81*, 604–623. [CrossRef]
36. Krugman, P.; Venables, A.J. Globalization and the Inequality of Nations. *Q. J. Econ.* **1995**, *110*, 857–880. [CrossRef]
37. Luo, Y.S.; Mensah, C.N.; Lu, Z.N.; Wu, C. Environmental regulation and green total factor productivity in China: A perspective of Porter's and Compliance Hypothesis. *Ecol. Indic.* **2022**, *145*, 109744. [CrossRef]
38. Okabe, T.; Kam, T. Regional economic growth disparities: A political economy perspective. *Eur. J. Polit. Econ.* **2017**, *46*, 26–39. [CrossRef]
39. Gu, B.M.; Liu, J.G.; Ji, Q. The effect of social sphere digitalization on green total factor productivity in China: Evidence from a dynamic spatial Durbin model. *J. Environ. Manag.* **2022**, *320*, 115946. [CrossRef]
40. Romer, P.M. Increasing Returns and Long-Run Growth. *J. Polit. Econ.* **1986**, *94*, 1002–1037. [CrossRef]
41. Martin, P.; Ottaviano, G.I. Growing locations: Industry location in a model of endogenous growth. *Eur. Econ. Rev.* **1999**, *43*, 281–302. [CrossRef]
42. Zhang, L.; Wang, Q.Y.; Zhang, M. Environmental regulation and CO_2 emissions: Based on strategic interaction of environmental governance. *Ecol. Complex.* **2021**, *45*, 100893. [CrossRef]
43. Cuberes, D.; Desmet, K.; Rappaport, J. Urban growth shadows. *J. Urban Econ.* **2021**, *123*, 1–48. [CrossRef]
44. Hsieh, C.T.; Klenow, P.J. Misallocation and Manufacturing Tfp in China and India. *Q. J. Econ.* **2009**, *124*, 1403–1448. [CrossRef]
45. Wang, B.; Wu, Y.; Yan, P. Regional environmental efficiency and environmental total factor productivity growth in China. *Econ. Res.* **2010**, *45*, 95–109.
46. Tone, K. A slacks-based measure of efficiency in data envelopment analysis. *Eur. J. Oper. Res.* **2001**, *130*, 498–509. [CrossRef]
47. Fukuyama, H.; Weber, W.L. A directional slacks-based measure of technical inefficiency. *Socio-Econ. Plan. Sci.* **2009**, *43*, 274–287. [CrossRef]
48. Young, A. Gold into base metals: Productivity growth in the People's Republic of China during the reform period. *J. Polit. Econ.* **2003**, *111*, 1220–1261. [CrossRef]
49. Chen, Z.Q.; Yu, B.L.; Yang, C.S.; Zhou, Y.Y.; Yao, S.J.; Qian, X.J.; Wang, C.X.; Wu, B.; Wu, J.P. An extended time series (2000–2018) of global NPP-VIIRS-like nighttime light data from a cross-sensor calibration. *Earth Syst. Sci. Data* **2021**, *13*, 889–906. [CrossRef]
50. Amaral, S.; Câmara, G.; Monteiro AM, V.; Quintanilha, J.A.; Elvidge, C.D. Estimating population and energy consumption in Brazilian Amazonia using DMSP night-time satellite data. *Comput. Environ. Urban Syst.* **2005**, *29*, 179–195. [CrossRef]
51. Chand, T.K.; Badarinath KV, S.; Elvidge, C.D.; Tuttle, B.T. Spatial characterization of electrical power consumption patterns over india using temporal dmsp-ols night-time satellite data. *Int. J. Remote Sens.* **2009**, *30*, 647–661. [CrossRef]
52. Elvidge, C.D.; Imhoff, M.L.; Baugh, K.E.; Hobson, V.R.; Nelson, I.; Safran, J.; Dietz, J.B.; Tuttle, B.T. Night-time lights of the world: 1994–1995. *ISPRS J. Photogramm.* **2001**, *56*, 81–99. [CrossRef]
53. Wu, J.S.; Niu, Y.; Peng, J.; Wang, Z.; Huang, X.L. Energy consumption dynamics in Chinese prefecture-level cities from 1995 to 2009 based on DMSP/OLS nighttime lighting data. *Geogr. Res.* **2014**, *33*, 625–634.
54. Fu, L. An empirical study on the relationship between industrial structure advancement and economic growth in China. *Stat. Res.* **2010**, *27*, 79–81.
55. Shao, S.; Zhang, Y.; Tian, Z.H.; Li, D.; Yang, L.L. The regional Dutch disease effect within China: A spatial econometric investigation. *Energ. Econ.* **2020**, *88*, 104766. [CrossRef]
56. Lesage, J.; Pace, R.K. *Introduction to Spatial Econometrics*; Chapman & Hall/CRC Press: Boca Raton, FL, USA, 2009.
57. Zeng, Y.; Feng, H.; Liu, J. Has the agglomeration of productive service industries improved the quality of urban economic growth? *Res. Quant. Econ. Tech. Econ.* **2019**, *36*, 83–100.
58. Vega, S.H.; Elhorst, J.P. The Slx Model. *J. Reg. Sci.* **2015**, *55*, 339–363. [CrossRef]
59. Jiang, T. Mediating and moderating effects in empirical studies of causal inference. *China Ind. Econ.* **2022**, *5*, 100–120.
60. Chen, Y.; Fan, Z.Y.; Gu, X.M.; Zhou, L.A. Arrival of Young Talent: The Send-Down Movement and Rural Education in China. *Am. Econ. Rev.* **2020**, *110*, 3393–3430. [CrossRef]
61. Yan, D.; Sun, W.; Sun, X. Study on the evolution of spatio-temporal population pattern and driving factors in the Yangtze River Delta. *Geoscience* **2020**, *40*, 1285–1292.

Disclaimer/Publisher's Note: The statements, opinions and data contained in all publications are solely those of the individual author(s) and contributor(s) and not of MDPI and/or the editor(s). MDPI and/or the editor(s) disclaim responsibility for any injury to people or property resulting from any ideas, methods, instructions or products referred to in the content.

Article

Analysis of the Impact of Land Use Change on Grain Production in Jiangsu Province, China

Xufeng Cao, Jiqin Han * and Xueying Li

College of Economics and Management, Nanjing Agricultural University, Nanjing 210095, China; 2013206009@njau.edu.cn (X.C.); 2021206010@stu.njau.edu.cn (X.L.)
* Correspondence: jhan@njau.edu.cn

Abstract: Located in the Yangtze River Delta region, Jiangsu Province has become the major grain production area of China and plays an important role in ensuring national food security. With rapid economic development and urbanization, the amount of cultivated land has decreased, which greatly affects food security. Based on the statistical data of grain production in Jiangsu Province since 2000 and the remote sensing data of 2000, 2010, and 2020, this paper used the stochastic frontier production function to calculate the output elasticity of various factors and the technical efficiency of grain production. The agglomeration effect of food production was investigated by using spatial correlation analysis. Finally, regression analysis was applied to examine the impact of land use change on grain yield and the technical efficiency of production. The results show that the grain-sown area is the decisive factor for the increase in grain output in Jiangsu Province. The technical efficiency of grain production in the province has been maintained at a relatively high level since 2000, showing a fluctuating upward trend, and the efficiency value in southern Jiangsu Province is greater than that in central and northern Jiangsu. The analysis of the spatial distribution characteristics of grain production technical efficiency shows that grain production has an agglomeration effect. The regression results showed that the complexity of land use and the density of the cultivated land patch were negatively correlated with grain yield and grain production technical efficiency, while the location of cultivated land was positively correlated with grain yield and grain production technical efficiency. The conclusion of this paper has important policy significance for promoting food production and ensuring food security.

Keywords: land use; grain production; food security; technical efficiency; influencing factors

1. Introduction

The Yangtze River Delta region has been a major grain production area in China since ancient times. With the development of industrialization and urbanization, this region has become an important manufacturing base of China. Since 2000, there has been an increase in population and the urban agglomeration scale in the Yangtze River Delta. The traditional grain production space has been greatly squeezed, and the per capita cultivated land area has been decreasing. With the rising living standard and migration of rural population to cities, the overall demand for food has significantly increased. The imbalance between supply and demand in food production has threatened China's food security [1].

Jiangsu province is an important economically developed area of the Yangtze River Delta region. In 2022, the GDP of the province reached CNY 12.29 trillion, accounting for 10.2% of national GDP. Meanwhile, it is an important grain production area of China. In 2022, Jiangsu province's grain output reached 37.69 million tons, accounting for 5.49% of China's total food production. However, with the rapid economic development of Jiangsu Province, a lot of cultivated land has been transformed into construction land, and the cultivated land area of the province has decreased from 5,016,300 hectares in 2000 to 4,075,900 hectares in 2020, a reduction of 18.7%. In addition to the decrease in the total

Citation: Cao, X.; Han, J.; Li, X. Analysis of the Impact of Land Use Change on Grain Production in Jiangsu Province, China. *Land* **2024**, *13*, 20. https://doi.org/10.3390/land13010020

Academic Editors: Wei Sun, Zhaoyuan Yu, Kun Yu, Weiyang Zhang and Jiawei Wu

Received: 19 November 2023
Revised: 16 December 2023
Accepted: 19 December 2023
Published: 22 December 2023

Copyright: © 2023 by the authors. Licensee MDPI, Basel, Switzerland. This article is an open access article distributed under the terms and conditions of the Creative Commons Attribution (CC BY) license (https://creativecommons.org/licenses/by/4.0/).

amount of cultivated land, the regional distribution of cultivated land has also changed, among which the total grain output and cultivated land area have shown a significant decline trend in Southern Jiangsu Province due to its higher urbanization rate. Therefore, food security is top of the agenda of all stakeholders of the food chain.

Increasing factor input to increase grain yield per unit area is an important measure to alleviate the contradiction between grain supply and demand caused by the decrease in cultivated land [2]. However, the input of factors cannot increase without limit. With the continuous input of factors, the marginal output of factors will continue to decline until zero, according to the law of diminishing marginal utility. Therefore, improving the technical efficiency of grain production has become an important way to ensure the increase in grain output [3].

The calculation of the technical efficiency of grain production is generally divided into non-parametric methods and parametric methods. Data envelopment analysis (DEA) is a more commonly used non-parametric method. Zhang et al. (2018) used the DEA method and Tobit regression model to examine grain production efficiency and its influencing factors in 13 major grain production areas of China from 2006 to 2016 [4]. Yadava et al. (2021) used the DEA method to calculate the technical efficiency of fertilizer use in India [5]. Salam et al. (2022) used the DEA method to calculate the technical efficiency of grain production in Punjab [6], and Berk et al. (2022) used the DEA method to calculate the resource utilization efficiency in corn production in Turkey [7]. The DEA method can only assess relative efficiency and ignores the effect of random errors. The advantage of SFA over DEA in terms of parametric methods is that it considers the influence of random error on the results. Determining the production function form in advance and then studying the production process can improve the accuracy of calculating technical efficiency and also allows for the analysis of the correlation between efficiency and influencing factors. Ghosh and Mazumdar (2018) calculated rice production efficiency in India using stochastic frontier production function [8]. Eguyen et al. (2003) analyzed the technical efficiency of rice production in the Mekong Delta of Vietnam based on the stochastic frontier function [9]. The stochastic frontier production function was applied to study the relationship between global warming and grain production efficiency [10] and it is also used to calculate the mechanical efficiency of rice production in China [11] and the technical efficiency of rice fertilizer use in China [12]. Alem (2021) analyzed the production performance of Norwegian grain production areas using a modified stochastic frontier production function [13]. Nathan et al. (2021) used the stochastic frontier production function to calculate the relationship between precision agriculture technology adoption and technical efficiency in the United States [14]. Chandel et al. (2022) used the stochastic frontier production function to calculate the technical efficiency of rice production in the Ganges Plain [15].

Regarding the influencing factors of grain production efficiency, Wang et al. (2011) measured and decomposed the grain production productivity of 138 counties in Hebei Province [16]. Their findings indicated that the improvement in agricultural technology, farmers' income, land consolidation, and other factors would improve grain production efficiency. With the rapid advancement of urbanization, the decrease in cultivated land resources and the increase in food demand have gradually become important constraints in the process of development. Therefore, scholars at home and abroad have conducted a wealth of studies on the impact of land use change on food production. First, land use change affects the grain-sown area, which in turn make an impact on grain production. A few studies have been conducted on changes in cultivated land and its impact on grain production in some areas of China, e.g., Chongqing, Inner Mongolia, Jiangsu, and Huang-Huai-hai Plain [17–20]. The results show that there is a causal relationship between land sown area and total grain production. Adjei et al. (2020) conducted a study on land use change and how this has an impact on food production in Ghana [21]. Liu et al. (2021) studied the overall characteristics of grain production changes in China and found that the change in sown area was the direct cause of grain crop yield changes at the national and regional scales. By exploring the impact of cultivated land use area on food security in

Baltic countries [22], Ambros and Granvik (2020) proposed that the blind construction of large-scale farms was of great harm to food security [23]. Bhermana et al. (2011) suggested the proposal of the intensification of cultivated land planning. Second, land use changes have an impact on grain production efficiency [24]. Based on the data of different land use types and statistical yearbook data of Henan Province from 2000 to 2018, Guo (2021) studied the spatial–temporal change in land use in Henan Province and its impact on grain production efficiency [3]. Rahman and Rahman (2009) explored the impact of land fragmentation on grain production efficiency in Bangladesh [25]. Manjunatha et al. (2013) studied the production efficiency of groundwater-irrigated farms in southern India and found that small farms had a relatively high resource utilization efficiency, while large farms faced resource waste due to land fragmentation and had a low production efficiency [26].

The current literature on the relationship between land use and grain production mainly focuses on the micro-scale, especially on the input of labor, agricultural machinery, and pesticides and fertilizers. The research data have mainly been obtained by sampling surveys. There are relatively few studies on the impact of land resources on grain production efficiency through remote sensing data. The mechanism of the effects of land use change on grain production efficiency still needs further investigation and understanding. Therefore, it is necessary to construct a systematic study on the impact of land use change on grain yield and efficiency. Based on the panel data of 53 regions in Jiangsu Province from 2000 to 2020, this paper estimated the grain production efficiency of this Province. Afterward, the key factors of the impact of land use change on grain production efficiency and yield were analyzed by using remote sensing images from 2000, 2010, and 2020 to interpret land use data. Finally, suggestions on land management are put forward so as to improve the technical efficiency of grain production.

2. Theoretical Research Framework

According to the production function theory, grain output is related to capital, labor force, and land input into grain production. For example, in Formula (1), Y represents total grain output, L represents capital, K represents labor force, and N represents land.

$$Y = f(L, K, N) \tag{1}$$

The capital input of general food production is replaced by actual material input, such as agricultural machinery and fertilizer, and the land input of food production is measured by the grain-sown area. Increasing agricultural machinery, fertilizer, labor force, and grain-sown area can all increase grain yield Y, but the limited cultivated land space in Jiangsu Province limits the maximum grain-sown area. Meanwhile, the marginal theory tells us that in the case of a certain sown area, increasing the input of agricultural machinery, fertilizer, and labor will have a diminishing marginal output effect, and the effect on the increase in grain output will become smaller. Therefore, the feasible way is to achieve the purpose of increasing grain production by improving technical efficiency.

On the one hand, the question is how to measure the existing technical efficiency, and how much technical inefficiency input is present in production? In this paper, the concept of total factor productivity is introduced, and the technical efficiency of grain production is measured by Solow residuals. By increasing the residuals in Formula (1) and selecting the appropriate grain production function, the technical efficiency of grain production can be obtained by estimating the residuals. On the other hand, what is the impact of land use change on technical efficiency? Industrialization and urbanization have a strong spatial competition for grain planting, which is unlikely to increase the existing cultivated land area and grain production area. However, through the adjustment and optimization of the spatial layout of cultivated land and grain production, it is possible to increase grain yield and technical efficiency.

According to the theory of spatial economics, location has a very important impact on resource allocation and production efficiency. Then, the change in the spatial layout of cultivated land will also have an important impact on the technical efficiency of grain production. With the continuous expansion of cities, cultivated land has become further away from suburban areas. The rising population and improvement in transportation networks has increased the fragmentation of cultivated land. The development of the manufacturing industry has also increased the intensity of land development and the complexity of land use. Continuous changes in the location, morphology, and surrounding environment of cultivated land have had a significant impact on food production.

This paper aims to establish a grain production function model of Jiangsu Province to verify that the sown area is the determining factor of grain output under the current agricultural development stage of Jiangsu Province. After calculating the technical efficiency of grain production, we analyze the effects of the spatial morphology of cultivated land on grain yield and technical efficiency. Finally, the spatial optimization scheme is proposed to streamline the spatial layout of grain production and improve grain productivity.

3. Methods and Data

3.1. Research Method

3.1.1. The Calculation Method of Grain Production Technical Efficiency

Stochastic frontier analysis is one of the most commonly used methods for measuring technical efficiency. The stochastic frontier analysis method was first proposed by Aigner [27] in 1977. After improvement and refinement by Battese and Coelli et al. [28], this model was widely adopted in subsequent research and has become the mainstream method for studying technical efficiency. In order to improve the accuracy of the model and better reflect the combined impact of different input factors on grain production efficiency in the production function, this study applied the translog stochastic frontier production function model for the analysis [29], as follows:

$$\ln(Y_{it}) = \beta_0 + \sum_j \beta_j \ln X_{jit} + \frac{1}{2}\sum_j \sum_l \beta_{jl} \ln X_{jit} \ln X_{lit} + \beta_t t \\ + \beta_{tt} t^2 + \sum_j \beta_{jt} t \ln X_{jit} + v_{it} - u_{it} \qquad (2)$$

$$TE_{it} = \exp(-u_{it}) \qquad (3)$$

Y represents the grain output of region i in year t. β_0 is the intercept term. $\beta_j, \beta_t, \beta_{jl}, \beta_{tt}, \beta_{jt}$ are the parameters to be estimated. t represents the year. X is the input factor. v and u are the error terms of the model, where v represents the random error term subject to normal distribution, u represents the technical inefficiency term that follows the truncated normal distribution, and v and u are independent of each other. TE_{it} represents the technical efficiency level of grain production in region i in year t. Input factors $X_{jit}(j = 1,2,3,4)$ are sown area, labor force, agricultural machinery power, and fertilizer application amount.

In order to overcome the shortcomings of a single model, Hansen and Racine [30] proposed the cutter model averaging method, which determines the weight by minimizing the cross-validation criterion, and then assigns different cutter weights to different models so as to obtain more comprehensive and comprehensive estimation results.

Assuming there are m candidate models, the knife cut the fit value of each candidate model's explanatory variable $\hat{y}^m = (\hat{y}_1^m, \cdots, \hat{y}_n^m)$, for which $\hat{y}_i^m$ is the fair value of the explanatory variable after excluding the i-th sample. The weight w_m of the knife cut model averaging method is a set of nonnegative vectors and sum to 1. The calculation formula is as follows:

$$w^* = \underset{w=(w_1,\cdots,w_M)\in\Omega_M}{\operatorname{argmin}} CV_n(w) = \frac{1}{n}\hat{e}(w)'\hat{e}(w) \\ = \frac{1}{n}\left(y - \sum_{m=1}^M w_m \hat{y}^m\right)'\left(y - \sum_{m=1}^M w_m \hat{y}^m\right) \qquad (4)$$

Among them, w^* is the cutting weight, $\hat{e}(w)$ is the weighted average residual, and $\sum_{m=1}^{M} w_m \hat{y}^m$ is the weighted average of the knife cut fit values. After obtaining the cutting weight w^*, the regression model of this study is as follows:

$$y_{it} = \sum_{m=1}^{M} w_m^* \hat{f}_m(x_{it}) = \sum_{m=1}^{M} w_m^* (\hat{\beta}_{0it}^m + \sum_{j=1}^{4} \hat{\beta}_{x_j}^m x_{jit} + \frac{1}{2} \sum_{j=1}^{4} \sum_{l=1}^{4} \hat{\beta}_{x_{jl}}^m x_{jit} x_{lit} \\ + \hat{\beta}_t^m t + \hat{\beta}_{tt}^m t^2 + \sum_{j=1}^{4} \hat{\beta}_{x_{jt}}^m t x_{jit} + \hat{v}_{it}^m - \hat{u}_{it}^m \quad (5)$$

Among them, w_m^* is the cutting weight of the m-th model, $\hat{f}_m(x_{it})$ is the regression result of the m-th model, and x_{jit} (j = 1,2,3,4) represents the sown area, labor force, agricultural machinery power, and fertilizer application amount. After deduction, the elasticity coefficient of the grain-sown area, labor force, agricultural machinery power, and fertilizer application amount of the final model is obtained as follows:

$$B_{x_j} = \sum_{m=1}^{M} w_m^* (\hat{\beta}_{x_j}^m + \sum_{l=1}^{4} \hat{\beta}_{x_{jl}}^m x_{lit} + \hat{\beta}_{x_{jt}}^m t)(j = 1,2,3,4) \quad (6)$$

3.1.2. Spatial Correlation Analysis of Technical Efficiency in Grain Production

Spatial econometric analysis is an important component of spatial statistics and an effective means of understanding spatial patterns [31]. For spatial econometric regression models, when the results are calculated, it is necessary to analyze the significance of their spatial correlation to see whether it is significantly different from zero. In the spatial error model, the common methods for the spatial autocorrelation test of random disturbance terms include Moran's I test, likelihood ratio (LR) test, Lagrange coefficient (LM) test, Wald test, etc. The difference is that Moran's I test is based on least squares estimates, while the likelihood ratio (LR) test, Lagrange coefficient (LM) test, and Wald test are based on maximum likelihood estimates.

Moran's I statistic for spatial autocorrelation is:

$$I = \frac{\sum_{i=1}^{n} \sum_{j=1}^{n} \omega_{ij} (x_i - \bar{x})(x_j - \bar{x})}{\frac{\sum_{i=1}^{n} (x_i - \bar{x})^2}{n} \sum_{i=1}^{n} \sum_{j=1}^{n} \omega_{ij}} \quad (7)$$

In the equation, x_j represents the grain production efficiency of regional units i and j, and ω_{ij} is the spatial weight value between i and j. This paper uses a Queen's second-order adjacency weight matrix, where Moran's I is progressive normal distribution. $\bar{x}$ is the average grain production efficiency in Jiangsu Province. The Moran's I index has a range of values between −1 and 1. At the given significance level, Moran's I index is greater than 0, indicating that there is a positive spatial autocorrelation of grain production efficiency, showing a clustering trend. Moran's I is less than 0, indicating that there is a negative spatial autocorrelation of grain production efficiency, showing a discrete trend. When Moran's I = 0, there is no spatial autocorrelation.

Moran's I spatial autocorrelation coefficient only reflects the spatial distribution pattern of the research phenomenon in the entire study area and cannot obtain the location of the aggregation area of the research phenomenon from it. To explore this phenomenon, Getis–Ord G_i^* statistics were used for local spatial autocorrelation analysis. In spatial clustering analysis, local clusters with higher values are called hotspots, while local clusters formed by lower values are called cold spots.

$$G_i^*(d) = \frac{\sum_{j=1}^{n} \omega_{ij}(d) x_j}{\sum_{j=1}^{n} x_j} \qquad (8)$$

In the formula, $\omega_{ij}(d)$ represents the spatial weight matrix, and x_j represents the sample values of region j. The larger the absolute value of G_i^* is, the more concentrated the grain production efficiency is, that is, the hot spot area is formed. When G_i^* is positive, this region is a positive hotspot region, and when G_i^* is negative, this region is a negative hotspot region. When the value of G_i^* is close to 0, it indicates that there is no aggregation of grain production efficiency.

3.1.3. Analysis of Influencing Factors of Technical Efficiency of Grain Production

The Tobit model is a restricted dependent variable regression model. Compared to the traditional OLS regression and discrete models, it can effectively avoid bias in the regression estimation process, and thus it has been widely used [32]. The explained variable in this paper is the technical efficiency of grain production. As the technical efficiency of grain production calculated by using the stochastic frontier production function ranges from 0 to 1, it is a typical constrained dependent variable. If OLS is used to regression the model, the results may be biased, so the Tobit regression model is chosen for analysis. The specific model is as follows:

$$TE_i = \begin{cases} \sigma_0 + \sum_k \sigma_k Z_{ki} + \xi_i, \delta_0 + \sum_k \delta_k Z_{ki} + \xi_i > 0 \\ 0, \delta_0 + \sum_k \delta_k Z_{ki} + \xi_i \leq 0 \end{cases} \qquad (9)$$

In the formula, TE represents the technical efficiency level of grain production. Z_{ki} is the independent variable that affects the efficiency of grain production. δ_0 and δ_k are the estimated parameters of the model. ξ_i is the model's error term, which belongs to a normal distribution.

Z_1 Land use diversity index [33,34] represents the impact of land use diversity, calculated using the Shannon index. The specific formula is as follows:

$$Z_1 = 1 - \sum d_i^2 / \left(\sum d_i\right)^2 \qquad (10)$$

In the formula, d_i represents the area of the i-th land use type in the study area, reflecting the complexity and utilization degree of land use types. The value ranges from 0 to 1. The larger the value, the higher the degree of land use diversity, and the lower the degree of land use diversity.

Z_2 Cultivated land patch density represents the degree of fragmentation of cultivated land, calculated as the ratio of the number of cultivated land patches to the area of cultivated land, i.e., the number of patches per unit area of cultivated land. The number of cultivated land patches is extracted from remote sensing image interpretation data using ArcGIS10.2 software.

Z_3 Cultivated land area ratio refers to the ratio of the cultivated land area to the total area in each region, reflecting the size of cultivated land in each region.

Z_4 Cultivated land distance refers to the linear distance from the geometric center of cultivated land to the administrative center in each region, reflecting the spatial positioning of cultivated land in the city.

Z_5 Land development intensity refers to the proportion of construction land area in the total area of a region. The high development intensity means that the proportion of construction land is high.

3.2. Data Source

Variables for calculating the technical efficiency of grain production include grain output, grain-sown area, labor force, agricultural machinery power, and fertilizer application amount (see Table 1). The data mainly come from the Jiangsu Provincial Statistical Yearbook (2000–2020). For data statistics and continuity convenience, statistical analysis was conducted based on the administrative regions divided by the 2020 Jiangsu Provincial Statistical Yearbook. Data from 53 sample districts were obtained. Because the statistical yearbook only includes statistics on the total power of agricultural machinery and the net amount of fertilizer application, it is not subdivided into various indicator values for grain production and use, except for the grain-sown area. Therefore, the data need to be processed accordingly. Drawing on relevant research and using the weight coefficient method [35] to separate the input factors of grain production, assuming that the material inputs per unit area of grain planting and other agricultural production are equal and that the labor force input and output are also equal, then:

Table 1. Statistical description of the main input variables for grain production efficiency.

Variable	Mean Value	Standard Deviation
Grain-sown area (1000 hectares)	404.707	240.468
Labor force (10,000 people)	259.970	114.927
Power of agricultural machinery (10,000 kilowatts)	306.853	175.893
Fertilizer application amount (10,000 tons)	25.080	17.880

Labor force = (Total agricultural output value/Total output value of agriculture, forestry, animal husbandry, and fishery) × (Grain-sown area/Total area of crop sown) × (Number of labor force in agriculture, forestry, animal husbandry, and fishery).

Power of agricultural machinery = (Grain-sown area/Total area of crop sown) × (Total power of agricultural machinery).

Fertilizer application amount = (Grain-sown area/Total area of crop sown) × (the net amount of chemical fertilizer applied).

According to the land use data interpreted from remote sensing images, there were 280,086, 216,224, and 218,805 land use patches in 2000, 2010, and 2020, respectively. Among them, the number of cultivated land patches was 39,398, 24,525, and 42,234. Based on this, the land use data of various districts and counties in Jiangsu Province were compiled and combined with partial statistical yearbook data, and the land use diversity index and cultivated land patch density were calculated.

4. Results

4.1. Output Elasticity of Grain Production Factors

The Stata 15 software was used to estimate three types of panel random frontier production functions, and the instruction jackknife was used to calculate the knife cut fit value $\hat{y}^m$ of the explanatory variables for the three models. Then, the respective weights were obtained according to the cross-validation criterion, as shown in Table 2.

Based on the estimated coefficients and cutting weights of different types of random frontier production functions, the elasticity coefficient is calculated. The calculation results are shown in Table 3.

Table 2. Estimation results of the production function of different types of stochastic frontiers.

Variable	BC92 Model	BC95 Model	Kumb90 Model
Grain-sown area	1.002	0.961	0.989
Labor force	0.035	0.029	−0.015
Power of agricultural machinery	0.052	0.0013	0.056
Fertilizer application amount	−0.036	−0.0176	−0.005
Technical efficiency	0.869	0.922	0.851
Knife cutting weight	0.335	0.326	0.339

Note: BC92 model is a stochastic frontier production function model proposed by Battese and Coelli in 1992. The expression of its inefficiency term is as follows: $u_{it} = \eta_t u_i = \{\exp[-\eta(t-T)]\}u_i$. BC95 model is a stochastic frontier production function model proposed by Battese and Coelli in 1995. The expression of its inefficiency term is as follows: $u_{it} = z_{it}\delta + w_{it}$. Kumb90 model is a stochastic frontier production function model proposed by Kumbhakar in 1990. The expression of its inefficiency term is as follows: $u_{it} = \gamma(t)u_i = \left[1 + \exp\left(bt + ct^2\right)\right]^{-1} + u_i$.

Table 3. Calculation of elastic coefficient values using the average method of cutting model.

Variable	Grain-Sown Area	Labor Force	Power of Agricultural Machinery	Fertilizer Application Amount
Elastic coefficient	0.984	0.016	0.037	−0.019

From the above table, it can be seen that the output elasticity of the grain-sown area, labor force, and power of agricultural machinery is positive. The output elasticity of the fertilizer application amount is negative. The results showed that increasing the input of the grain-sown area, labor force, and agricultural machinery power could increase grain yield, but increasing the input of fertilizer application would decrease grain yield. The yield elasticity of the grain-sown area was the largest, reaching 0.984, indicating that under the given technical conditions, keeping other input factors unchanged and increasing the sown area by an additional 1%, the grain output of the province would increase by 0.984%. This reflects the critical importance of land factors for food security in Jiangsu Province.

4.2. Technical Efficiency of Food Production

According to the calculation results of the Stata 15.0 software, the grain technical efficiency curve of each city in Jiangsu Province was drawn, as shown in Figure 1.

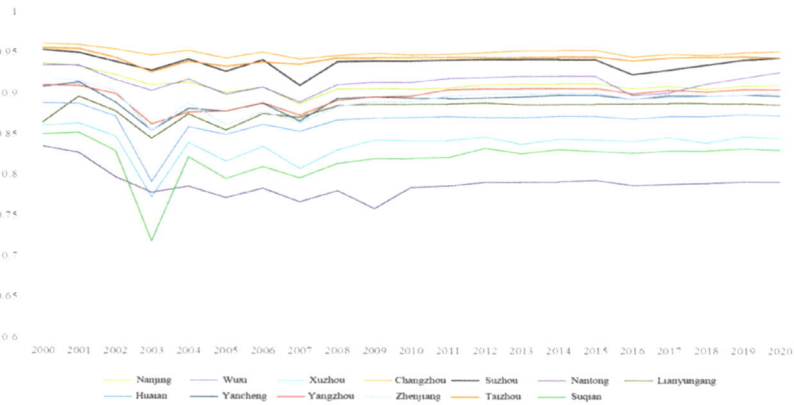

Figure 1. Grain production efficiency curve in Jiangsu Province.

It is calculated that the average technical efficiency of grain production is 0.880 during 2000–2020 in Jiangsu Province. On the whole, the production efficiency showed a rising trend in the fluctuation. The efficiency of grain production in various cities of Jiangsu

Province varies depending on the region. The efficiency of grain production in southern Jiangsu is higher, followed by central Jiangsu, and it is the lowest in northern Jiangsu.

Specifically, the average level of grain production efficiency in the southern region of Jiangsu is 0.921. Among the five cities, Changzhou has the highest average level of grain production efficiency at 0.950, while Zhenjiang has the lowest average level of grain production efficiency at 0.892.

The average level of grain production efficiency in the central Jiangsu region is 0.876. Among the three cities, Taizhou City has the highest average level of grain production efficiency at 0.943. Nantong City has a relatively low average level of grain production efficiency at only 0.789. It can be seen that there is a significant difference in the efficiency of grain production among different cities in the central Jiangsu region.

The average level of grain production efficiency in the northern Jiangsu region is 0.859. Among the five cities, Yancheng City has the highest average level of grain production efficiency at 0.891.

The three time nodes of 2000, 2010, and 2020 are selected. As can be seen from Figures 2 and 3, the spatial distribution of the grain production efficiency in Jiangsu Province has obvious changes.

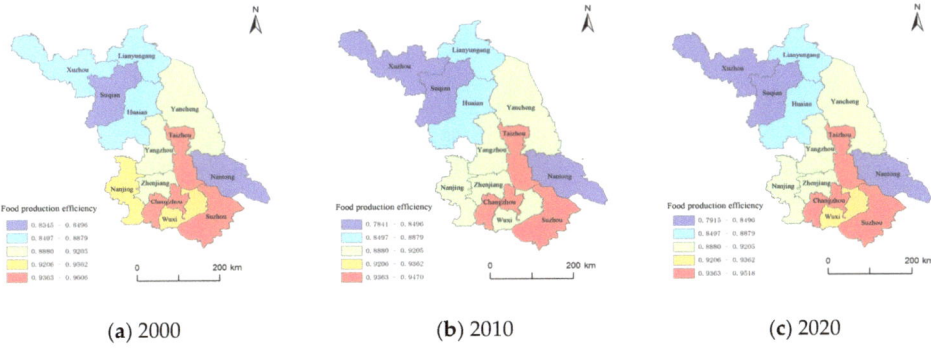

(**a**) 2000 (**b**) 2010 (**c**) 2020

Figure 2. Spatial distribution of grain production efficiency in Jiangsu Province in 2000, 2010, and 2020 (city level).

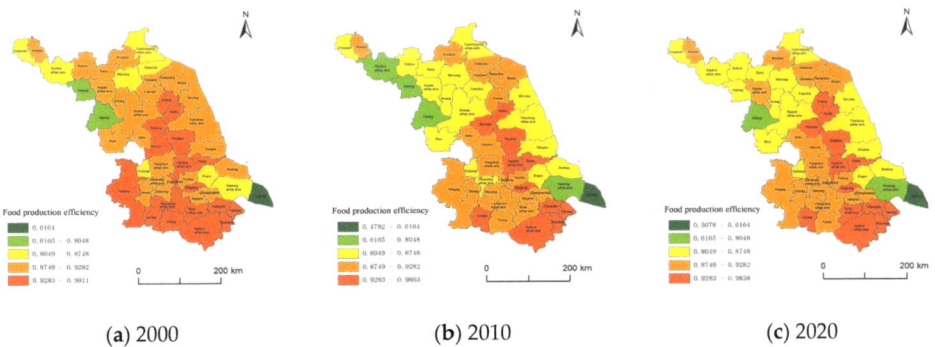

(**a**) 2000 (**b**) 2010 (**c**) 2020

Figure 3. Spatial distribution of grain production efficiency in Jiangsu Province in 2000, 2010, and 2020 (county level).

In order to further explore grain production efficiency at the county level, this paper analyzes the spatial characteristics of grain production efficiency in 2000, 2010, and 2020 from the perspective of the county level, as shown in Figure 3. From 2000 to 2010, the spatial distribution of the grain production efficiency mainly showed a decrease in grain production efficiency. From 2010 to 2020, except for some areas such as Xuzhou urban area,

Suqian urban area, Funing County, and Suining County, the rest of the region basically maintained a stable recovery of grain production efficiency.

4.3. Spatial Correlation of Technical Efficiency

The ArcGIS10.2 software was used to calculate the global Moran's I index of the spatial distribution of the grain production efficiency in Jiangsu Province in 2000, 2010, and 2020 from the city level and county level, respectively (Table 4), and all of the Moran's I indexes were positive. The results showed that the grain production efficiency in Jiangsu Province showed a positive spatial autocorrelation. The spatial agglomeration of grain production in Jiangsu Province showed that cities with higher (lower) grain production tended to cluster near each other.

Table 4. Global Moran index of grain production in Jiangsu Province in 2000, 2010, and 2020.

	Time	Moran's I	z	p
At the city level	2000	0.379	2.372	0.017
	2010	0.155	1.278	0.201
	2020	0.231	1.681	0.092
At the county level	2000	0.211	2.710	0.006
	2010	0.184	2.581	0.009
	2020	0.221	3.047	0.002

The Getis–Ord G*i index was used to calculate the grain production efficiency in Jiangsu Province in 2000, 2010, and 2020, and the spatial agglomeration distribution pattern of grain production in Jiangsu Province is shown in Figures 4 and 5.

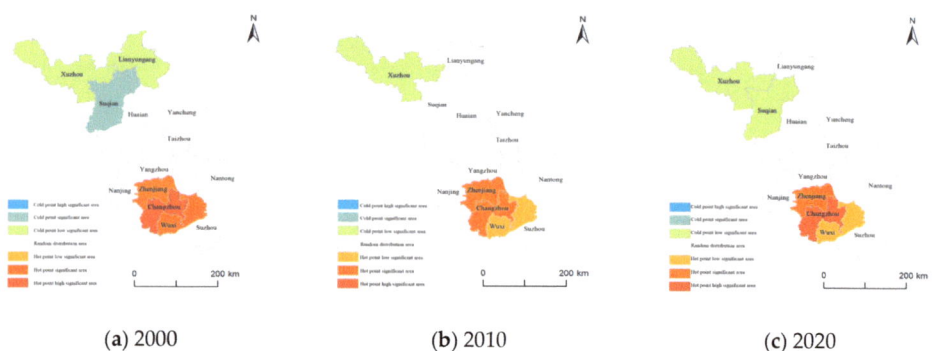

(a) 2000 (b) 2010 (c) 2020

Figure 4. Evolution of cold hot spots in grain production in Jiangsu Province in 2000, 2010, and 2020 (city level).

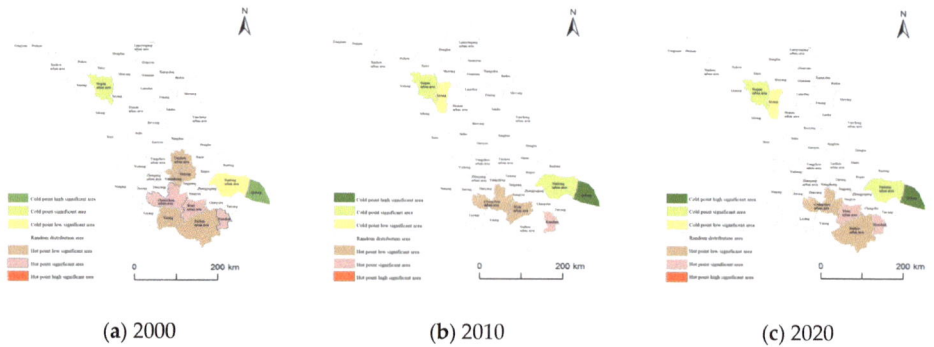

(a) 2000 (b) 2010 (c) 2020

Figure 5. Evolution of cold hot spots in grain production in Jiangsu Province in 2000, 2010, and 2020 (districts and counties).

From Figure 4, the cold points of the grain production efficiency in 2000 are in Xuzhou, Suqian, and Lianyungang cities in northern Jiangsu. In 2020, the cold points of the grain production efficiency are only in Xuzhou and Suqian, which belong to the regions with low and significant cold points. The hot spots of the grain production efficiency are mainly concentrated in southern Jiangsu, including Zhenjiang, Changzhou, and Wuxi. From 2000 to 2020, Zhenjiang has always been a medium significant hotspot area, Changzhou has changed from a high significant hotspot area to a medium significant hotspot area and then to a high significant hotspot area again until 2020, while Wuxi has gradually changed from a medium significant hotspot area to a low significant hotspot area. There is a large difference in grain production efficiency between northern Jiangsu and southern Jiangsu. In northern Jiangsu, there is a significant spatial agglomeration feature of low grain production efficiency, while in southern Jiangsu, there is a significant agglomeration feature of high grain production efficiency.

From Figure 5, the distribution of hot and cold spots of grain production efficiency from 2000 to 2010 at the county level changed mainly from a random distribution area to a low significant cold point area in Siyang, from a low significant cold point area to a medium significant cold point area in Nantong, and from a medium significant hot point area to a low significant hot point area in Changzhou and Wuxi. From 2010 to 2020, no other changes were noticed except Suzhou urban area and Wuxi urban area. Both areas changed from hot low significant to hot high significant areas. From the distribution point of view, the cold point agglomeration of grain production efficiency is concentrated in Nantong city, Qidong city, Suqian city, and Siyang city, indicating that there is a spatial agglomeration of low grain production efficiency in these areas. The hotspots are concentrated in parts of Changzhou, Wuxi, and Suzhou. From the county level, there have not been any hotspots and significant areas in the province since 2000. There is no high value spatial agglomeration of food production efficiency.

4.4. Effects of Land Use Change on Food Production

Taking grain yield (Y) and technical efficiency (TE) as dependent variables, OLS and Tobit methods were used for the regression calculation, and the coefficients of the land use diversity index, cultivated land patch density, cultivated land area ratio, cultivated land distance, and land development intensity were obtained, as shown in Tables 5 and 6.

Table 5. Panel data estimation of the impact of land use change on grain yield in Jiangsu Province.

	Coefficient	Standard Deviation	t-Statistic
Land use diversity index	−4.597 ***	0.256	−17.97
Cultivated land patch density	−0.289 ***	0.024	−12.04
Cultivated land area ratio	0.230 ***	0.065	3.54
Cultivated land distance	0.102 ***	0.031	3.35
Land development intensity	−1.423 ***	0.099	−14.39
Constant	−4.597 ***	0.256	−17.97

Note: *** indicate significance levels at 1%.

Table 6. Panel data estimation of the impact of land use change on grain technology efficiency in Jiangsu Province.

	Coefficient	Standard Deviation	t-Statistic
Land use diversity index	−0.062	0.104	−0.60
Cultivated land patch density	−0.001 *	0.000	−1.74
Cultivated land area ratio	−0.195 ***	0.052	−3.77
Cultivated land distance	0.002	0.002	0.86
Land development intensity	−0.309 ***	0.064	−4.86
Constant	−0.062	0.104	−0.60

Note: *** and * indicate significance levels at 1% and 10%.

The regression results showed that the land use diversity index and cultivated land patch density had negative effects on grain yield and technical efficiency, cultivated land distance was positively correlated with grain yield and technical efficiency, and the improvement in land development intensity was negatively correlated with grain yield and technical efficiency.

5. Discussion

When food security is top of the agenda of different stakeholders, it is of great significance to examine the optimization of cultivated land use. Our study reveals the strong correlation between cultivated land area and grain production in Jiangsu Province through the analysis of the stochastic frontier production function of grain output and input factors. The technical efficiency of grain production is calculated at the city and county level of the province. Using the spatial data of land distribution creatively, the impacts of land use change on grain yield and efficiency in the province are analyzed. According to the above research results, the output elasticity of grain production input factors, the spatial distribution of the technical efficiency of grain production, the agglomeration effect, and the influencing factors of land use change on grain production will be discussed.

According to the calculation results of the output elasticity of the production factors, it can be seen that land factors play a decisive role in the current stage of grain production in Jiangsu Province. This result is basically consistent with the finding of Li (2017) [36]. In his research, the output elasticity of the grain-sown area is the largest, reaching 0.75, while labor force and agricultural machinery account for 0.07 and 0.02, respectively. The same value was found in this paper. The output elasticity of labor force is 0.016, indicating that an additional 1% increase in labor force will increase food production by 0.016%, keeping other input factors and technical conditions unchanged. On the one hand, this conclusion is similar to the research conclusion of Zeng et al. (2012) [32], that is, farmers are the main laborers in agricultural production; on the other hand, the coefficient of 0.016 indicates that the marginal income of rural labor is already low under the current conditions, and the input of a large number of labor forces does not contribute much to the growth of food production. The output elasticity of farm machinery power is 0.037, indicating that an additional 1% increase in total farm machinery power will increase grain output by 0.037%. This value is small, and the contribution of increasing this input factor to output growth is small. The main reason for this may be that in order to avoid being affected by natural disasters, in busy farming seasons such as spring ploughing and autumn harvest, farmers may invest multiple machines to work at the same time to complete the seeding and harvesting process as soon as possible, which may lead to the excessive input of machinery, as found in the study of Guan (2006) [37]. Compared with the national average, the output elasticity of the sown area in Jiangsu Province accounts for a more prominent proportion, reaching 0.984, which is about 0.23 higher than the national level. As a developed region in the eastern coastal region, this province has a high level of agricultural mechanization, a large amount of machinery input, a dense population, and abundant labor resources. Therefore, the marginal diminishing effect of agricultural machinery and labor input is more obvious. In this way, the factor of sown area becomes more important. What is more different is that, compared with the national fertilizer output elasticity of 0.18, the fertilizer output elasticity of Jiangsu Province is −0.19, indicating that an additional 1% increase in fertilizer application will reduce grain output by 0.019%. This is consistent with the conclusion put forward by Wang (2018) [38], that is, the application rate of chemical fertilizer has shown a decreasing marginal utility. An excessive amount of chemical fertilizer application will cause soil quality degradation and pollution [39], which will endanger the further improvement in grain yield.

According to the efficiency calculation and spatial distribution in the past 20 years, the technical efficiency of grain production in southern Jiangsu is generally higher than that in central and northern Jiangsu, which is consistent with Zhang Qian's research findings [40]. In 2020, the average grain production efficiency of cities in southern Jiangsu, central Jiangsu, and northern Jiangsu was 0.927, 0.880, and 0.866, respectively. There are historical reasons, natural reasons, and economic and social reasons for the formation of the different grain production efficiency. Historically, south Jiangsu has always been the main grain-producing area of our country, known as the "granary", accumulating rich experience of intensive farming. In terms of natural conditions, the southern region of Jiangsu has a dense river network, convenient irrigation, abundant rainfall, and fertile soil, while the conditions in the central and northern regions of Jiangsu are relatively poor; in particular, the soil salinization in the northern region of Jiangsu has a certain impact. From the perspective of economic and social development, southern Jiangsu has a developed economy, advanced technology, and convenient transportation, while central and northern Jiangsu are relatively backward. The sown area of grain crops in central and northern Jiangsu is more than that in southern Jiangsu, and the labor force engaged in grain production is also more than that in southern Jiangsu, but the grain production efficiency is relatively low. Therefore, to improve the efficiency of grain production in southern Jiangsu, the most important thing is to ensure that there is enough cultivated land. Specifically, it is necessary to strictly observe the red line of cultivated land, increase the development of cultivated land, and innovate the mode of urban development. For northern Jiangsu, it is necessary to reduce the investment of redundant factors in grain production, increase the investment of capital and scientific research, accelerate the construction of agricultural infrastructure, and attract talents to accelerate the development of modern agriculture.

The cold spots of the grain production efficiency are mainly concentrated in the north of Jiangsu, and the hot spots of the grain production efficiency are mainly concentrated in the south of Jiangsu. Through the analysis of spatial aggregation characteristics, it can be seen that the efficiency of the counties with strong traditional grain production has been consolidated and improved, and the high-value agglomeration areas have also been transferred to these counties. We should focus on breaking through the technical bottleneck of improving the agricultural production technical efficiency in northern Jiangsu. Since 2000, the growth rate of the grain production efficiency in northern Jiangsu has been slow and at a low value. The government should provide relevant policy guidance for grain production in northern Jiangsu and help these areas break through the bottleneck of improving the grain production technical efficiency by further optimizing the allocation of agricultural production resources and improving the level of production management.

The regression results of the impact of land use change on grain production show that the improvement in the land use diversity index in Jiangsu province has a negative impact on grain production. With the growth of population and the rapid development of social economy, the change in land use in Jiangsu Province has intensified. Land has gradually been enriched from a relatively single use to multiple uses, such as industry, transportation, entertainment, and residential, and the proportion of various uses has increased, except as cultivated land. Land use structure has become more complex, and the diversity index of land use has also been increasing. The effect on grain yield and efficiency is also significant. The model estimation results show that grain yield and efficiency decrease as the land use diversity index increases. The cultivated land patch density has a negative impact on grain yield and efficiency, which is similar to the conclusion of Ntihinyurwa et al. (2021) [41]. With the Mosaic occupation of cultivated land by other land types, especially the continuous expansion of rural settlements and urban construction land, and the rapid development of the transportation network, the fragmentation of cultivated land continues to increase, hindering the promotion of mechanical equipment and large-scale agricultural technology, which is not conducive to the intensive production of grain and is bound to affect grain production. Berk (2022) also concluded that agricultural land fragmentation may lead to inefficiency [7]. The distance of cultivated land has a positive correlation with grain output, indicating that in recent years, with the shift of the center of gravity of cultivated land to the urban periphery, the distance between it and the administrative center has increased, and the probability of land being used for urbanization construction has become smaller and smaller. In addition to the definition of "three districts and three lines", most of these areas are outside the urban development boundary, vigorously developing agriculture has become the leading direction, and the grain-producing areas are basically stable, which is conducive to the increase in grain output. The improvement in the land development intensity is negatively correlated with grain output, indicating that the intensity of land development is increasing with the improvement in the urbanization level. The crowding out of a large amount of cultivated land by construction land will inevitably lead to a prominent contradiction between human and land, resulting in the decline in grain output and efficiency.

In this paper, the effects of grain production efficiency and land use change on grain production efficiency in Jiangsu Province over the last 20 years were discussed, and some useful conclusions were obtained. The deficiency of this paper is that the overall situation of food production has been studied, and the output structure of food production has not been studied. Subdividing the production of major food crops, such as the impact of land use change on rice, wheat, etc., is a direction that can be studied in the future.

6. Conclusions

This study uses the stochastic frontier production function to verify the important role of grain-sown area in grain production, calculates the technical efficiency of grain production by matching the statistical data of grain production in Jiangsu province with the remote sensing data of cultivated land layout, and analyzes the impact of land use change on grain production yield and technical efficiency from a spatial perspective. The main conclusions of this study are as follows:

Firstly, through the calculation and analysis of the stochastic frontier production function, the grain-sown area is the decisive factor for the increase in grain output in Jiangsu province. Under the current technology level, grain output will increase by 0.984% for every 1% increase in sown area. The increase in agricultural machinery power and labor force has very little impact on the increase in grain yield, and the impact of the fertilizer application rate on the increase in grain yield is even negative. This shows that under the current level, the most effective way to increase grain output and ensure food security in Jiangsu Province is to increase the sown area. The size of the grain-sown area is directly related to the cultivated land area, so it is very important to ensure that the cultivated land area does not decrease and meet the demand for grain sowing.

Secondly, since 2000, the technical efficiency of grain production in Jiangsu Province has been maintained at a relatively high level, showing a fluctuating upward trend. In terms of the whole province, the grain production efficiency in southern Jiangsu is generally higher than that in northern and central Jiangsu. The southern region of Jiangsu relies on advanced agricultural technology to drive grain production, while the economic development level of the northern region of Jiangsu is relatively low, and the promotion of modern agricultural technology and agricultural infrastructure construction is not as good as that of the southern region of Jiangsu. In the future, it is necessary to continuously improve the level of modern agricultural technology in northern Jiangsu and make good use of the cultivated land space in northern Jiangsu.

Thirdly, the analysis of the spatial distribution characteristics of the technical efficiency of grain production in Jiangsu Province shows that there is a positive spatial correlation of grain production in Jiangsu province, and grain production presents agglomeration benefits. In the past 20 years, the distribution of cold hot spots of grain production efficiency in Jiangsu province has not changed much, the cold spots of grain production technical efficiency are mainly concentrated in the north of Jiangsu, and the hot spots of grain production technical efficiency are mainly concentrated in the south of Jiangsu. Through the analysis of spatial aggregation characteristics, it can be seen that the efficiency of the counties with strong traditional grain production has been consolidated and improved, and the high-value agglomeration areas have also been transferred to these counties.

Fourthly, through the analysis of the impact of land use change on the total grain output and the technical efficiency of grain production in the province, it is found that the increase in the complexity of land use, especially the increase in the proportion of construction land, has a negative impact on both grain output and the technical efficiency of grain production. From the perspective of the location of cultivated land, the farther cultivated land is away from the city, the more conducive it is to the improvement in grain output and grain production efficiency. From the perspective of cultivated land shape, the larger the density of the cultivated land patch and the more fragmented the cultivated land are, the more unfavorable the increase in grain yield and technical efficiency of grain production is.

In short, in order to ensure food security, Jiangsu province should strengthen the protection of cultivated land to ensure that the due grain-sown area is not reduced. Second, we should reduce the fragmentation of cultivated land in the province and improve the technical efficiency of grain production. Third, we should break the unbalanced development of the region and focus on improving the technical efficiency of northern Jiangsu.

Author Contributions: Conceptualization, X.C. and J.H.; writing—original draft, X.C.; writing—review and editing, J.H. and X.L.; data analysis, X.C. funding acquisition, J.H. All authors have read and agreed to the published version of the manuscript.

Funding: This research was funded by the Ministry of Science and Technology of China "Discovery, Creation and Modern Production Technology Demonstration of Genetic Resources of Major Crops" (No. 2020YFE0202900).

Data Availability Statement: The data presented in this study are available upon request from the first author. The data are not publicly available due to data publisher regulations.

Conflicts of Interest: The authors declare no conflict of interest.

References

1. Chen, Y. Modeling and Analysis of Land Use Change and Its Impact on Grain Production in China. *China Land Sci.* **2000**, *4*, 22–26.
2. Li, H. Land Resource Allocation, Utilization Efficiency, and Regional Economic Growth. Ph.D. Thesis, Hunan University, Changsha, China, 2016.
3. Guo, L. Land Use Change in Henan Province and Its Impact on Grain Production Efficiency. Master's Thesis, Henan University, Zhengzhou, China, 2021.
4. Zhang, Q.; Zhang, F.; Chen, X. Research on the Calculation of Production Efficiency in China's Major Grain Production Areas. *Price Theory Pract.* **2018**, *9*, 155–158.
5. Yadava, A.K.; Komaraiah, J.B. Technical Efficiency of Chemical Fertilizers Use and Agricultural Yield: Evidence from India. *Iran. Econ. Rev.* **2021**, *27*, 121–134. [CrossRef]
6. Salam, A.; Hameed, A.; Adr, A.D.R.; Sonobe, T. Technical Efficiency in Production of Major Food Grains in Punjab, Pakistan. *Asian Dev. Rev.* **2022**, *39*, 201–222. [CrossRef]
7. Berk, A.O.; Güney, L.; Sangün. Measurement of resource use efficiency in corn production: A two-stage data envelopment analysis approach in Turkey. *Ciência Rural.* **2022**, *52*, e20210022. [CrossRef]
8. Ghosh, C.; Mazumdar, D. Factors Affecting Cost Efficiencies of Rice Production of Indian States: A Stochastic Frontier Approach. *Indian J. Agric. Econ.* **2018**, *73*, 165–182.
9. Hien, N.T.M.; Kawaguchi, T.; Suzuki, N. A Study on Technical Efficiency of Rice Production in The Mekong Delta-Vietnam by Stochastic Frontier Analysis. *Summ. Dep. Agric. Kyushu Univ.* **2003**, *48*, 325–357.
10. Wang, P.; Deng, X.; Jiang, S. Global warming, grain production and its efficiency: Case study of major grain production region. *Ecol. Indic.* **2019**, *105*, 563–570. [CrossRef]
11. Min, S.; Paudel, K.P.; Feng-Bo, C. Mechanization and efficiency in rice production in China. *J. Integr. Agric.* **2021**, *20*, 1996–2008.
12. Sun, Z.; Li, X. Technical Efficiency of Chemical Fertilizer Use and Its Influencing Factors in China's Rice Production. *Sustainability* **2021**, *13*, 1155. [CrossRef]
13. Alem, H. A Meta-frontier Analysis on the Performance of Grain-Producing Regions in Norway. *Economies* **2021**, *9*, e9010010. [CrossRef]
14. Delay, N.D.; Thompson, N.M.; Mintert, J.R. Precision agriculture technology adoption and technical efficiency. *J. Agric. Econ.* **2022**, *73*, 195–219. [CrossRef]
15. Chandel, R.B.S.; Khan, A.; Li, X.; Xia, X. Farm-Level Technical Efficiency and Its Determinants of Rice Production in Indo-Gangetic Plains: A Stochastic Frontier Model Approach. *Sustainability* **2022**, *14*, su14042267. [CrossRef]
16. Qian, W.; Xiao-Bin, J.; Yin-Kang, Z.; Shamuxi, A. Analysis of the Evolution of Spatial Aggregation Pattern of Food Production Efficiency at County Level: Taking Hebei Province as an Example. *Soil Water Conserv. Bull.* **2011**, *31*, 234–238.

17. Zou, S.; Liao, H.; Xiang, S. Analysis of changes in arable land use and food security in Chongqing. *J. Southwest Norm. Univ.* **2010**, *35*, 218–223.
18. Jia, J.; Chi, W.; Bu, R.; Yan, H. The impact of land use change on grain production capacity in Inner Mongolia. *West. Resour.* **2011**, *6*, 61–63.
19. Liu, W. Current Situation, Existing Problems, and Countermeasures of the Wheat Industry in Jiangsu Province. *China Seed Ind.* **2021**, *318*, 19–22.
20. Hong, S.; Hao, J.; Zhou, N.; Chen, L. Changes in arable land and its impact on changes in grain production patterns in the Huanghuaihai Plain. *J. Agric. Eng.* **2014**, *30*, 268–277.
21. Adjei, V.; Anlimachie, M.A.; Ativi, E.E. Understanding the Nexus between Climate Change, the Shift in Land Use toward Cashew Production and Rural Food Security in Ghana; the Experiences of Farmers in the Transition Zone of Ghana. *J. Atmos. Sci. Res.* **2020**, *3*, 9. [CrossRef]
22. Liu, Z.; Zhong, H.; Li, Y.; Wen, Q. Characteristics of China's grain production and its impact on regional grain supply and demand pattern in recent 20 years. *J. Nat. Resour.* **2021**, *36*, 1413–1425.
23. Ambros, P.; Granvik, M. Trends in Agricultural Land in EU Countries of the Baltic Sea Region from the Perspective of Resilience and Food Security. *Sustainability* **2020**, *12*, 5851. [CrossRef]
24. Bhermana, A.; Agustini, S.; Irwandi, D.; Firmansyah, M.A. Spatial land use planning for developing sustainable food crop areas using land evaluation approach and GIS application (a case study of Pulang Pisau Regency, Central Kalimantan). *IOP Conf. Ser. Earth Environ. Sci.* **2021**, *648*, 012011. [CrossRef]
25. Rahman, S.; Rahman, M. Impact of land fragmentation and resource ownership on productivity and efficiency: The case of rice producers in Bangladesh. *Land Use Policy* **2009**, *26*, 95–103. [CrossRef]
26. Manjunatha, A.V.; Anik, A.R.; Speelman, S.; Nuppenau, E.A. Impact of land fragmentation, farm size, land ownership, and crop diversity on profit and efficiency of irrigated farms in India. *Land Use Policy* **2013**, *31*, 397–405. [CrossRef]
27. Aigner, D.; Lovell, C.A.K.; Schmidt, P. Formulation and Estimation of Stochastic Frontier Production Function Models. *J. Econom.* **1977**, *6*, 21–37. [CrossRef]
28. Battese, G.E.; Coelli, T.J. Frontier production functions, technical efficiency and panel data: With application to paddyfarmers in India. *J. Product. Anal.* **1992**, *3*, 153–169. [CrossRef]
29. Wang, S.; Zhou, Y.; Tian, X. Measurement of the Efficiency of Pig Breeding Technology in China. *Stat. Decis.* **2022**, *38*, 65–69.
30. Hansen, B.E.; Racine, J.S. Jackknife Model Averaging. *J. Econom.* **2012**, *167*, 38–46. [CrossRef]
31. Yang, Y.; Deng, X.; Li, Z.; Wu, F.; Li, X. The impact of land use change on food production efficiency in the North China Plain from 2000 to 2015. *Geogr. Res.* **2017**, *36*, 2171–2183.
32. Zeng, F.; Gao, M. Accounting for Grain Production Efficiency in China and Analysis of Its Influencing Factors: An Empirical Study Based on the SBM-Tobit Model Two Step Method. *Agric. Technol. Econ.* **2012**, *7*, 63–70.
33. Simpson, E.H. Measurement of diversity. *Nature* **1949**, *163*, 688. [CrossRef]
34. Comer, D.; Greene, J.S. The development and application of a land use diversity index for Oklahoma City, OK. *Appl. Geogr.* **2015**, *60*, 46–57. [CrossRef]
35. Fan, L. Technical Efficiency, Technological Progress, and Grain Productivity Growth. *Econ. J.* **2016**, *33*, 31–36.
36. Li, T. Research on the Impact of Structural Adjustment and Technological Progress on Grain Production in China. Ph.D. Thesis, Nanjing Agricultural University, Nanjing, China, 2017.
37. Guan, J.C.; Yam, R.C.M.; Mok, C.K.; Ma, N. A study of the relationship between competitiveness and technological innovation capability based on DEA models. *Eur. J. Oper. Res.* **2006**, *170*, 971–986. [CrossRef]
38. Wang, J.C.; Mu, Y. Change of grain production in main producing areas and its influencing factors: Based on county data of Henan Province. *Agric. Outlook* **2018**, *18*, 50–57.
39. Wang, X.; Cai, D.; Grant, C.; Hoogmoed, W.B.; Oenema, O. Factors controlling regional grain yield in China over the last 20 years. *Agron. Sustain. Dev.* **2015**, *35*, 1127–1138. [CrossRef]
40. Zhang, Q. A Regional Comparative Study on Grain Production Efficiency and Capacity in Jiangsu Province. Master's Thesis, Yangzhou University, Yangzhou, China, 2016.
41. Ntihinyurwa, P.D.; de Vries, W.T. Farmland Fragmentation, Farmland Consolidation and Food Security: Relationships, Research Lapses and Future Perspectives. *Land* **2021**, *10*, 129. [CrossRef]

Disclaimer/Publisher's Note: The statements, opinions and data contained in all publications are solely those of the individual author(s) and contributor(s) and not of MDPI and/or the editor(s). MDPI and/or the editor(s) disclaim responsibility for any injury to people or property resulting from any ideas, methods, instructions or products referred to in the content.

Article

Evolutionary Game Analysis of Ecological Governance Strategies in the Yangtze River Delta Region, China

Qing Wang * and Chunmei Mao

College of Public Administration, Hohai University, Nanjing 211100, China; 190213120007@hhu.edu.cn
* Correspondence: wqing@hhu.edu.cn

Abstract: Under integrated ecological and green development in the Yangtze River Delta, the regional ecology is adversely affected by ineffective synergistic governance. Regional environmental governance is a collaborative process involving multiple stakeholders and mutual engagement, with each participant pursuing their interests and common goals simultaneously. This study employed stakeholder theory. A tripartite evolutionary game model of the public, enterprises, and local governments was constructed to analyze the behavioral strategies and influencing factors for the parties involved, and the impacts of key factors on the stability of the evolutionary game system were evaluated. The results indicate that ecological environmental governance in the Yangtze River Delta region is a complex and evolving system involving multiple stakeholders, within which system stability is influenced by stakeholders' behavioral strategies. The interests of each party are affected by the cost of public involvement in ecological environment governance and the benefits and subsidies that enterprises receive for active environmental governance. The costs and penalties paid by local governments for lax regulations impact their behavioral strategies. This study provides policy recommendations for ecological governance in the study region, including the government–enterprise co-construction of liquid regulatory funds, government–enterprise–public partnerships in low-cost regulatory models, and the sharing of high-quality regulatory outcomes.

Keywords: multiple stakeholders; evolutionary game; behavioral strategies; numerical simulation; ecological environment governance; collaborative framework

Citation: Wang, Q.; Mao, C. Evolutionary Game Analysis of Ecological Governance Strategies in the Yangtze River Delta Region, China. *Land* **2024**, *13*, 212. https://doi.org/10.3390/land13020212

Academic Editor: Brian D. Fath

Received: 19 January 2024
Revised: 2 February 2024
Accepted: 7 February 2024
Published: 8 February 2024

Copyright: © 2024 by the authors. Licensee MDPI, Basel, Switzerland. This article is an open access article distributed under the terms and conditions of the Creative Commons Attribution (CC BY) license (https://creativecommons.org/licenses/by/4.0/).

1. Introduction

In recent years, positive outcomes have been attained through integrated ecological and green development of the Yangtze River Delta region, inspired by Xi Jinping's "Thought on Socialism with Chinese Characteristics for a New Era". However, due to the region's excellent integrated growth and the acceleration of the new urbanization process, cross-border pollution has become an urgent bottleneck to further development in Shanghai, Jiangsu, Zhejiang, and Anhui (hereafter referred to as "three provinces and one city"). Against this background, in January 2021, the Office of the Leading Group for Integrated Development of the Yangtze River Delta released the "Plan for Joint Ecological Environment Protection in the Yangtze River Delta Region", which explicitly advocates for the promotion of collaborative environmental governance and strengthening of the foundation for green development. However, collaborative governance of the ecological environment entails the participation of multiple parties, and the divergent interests and conflicting behavioral orientations of these parties pose a predicament for collective action in the Yangtze River Delta region. Therefore, it is particularly necessary to explore the strategic choices of various stakeholders and their influencing factors in the environmental governance of the Yangtze River Delta region.

Considerable academic research has been conducted on regional ecological and environmental governance strategies. In particular, many studies have reported on the current state, features, and efficiency of environmental governance in the Yangtze River Delta region [1–6]. Through both qualitative and quantitative approaches, researchers have proposed strategies for collaborative governance by multiple stakeholders. Among them, Mao et al. pointed out that the joint prevention and control of air pollution in the Yangtze River Delta region suffers from unresolved problems, such as a weak synergy of regional policies and low social participation [2]. Based on the case studies of Hangzhou and the Hefei metropolitan area, Suo et al. analyzed the core environmental pluralistic co-governance characteristics and reported that the government is the core body tasked with establishing a pluralistic synergistic environmental co-governance model [3]. Based on empirical data, other scholars have conducted systematic evaluations of the spatiotemporal characteristics of environmental pollution and assessed the effectiveness of regional environmental governance [4]. Further, they suggested that collaborative governance of the Yangtze River Delta requires the creation of a long-term dynamic mechanism that should include mechanisms to promote government responsibility and public participation [5,6]. Studies have focused on the significance of involving multiple actors. Most scholars have proposed macro-level countermeasures or collaborative governance led by local governments. However, different behavioral strategy choices among local governments have not yet been specifically explained at the micro level, and behavioral strategies and their interactions among multiple subjects, such as local governments, private enterprises, and the public, have not been studied in depth.

The most representative research on the multiparty joint governance of regional ecosystems is the synergistic governance model proposed by Ansell and Gash and Emerson et al. Ansell and Gash [7] developed the synergistic governance model comprising four components: starting conditions (S), catalytic leadership (F), institutional design (I), and synergistic process (C), with the synergistic process being the key element, while starting conditions, institutional design, and leadership provide supportive factors for synergistic governance. Emerson et al. [8] described a collaborative governance model comprised of three interacting components: "principled" participation, shared motivation, and the ability to act jointly. Based on this, scholars in China have developed various frameworks and models to analyze ecological and environmental governance [9–11]. These frameworks are based on the actual atmospheric or water environmental governance in the area. Among them, Wu et al. [9] proposed a "structure-process" analytical framework for collaborative air pollution management in the Yangtze River Delta region. They argued that a range of structural and process mechanisms can reduce transaction costs and cooperation risks and are beneficial for the functioning of the collaborative air pollution management model in this region. Other scholars have developed a "Dynamics-Structure-Process" analytical model and advocated for an inter-administrative environmental governance model named "Inter-Governmental Consultation" based on an analysis of governance costs and cooperation benefits [10]. Researchers have developed gravity models to demonstrate the spatial relationships of collaborative governance. These researchers have suggested a pathway to enhance collaborative governance of pollution and carbon reduction within China's three primary urban agglomerations in the Yangtze River Delta region [11]. Studies have been carried out to devise an analytical framework or model aimed at exploring the factors influencing ecological and environmental governance, as well as the mechanisms for nurturing synergistic approaches to ecological and environmental governance in the Yangtze River Delta region. However, previous research has not elucidated the fundamental determinants behind the adoption of behavioral strategies by multiple stakeholders and the implications of these determinants for environmental governance.

Evolutionary game theory is an approach that combines traditional game theory analysis with dynamic evolutionary process analysis, with an emphasis on "evolutionarily stable strategies" and "replication dynamics" [12–17]. By analyzing limited rationality and group behavior, evolutionary game theory reveals the interactions and behaviors of multiple stakeholders, which offers a novel approach of studying the behavioral strategies of multi-interest stakeholders and elucidating key underlying factors [18–45]. Academic research on the evolutionary game of regional ecological and environmental governance is divided into three primary areas. The first is evolutionary games between government agents [18–24]. Scholars have analyzed the problem of ecological environmental governance from the perspective of evolutionary games between the central and local governments or between local governments [18–20]. Numerous Chinese scholars have focused on ecological and environmental governance in the Beijing–Tianjin–Hebei region, investigating the impact of collaborative air pollution management in the region and identifying influencing factors [21–23]. In these studies, game models to explore the relationships between the central and local governments or among local governments have been used [18–24]. In the Yangtze River Delta region, relatively few evolutionary game studies have been conducted on ecological and environmental governance. Bo et al. investigated the evolutionary game theory of haze governance behavior in this region by constructing a game model with local governments [24]. The second area concerns evolutionary games between the government and businesses [25,26]. Some scholars have studied how government policies affect enterprises' strategic choices by applying this government–enterprise game model. Song et al. found that formal environmental regulation can effectively promote enterprise innovation, and Izabela et al. verified the interactive effects of government and firm strategies [25,26]. The third area studies evolutionary games between multiple interested parties [27–46]. Some scholars systematically analyzed the strategy selection process of three primary parties by constructing a three-party evolutionary game model [27–31,43–45]. They explored the mutual influence mechanism of the behavioral strategies of all parties' interests under environmental regulations. In the Beijing–Tianjin–Hebei region, scholars have conducted numerous studies on ecological management based on evolutionary games [21–23]. However, there is a lack of evolutionary game studies on ecological and environmental governance in the Yangtze River Delta region [27,41,42]. Extant studies on ecological environment governance that have employed the evolutionary game approach have only examined the game played by central and local governments or between the governments of the three provinces and one city. These studies did not include other relevant stakeholders such as enterprises and the public in the evolutionary game analysis framework and thus it does not represent the actual situation of regional ecological and green integration development.

In summary, a research framework for ecological governance in the Yangtze River Delta Region is as shown in Figure 1: (1) This study incorporates enterprises and the public into a collaborative governance mechanism to explore the behavioral strategies of multiple stakeholders involved in ecological governance in the Yangtze River Delta region. (2) Based on the assumption of limited rationality, this study constructs a tripartite evolutionary game model of "public-enterprise-local government" to explore the strategic choices made by each stakeholder in ecological environment governance and their impacts on each other. This study was conducted to clarify the effects of crucial factors on ecological environment governance in the Yangtze River Delta region. (3) By simulating and analyzing the differential impacts of various factors and exploring their underlying mechanisms, this study enabled the construction of a synergistic governance model of "government-enterprise co-construction of liquidity regulatory funds, government-enterprise-public partnership in low-cost regulatory modes, and sharing of high-quality regulatory outcomes". These efforts may boost the interests of stakeholders involved in ecological and environmental governance while providing theoretical support and policy recommendations to promote sustainable development in the Yangtze River Delta.

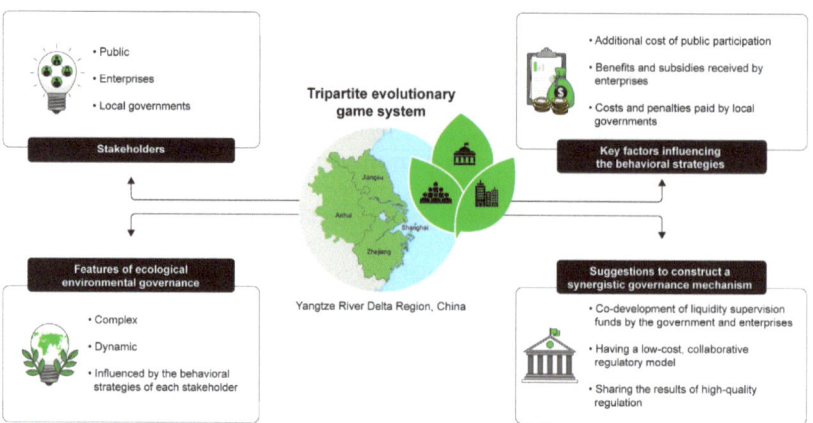

Figure 1. Research framework for ecological governance in the Yangtze River Delta region (prepared by the authors).

2. Materials and Methods

2.1. Study Area

In November 2018, at the First China International Import Expo (CIIE), Chinese President Xi Jinping announced his support for upgrading the integrated development of the Yangtze River Delta to a national strategy. The scope of the regional plan includes the entire area of Shanghai, Jiangsu, Zhejiang, and Anhui provinces (an area of 358,000 kilometers^2). The state of the ecological environment, which is the basis for human survival and development, determines a region's capacity for sustainable development. The Yangtze River Delta region, with its rich ecosystem types, deep ecological culture, and great ecological carrying capacity, is an important place for Xi Jinping's idea of ecological civilization to germinate. It is also a pioneer zone for building a beautiful China. At present, the integration of the Yangtze River Delta region is at a stage of higher quality development. The ecological environment has become an indispensable dimension for assessing the higher quality integrated development of the Yangtze River Delta.

Recently, the local governments of the three provinces and one city implemented the outline of the plan for the integrated development of the Yangtze River Delta region, clarified their respective roles and positions in the ecological environment, and made concerted efforts to build a beautiful and green Yangtze River Delta. Shanghai has sufficient economic, technological, and organizational advantages to play a leading role. Jiangsu Province serves as a model and has joined forces with Zhejiang Province and Shanghai to build a demonstration zone for integrated eco-green development. Zhejiang Province occupies an important position in the country in terms of water resources, including marine as well as other biological resources, and is the "Great Garden", as well as a practitioner of the concept of "two mountains". With good ecological resources and strong environmental protection, Anhui Province plays an important barrier protection role in Yangtze River Delta ecology. To better solve the ecological and environmental pollution problems, the three provinces and one city launched a set of governance practices. For example, Shanghai Municipality, Jiangsu Province, and Zhejiang Province jointly signed the first declaration on regional environmental cooperation in China in Hangzhou, which set out the need to strengthen cooperation across regional boundaries to solve environmental problems. Anhui Province has taken the lead in implementing the reform of the ecological forest management system and construction of the "atmospheric ecological compensation" model.

2.2. Stakeholders in Ecological and Environmental Governance in the Yangtze River Delta Region

Due to the public nature, complexity, and cross-border characteristics of regional ecological and environmental issues, regional ecological governance requires not only strong input from the local governments of the three provinces and one city but also cooperation among various stakeholders. According to Freeman's stakeholder theory, which provides a theoretical framework for analyzing the behavior of ecological and environmental governance stakeholders in the Yangtze River Delta region, the ecological and environmental governance stakeholders' role in the Yangtze River Delta region has evolved from "passive influence" to "active participation" to "collaborative governance" [46,47]. Combining the degree of closeness of the relationship between stakeholders and ecological environmental governance, stakeholders are defined as the public, private enterprises, and local governments. The interests and interrelationships of the different game subjects were analyzed to identify logical relationships between the tripartite game subjects of ecological environmental governance in the Yangtze River Delta region, as shown in Figure 2.

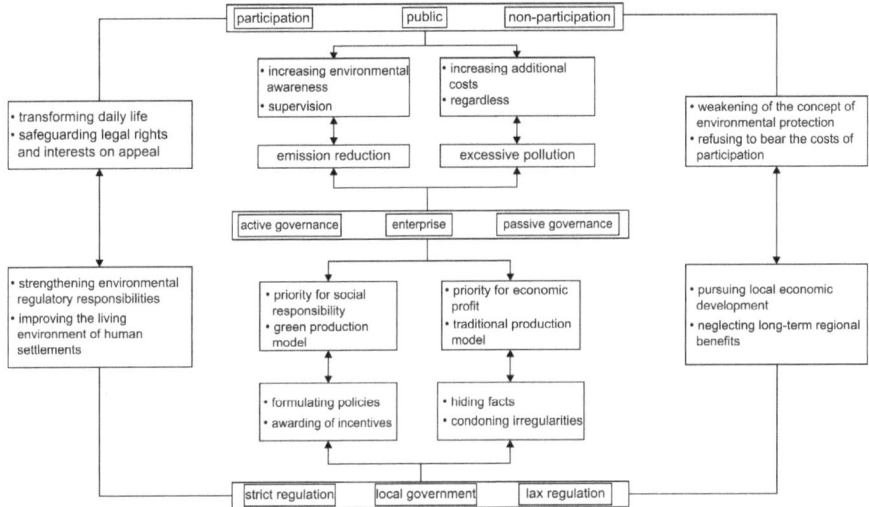

Figure 2. Logical relationship diagram of the three-party evolutionary game model (prepared by the authors).

In the regional governance of the ecological environment, the public, which is primarily affected by pollution in the region, experienced a noticeable delay in participating in the governance process [48]. The government–business model of governance alone cannot achieve the best strategy for environmental protection; the effectiveness of ecological and environmental governance is significantly influenced by the extent of public participation [49]. When the public takes the initiative to participate in eco-environmental governance, they actively respond to the government's call for environmental protection by changing their lifestyles through green travel and green consumption. When the living environment and physical and mental health are damaged by corporate production, the public can choose to report corporate emissions or appeal to defend their rights, and the local government may reward their behavior. Therefore, the public's behavior can push enterprises to change their production concepts from the bottom up and strengthen the government's adherence to regulatory responsibilities.

As enterprises are the main producers of ecological pollution, their behavioral strategies play a crucial role in ecological governance. In the context of the government's efforts to promote ecological governance, enterprises can earn subsidies from the local government and garner public recognition, while also establishing an admirable corporate image, if they prioritize social responsibility and opt for proactive governance. However, they must bear in mind that this requires them to incur a specific sum in green production costs. If a business seeks to maximize financial gains and is unwilling to pay elevated costs for transitioning to new production methods, even if this leads to a temporary boost in revenue, it may face reprimands from local authorities and public scrutiny. Such consequences are not conducive to the sustainable growth and development of an enterprise.

Local governments in the three provinces and one city act as regulators of regional ecological and environmental governance. To create a green and beautiful Yangtze River Delta regional habitat, local authorities encourage innovation in eco-friendly practices by enterprises and offer policy incentives and subsidies to the public to promote environmental protection. At the same time, when the public chooses not to participate and enterprises choose to produce in violation of negative emission goals, local governments need to intervene in a timely manner and take administrative measures such as deadline rectification. As the quality of life improves and the image of the city is enhanced, the social credibility and influence of the local government will be significantly enhanced, but the higher costs incurred by strict regulations will also place financial pressure on that government. When local governments relax their requirements on the public and enterprises to seek local economic growth, they may conceal ecological and environmental pollution and condone illegal discharge by enterprises [50]. Consequently, local governments incur adverse societal impacts and the loss of public confidence.

2.3. Three-Party Evolutionary Game Model Construction

2.3.1. Model Assumptions

Based on the interests of each subject in the regional environmental governance, this study operated under the following assumptions combined with the actual situation of ecological and environmental governance.

Hypothesis 1. *There are three types of interests in the regional environmental governance system: public, enterprise, and local governments. All three have limited rationality.*

Hypothesis 2. *There are two types of behavioral strategy choices for all three parties. Among them, the strategic space for the public is (participation, nonparticipation), the strategic space for enterprises is (active governance and passive governance), and the strategic space for local governments is (strict regulation and lax regulation). The probability that the public chooses to participate is x ($0 < x < 1$), and the probability that they choose not to participate is $1 - x$. The probability that a firm chooses positive governance is y ($0 < y < 1$), and the probability that it chooses negative governance is $1 - y$. The probability that a local government chooses strict regulation is z ($0 < z < 1$), and the probability that it chooses lax regulation is $1 - z$.*

Hypothesis 3. *When the public chooses to participate, the public needs to invest a certain amount of time, transport, consumption, and other basic costs (C_1) to adapt to the green lifestyle, and, accordingly, the public obtains benefits such as environmental beautification, physical and mental pleasure, and so on (R_1). If enterprises pursue only economic benefits and neglect social responsibility, the public will experience losses such as pollution of their living environment and poor health conditions (S_3). However, under the government's policy of strict regulation, the public will actively report enterprises and file complaints to protect their rights, and enterprises will incur additional costs (F_1). If firms are socially responsible, the public gains through activities such as corporate compensation and local government incentives (S_2). Conversely, when the public chooses nonparticipation, it will not gain anything, but ecological degradation will result in losses of public physical and mental health (S_1).*

Hypothesis 4. *As far as the enterprises are concerned, when an enterprise chooses an active governance strategy, the enterprise responds to the government's call for green production and pays the costs of upgrading equipment, training personnel, and introducing innovative technologies such as transforming production methods (C_2). Compared to traditional production methods, firms lose a portion of their operating profits (S_4) to fulfill their social responsibilities; they also gain a portion of their operating profits and image enhancement, among other gains (R_2), and local governments subsidize these positive corporate governance behaviors. Conversely, when a firm chooses a negative* governance strategy, it pays lower production method transformation costs (C_3) ($C_2 > C_3 > 0$) to obtain more operating profit (R_3) ($R_2 > R_3 > 0$). At this point, the public will report negative corporate governance behaviors, the local government will penalize negative corporate governance behaviors (F_2), and the firm will incur losses such as public compensation and negative public relations (S_5).

Hypothesis 5. *When the local government chooses a strict regulatory strategy, it is costly (C_4) to guide and incentivize the public and enterprises to carry out ecological and environmental management. Correspondingly, local governments obtain benefits (R_4) such as performance attainment, city image improvement, and social influence enhancement. Local governments that choose a lax regulatory strategy have a "free-rider" attitude and pay lower regulatory costs (C_5). At this point, if the enterprise actively manages the ecological environment, the local government obtains higher social benefits (A_1), and if the enterprise chooses to manage the ecological environment negatively, the local government is penalized because of the damage to the city's image (A_2). If the public participates in ecological environmental governance, the local government may be punished because of negative social opinion (B_1). If the public does not participate in ecological environmental governance, the local government will obtain fewer social benefits (B_2) than if they did participate.*

2.3.2. Benefits Matrix for Subjects in the Three-Party Evolutionary Game

According to the basic assumptions and parameter settings of the above model, a subject–benefit matrix of the tripartite evolutionary game was constructed as shown in Table 1.

Table 1. Benefits matrix for subjects of the three-party evolutionary game (prepared by the authors).

Combination of Strategies	Benefit Function		
	Public	Enterprise	Local Government
(participation, active governance, strict regulation)	$-C_1 + R_1 - F_1 - S_2$	$-C_2 + R_2 + T - S_4$	$-C_4 + R_4$
(participation, active governance, lax regulation)	$-C_1 + R_1 - S_2$	$-C_2 + R_2 - S_4$	$-C_5 + A_1 - B_1$
(participation, passive governance, strict regulation)	$-C_1 + R_1 - F_1 - S_3$	$-C_3 + R_3 - F_2 - S_5$	$-C_4 + R_4$
(participation, passive governance, lax regulation)	$-C_1 + R_1 - S_3$	$-C_3 + R_3 - S_5$	$-C_5 - A_1 - B_1$
(nonparticipation, active governance, strict regulation)	$-S_1$	$-C_2 + R_2 + T$	$-C_4 + R_4$
(nonparticipation, active governance, lax regulation)	$-S_1$	$-C_2 + R_2$	$-C_5 + A_1 + B_1$
(nonparticipation, passive governance, strict regulation)	$-S_1$	$-C_3 + R_3 - F_2$	$-C_4 + R_4$
(nonparticipation, passive governance, lax regulation)	$-S_1$	$-C_3 + R_3$	$-C_5 - A_2 + B_2$

Notes: C_1 represents the costs incurred when the public participates in governance; C_2 represents the costs of positive governance by enterprises; C_3 represents the costs of negative governance by enterprises; C_4 represents the costs of strict regulation by local government; C_5 represents the costs of lax regulation by local government; S_1 represents the losses when the public does not participate in governance; S_2 represents the losses to the public when enterprises are actively governed; S_3 represents the losses to the public when enterprises are negatively governed; S_4 represents the profits lost when enterprises are actively governed; S_5 represents the losses suffered when enterprises are negatively governed; R_1 represents the benefits received when the public participates in governance; R_2 represents the benefits of active governance by enterprises; R_3 represents the profits of negative governance by enterprises; R_4 represents the benefits of strict regulation by local governments; A_1 represents the social benefits that local governments receive when they are loosely regulated and when enterprises are positively governed; A_2 represents the penalties incurred by local governments when they are lax in regulation and are negatively governed by enterprises; B_1 represents the penalties incurred by local governments when they are laxly regulated and the public is participative in governance; B_2 represents the social benefits received by local governments when they are lax in regulation and the public is not participative in governance; F_1 represents the additional cost of public participation in governance when local governments strictly regulate; F_2 represents the penalties for negative enterprise governance when local governments strictly regulate; T represents the subsidy received by enterprises for active governance when local governments strictly regulate.

3. Results

3.1. Analysis of the Three-Party Evolutionary Game Model

3.1.1. Dynamic Equations of Replication of the Subject of the Three-Party Evolutionary Game

According to the payoff matrix of subjects in the three-party evolutionary game shown in Table 1, the expected payoffs for the public choosing the participation strategy U_{11}, the expected payoffs of choosing the nonparticipation strategy U_{12}, and the average expected payoffs U_1 are

$$U_{11} = y*z*(-C_1+R_1-F_1-S_2)+y*(1-z)*(-C_1+R_1-S_2)+(1-y)*z*(-C_1+R_1-F_1-S_3)+(1-y)*(1-z)*(-C_1+R_1-S_3) \quad (1)$$

$$U_{12} = y*z*(-S_1)+y*(1-z)*(-S_1)+(1-y)*z*(-S_1)+(1-y)*(1-z)*(-S_1) \quad (2)$$

$$U_1 = xU_{11}+(1-x)U_{12} \quad (3)$$

At this point, the replication dynamic equation for public strategy choice is

$$F(x) = dx/dt = x(1-x)(U_{11}-U_{12}) = x*(x-1)*(C_1-R_1-S_1+S_3+F_1*z+S_2*y-S_3*y) \quad (4)$$

Similarly, the enterprise's expected return U_{21} from choosing a positive governance strategy, the expected return U_{22} from choosing a negative governance strategy, and the average expected return U_2 are

$$U_{21} = x*z*(-C_2+R_2+T-S_4)+x*(1-z)*(-C_2+R_2-S_4)+(1-x)*z*(-C_2+R_2+T)+(1-x)*(1-z)*(-C_2+R_2) \quad (5)$$

$$U_{22} = x*z*(-C_3+R_3-F_2-S_5)+x*(1-z)*(-C_3+R_3-S_5)+(1-x)*z*(-C_3+R_3-F_2)+(1-x)*(1-z)*(-C_3+R_3) \quad (6)$$

$$U_2 = yU_{21}+(1-y)U_{22} \quad (7)$$

At this point, the replication dynamic equation for the enterprises' strategy choice is

$$F(y) = dy/dt = y(1-y)(U_{21}-U_{22}) = -y*(y-1)*(C_3-C_2+R_2-R_3+F_2*z-S_4*x+S_5*x+T*z) \quad (8)$$

The government's expected return U_{31} from choosing a strict regulatory strategy, the expected return U_{32} from choosing a lax regulatory strategy, and the average return expectation U_3 are

$$U_{31} = x*y*(-C_4+R_4)+x*(1-y)*(-C_4+R_4)+(1-x)*y*(-C_4+R_4)+(1-x)*(1-y)*(-C_4+R_4) \quad (9)$$

$$U_{32} = x*y*(-C_5+A_1-B_1)+x*(1-y)*(-C_5-A_2-B_1)+(1-x)*y*(-C_5+A_1+B_2)+(1-x)*(1-y)*(-C_5-A_2+B_2) \quad (10)$$

$$U_3 = zU_{31}+(1-z)U_{32} \quad (11)$$

At this point, the replication dynamic equation for the government strategy choice is

$$F(z) = dz/dt = z(1-z)(U_{31}-U_{32}) = -z*(z-1)*(A_2-B_2-C_4+C_5+R_4-A_1*y+B_1*x-A_2*y+B_2*x) \quad (12)$$

3.1.2. Stability Analysis of the Equilibrium Point in the Three-Party Evolutionary Game

Based on the replicated dynamic equations for public, business, and government strategy choices, a Jacobi matrix of a three-party evolutionary game system was constructed as follows:

$$J = \begin{bmatrix} J_1 & J_2 & J_3 \\ J_4 & J_5 & J_6 \\ J_7 & J_8 & J_9 \end{bmatrix} = \begin{bmatrix} \frac{\partial F(x)}{\partial x} & \frac{\partial F(x)}{\partial y} & \frac{\partial F(x)}{\partial z} \\ \frac{\partial F(y)}{\partial x} & \frac{\partial F(y)}{\partial y} & \frac{\partial F(y)}{\partial z} \\ \frac{\partial F(z)}{\partial x} & \frac{\partial F(z)}{\partial y} & \frac{\partial F(z)}{\partial z} \end{bmatrix} = \begin{bmatrix} (2x-1)*(C_1-R_1-S_1+S_3+F_1*z+S_2*y-S_3*y) & x*(x-1)*(S_2-S_3) & x*(x-1)*(F_1) \\ -y*(y-1)*(-S_4+S_5) & -(2y-1)*\binom{C_3-C_2+R_2-R_3+F_2*z}{-S_4*x+S_5*x+T*z} & -y*(y-1)*(F_2+T) \\ -z*(z-1)*(B_1+B_2) & -z*(z-1)*(-A_1-A_2) & -(2z-1)*\binom{A_2-B_2-C_4+C_5}{+R_4-A_1*y+B_1*x} \\ & & -A_2*y+B_2*x \end{bmatrix}$$

According to the equilibrium principle of replicating dynamic equations, the three-way evolutionary game between the public, firms, and government has an evolutionary stabilization strategy that serves as a pure Nash equilibrium strategy. This strategy includes eight pure strategy equilibria: (0, 0, 0), (1, 0, 0), (0, 1, 0), (0, 0, 1), (1, 1, 0), (1, 0, 1), (1, 0, 1), (0, 1, 1), (0, 1, 1), and (1, 1, 1). These equilibria were obtained by setting $F(x) = 0$, $F(y) = 0$, and $F(z) = 0$. These eight pure strategy equilibria were substituted into the Jacobi matrix, and the eigenvalues of each equilibrium were determined. The equilibrium points of the three-party evolutionary game system and the eigenvalues of the Jacobi matrix are shown in Table 2.

Table 2. Equilibrium points of the tripartite evolutionary game system and eigenvalues of the Jacobi matrix (prepared by the authors).

Equilibrium Point	Characteristic Value		
	λ_1	λ_2	λ_3
E_1 (0, 0, 0)	$R_1 - C_1 + S_1 - S_3$	$C_3 - C_2 + R_2 - R_3$	$A_2B_2 - C_4 + C_5 + R_4$
E_2 (1, 0, 0)	$R_1 - R_1 - S_1 + S_3$	$C_3 - C_2 + R_2 - R_3 - S_4 + S_5$	$A_2 + B_1 - C_4 + C_5 + R_4$
E_3 (0, 1, 0)	$R_1 - C_1 + S_1 - S_2$	$C_2 - C_3 - R_2 + R_3$	$C_5 - B_2 - C_4 - A_1 + R_4$
E_4 (0, 0, 1)	$R_1 - F_1 - C_1 + S_1 - S_3$	$C_3 - C_2 + F_2 + R_2 - R_3 + T$	$B_2 - A_2 + C_4 - C_5 - R_4$
E_5 (1, 1, 0)	$R_1 - R_1 - S_1 + S_2$	$C_2 - C_3 - R_2 + R_3 + S_4 - S_5$	$B_1 - A_1 - C_4 + C_5 + R_4$
E_6 (1, 0, 1)	$C_1 + F_1 - R_1 - S_1 + S_3$	$C_3 - C_2 + F_2 + R_2 - R_3 - S_4 + S_5 + T$	$C_4 - B_1 - A_2 - C_5 - R_4$
E_7 (0, 1, 1)	$R_1 - F_1 - C_1 + S_1 - S_2$	$C_2 - C_3 - F_2 - R_2 + R_3 - T$	$A_1 + B_2 + C_4 - C_5 - R_4$
E_8 (1, 1, 1)	$C_1 + F_1 - R_1 - S_1 + S_2$	$C_2 - C_3 - F_2 - R_2 + R_3 + S_4 - S_5 - T$	$A_1 - B_1 + C_4 - C_5 - R_4$

Notes: C_1 represents the costs incurred when the public participates in governance; C_2 represents the costs of positive governance by enterprises; C_3 represents the costs of negative governance by enterprises; C_4 represents the costs of strict regulation by local government; C_5 represents the costs of lax regulation by local government; S_1 represents the losses when the public does not participate in governance; S_2 represents the losses to the public when enterprises are actively governed; S_3 represents the losses to the public when enterprises are negatively governed; S_4 represents the profits lost when enterprises are actively governed; S_5 represents the losses suffered when enterprises are negatively governed; R_1 represents the benefits received when the public participates in governance; R_2 represents the benefits of active governance by enterprises; R_3 represents the profits of negative governance by enterprises; R_4 represents the benefits of strict regulation by local governments; A_1 represents the social benefits that local governments receive when they are loosely regulated and when enterprises are positively governed; A_2 represents the penalties incurred by local governments when they are lax in regulation and are negatively governed by enterprises; B_1 represents the penalties incurred by local governments when they are laxly regulated and the public is participative in governance; B_2 represents the social benefits received by local governments when they are lax in regulation and the public is not participative in governance; F_1 represents the additional cost of public participation in governance when local governments strictly regulate; F_2 represents the penalties for negative enterprise governance when local governments strictly regulate; T represents the subsidy received by enterprises for active governance when local governments strictly regulate.

According to the eigenvalue analysis of the Jacobi matrix, if all eigenvalues in the Jacobi matrix are less than zero, the equilibrium is the stable point of the evolutionary game system. On the basis of the eight equilibrium points indicated above, this study showed that there are six stable points: (1, 0, 0), (0, 1, 0), (0, 0, 1), (1, 0, 1), (0, 1, 1), and (1, 1, 1). This study explored the behavioral–strategic relationships between the subjects of the ecological and environmental governance evolution game in the Yangtze River Delta region under different stability scenarios based on the stability point.

Situation 1: E_2 (1, 0, 0) indicates (public participation, negative enterprise governance, and lax local government regulation). Awareness of public participation has increased, but public participation strategies have not led to changes in the strategic choices of other interest groups. Enterprises need to consider the loss of reputation due to public reporting. In the long-term evolutionary process, enterprises may initially choose a positive governance strategy, but when they find that the local government is lax in regulation, even if they are fined by the local government due to public reporting, enterprises may choose a negative governance strategy driven by profit maximization. Local governments need to consider penalties for public participation and reputational damage when choosing a lax regulatory strategy. However, benefits to local governments are still less than the costs of regulation in the case of public participation, and the relative net benefits are still less than zero in the case of strict regulation, which makes a lax regulatory strategy the optimal choice for local governments.

Situation 2: E_3 (0, 1, 0) indicates (public nonparticipation, active enterprise governance, and lax local government regulation). The public chose the nonparticipation strategy based on the active governance behavior of the enterprise and an awareness that the public would not gain anything even if they reported this behavior. The benefits gained by enterprises through green production and legal emissions outweigh the negative governance costs. The local government will increase the intensity of regulation considering the costs incurred by firms in active governance. However, finding that the relative net benefit is still less than zero when strict regulation is applied, the local government will choose to reduce the intensity of regulation, favoring a lax regulatory strategy.

Situation 3: E_4 (0, 0, 1) indicates (public nonparticipation, negative enterprise governance, and strict local government regulation). Local governments choose strict regulatory strategies when the relative net benefit of regulation is greater than zero. The benefit to the public from reporting negative enterprise governance behaviors increases, but the relative net benefit remains less than zero, and the public is biased toward a nonparticipation strategy. The subsidies and penalties received by enterprises will increase; however, the cost of active treatment may be high or the cost of illegal discharge may be low, for example, and enterprises will still choose a negative treatment strategy.

Situation 4: E_6 (1, 0, 1) indicates (public participation, passive enterprise governance, and strict local government regulations). In contrast to Situation 3, the public chooses to participate in ecological governance resulting in a relative net benefit greater than zero. With public reporting, firms need to consider losses such as public compensation and corporate image. Even after deducting losses from local government fines and public reporting, the relative net benefit when an enterprise governs negatively is still greater than zero and enterprises will still choose a negative governance strategy. Local governments need to consider the rewards for public reporting, as well as penalties associated with lax regulation, and will opt for a strict regulatory strategy if lax regulation results in significant losses.

Situation 5: E_7 (0, 1, 1) indicates (public nonparticipation, active enterprise governance, and strict local government regulation). The public choose not to participate in the strategy because positive corporate governance behavior does not negatively affect their living environment, and there is no gain from reporting it. Over the long term, enterprises may initially choose a negative governance strategy, but as local governments become more stringent in their regulations, enterprises will pay more fines while continuing to govern negatively and then receive more subsidies if they govern positively. Therefore, in the long term, enterprises gradually opt for an active governance strategy. A strict regulatory strategy is optimal for local governments considering both economic and social gains.

Situation 6: E_8 (1, 1, 1) indicates (public participation, active enterprise governance, and strict local government regulations). As the probability of enterprises choosing to govern actively and local governments choosing to regulate rigorously increases, the public's living environment improves and trust in enterprises and local governments deepens. By choosing an active governance strategy, enterprises can increase their production capacity through local government subsidies and technological upgrades, improve their image and visibility, and achieve high-quality development. Strict regulatory activities by local governments tend to normalize, regulatory systems tend to improve, regulatory costs are gradually reduced, and rewards and penalties for enterprises and the public gradually weaken. Ultimately, the regional environmental governance achieves a benign interactive situation of public participation, active enterprise governance, and strict supervision by local governments.

3.2. Simulation Analysis of the Three-Party Evolutionary Game

In the ecological and environmental governance of the Yangtze River Delta region, the strategic behavioral choices of the public, enterprises, and local governments are mutually influenced. To better visualize the progression of interactions in the tripartite game, and to examine influential factors pertaining to stakeholders' behavioral strategies, MATLAB R2023a software was employed to conduct numerical simulations.

Combined with the actual situation of ecological environment governance in the Yangtze River Delta region and expert discussions, this study set the initial probability of the subject's willingness to 0.5 uniformly, and the relevant parameters were assigned as follows: $C_1 = 2$, $R_1 = 5$, $S_1 = 3$, $F_1 = 5$, $S_2 = 2$, $S_3 = 1$, $C_2 = 3$, $R_2 = 3$, $T = 2$, $C_3 = 2$, $R_3 = 2$, $F_2 = 4$, $S_4 = 1$, $S_5 = 1$, $C_4 = 1$, $R_4 = 4$, $C_5 = 4$, $A_1 = 3$, $A_2 = 1$, $B_1 = 4$, and $B_2 = 2$. The specific data do not represent the actual time or amount of money but indicate the relative size of each parameter, which helps to show the dynamic game evolution process more objectively and clearly.

3.2.1. Initial Setup for Simulation of Evolutionary Game Systems

The vertical axis represents the behavioral strategy probability p of the three parties' subjects of interest, the horizontal axis represents the evolution speed of the behavioral strategy probability of each subject of interest, and the entire three-party evolution game process is viewed as a change in time t. The initial state simulation results for the three-party evolutionary game system are shown in Figure 3, which verifies that under the conditions of $R_1 - F_1 - C_1 + S_1 - S_2 < 0$, $C_2 - C_3 - F_2 - R_2 + R_3 - T < 0$, $A_1 + B_2 + C_4 - C_5 - R_4 < 0$. In it, x tends to be 0, y tends to be 1, and z tends to be 1. Finally, the ecological environmental governance of the Yangtze River Delta Region has reached a stable state in which the public is not involved, enterprises are actively governed, and local governments are strictly supervised, i.e., the three-party evolutionary game system stabilizes at the equilibrium point (0, 1, 1).

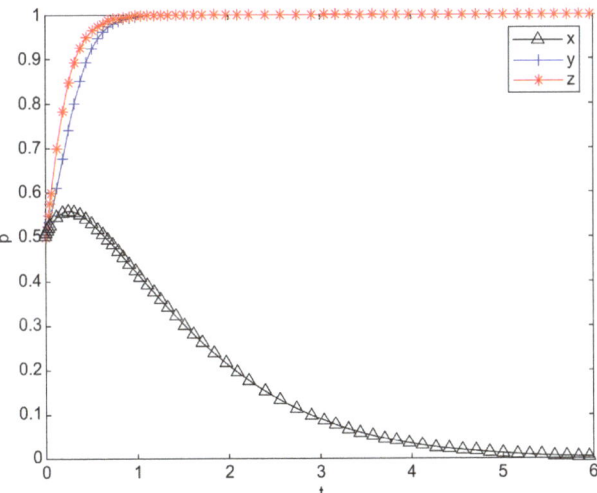

Figure 3. Simulation of the initial three-party evolutionary game strategy (prepared by the authors).

3.2.2. Effect of Public Parameters on the Stability of Evolutionary Game Systems

In the above evolutionary game system, the initial state of the evolutionary game is maintained at (0.5, 0.5, 0.5). The effect of the additional cost F_1 paid by the public to participate in ecological environment governance on the stability of the evolutionary game system was examined.

Figure 4 shows the simulation results when $F_1 = 3$ or 5 and the other parameters remain unchanged. Comparing these results with those shown in Figure 3, the context of strict regulation by the local government and the extra cost paid by the public to participate in ecological environmental governance not only affects trends in the public's strategy evolution but also affects trends in the strategy evolution of enterprises and local government. When the additional cost of public participation in ecological environmental governance ($F_1 = 3$) is low, the public tends to participate in the strategy, enterprises actively manage, and the local government strictly regulates. When the extra costs paid by the public to participate in ecological governance ($F_1 = 5$) are high, compared to $F_1 = 3$, the change in the public's strategy from participation to nonparticipation is high, probably due to the extra costs paid by the public, which are higher than the benefits gained under the nonparticipation strategy. The public will actively participate in ecological governance in the early stages; however, after the trade-offs emerge, they will no longer participate in ecological governance.

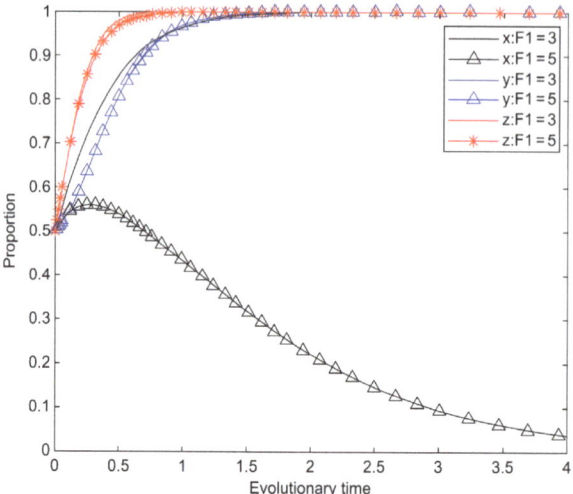

Figure 4. Effect of F_1 on the evolution of the system (prepared by the authors).

Local governments may experience a short-term delay in implementing a strict supervision strategy. This delay can be due to shift from a participation to a nonparticipation strategy by the public. The local government may perceive that the enterprises have achieved good results in ecological governance and therefore face less pressure from the public (i.e., opinions and complaints). Consequently, there would be a slight lag in supervising enterprises during the period. However, if the local government determines that the enterprise's ecological governance does not meet expectations, it will increase supervision and ultimately resort to strict measures. Although the enterprise operates under strict government supervision, the costs and benefits will remain unchanged. The enterprise would be committed to upholding environmental management standards with no significant alteration of its strategy from the initial state. Therefore, in the context of strict local government regulations, enterprises are compelled to implement active governance strategies regardless of whether the public opts to participate actively in the strategy. The above indicates that local governments play a crucial role in ecological governance.

3.2.3. Effect of Enterprise Parameters on the Stability of Evolutionary Game Systems

This section focuses on the impact of the benefit R_2 and local government subsidy T on the stability of the evolutionary game system when enterprises actively manage the ecological environment. For comparison, the initial state of the evolutionary game is maintained at (0.5, 0.5, 0.5).

Figure 5 shows the simulation results when $R_2 = 3$ or 5 and the other parameters remain unchanged. The level of benefits obtained from the enterprise's active governance strategy affects the enterprise's strategy evolution, as well as the strategy evolution of other stakeholders. As the benefits gained from enterprises' active governance strategies increase, the probability of enterprises engaging in active governance increases significantly, thus stabilizing active governance as a strategy for longer periods. The enterprise selects a favorable governance strategy to improve its benefits, which stimulates its participation in ecological governance. Consequently, the enterprise increases its ecological governance efforts, upgrades and reforms its equipment, and expands its production scale to achieve a beneficial system. This ultimately leads to enterprise development becoming increasingly positive. The public's participation in ecological governance efforts can lead to improving the quality of the living environment. However, some may choose not to participate due to their confidence in the local government and corporate ecological governance. It is vital to maintain a harmonious coexistence between man and nature. The local government has not increased its investment in ecological governance during the period, which implies that it is the initial stage of governance strategies in which a favorable social environment for businesses and the public is provided. The simulation results demonstrate that the objective of ecological governance is to attain high-quality development for enterprises. Only when enterprises achieve the objectives of low pollution, high output, and high income can they realize a favorable status of synergistic development among local government, the public, and enterprises. This will lay a solid foundation for the achievement of the dual-carbon goal.

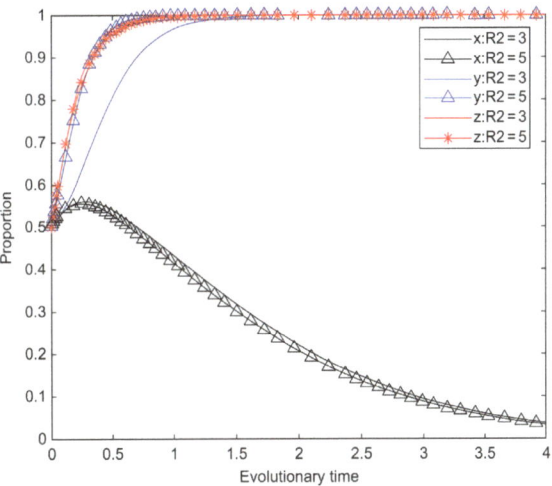

Figure 5. Effect of R_2 on system evolution (prepared by the authors).

Figure 6 shows the simulation results when T = 3 or 5 and the other parameters remain unchanged. In the context of strict local government regulations, changes in local government subsidies received through enterprises' active governance strategies have different impacts on the strategic evolution of stakeholders. When local governments increasingly provide subsidies to enterprises, those utilizing government subsidies can invest in ecological governance, equipment upgrading and transformation, and production expansion. Such activities can lead to a conscious increase in their own supervision and management, ensuring that enterprises actively adopt the strategy over the long term. Compared to the initial state, the local government's strategy has been slow in terms of progress. This is due to the increased costs of ecological governance imposed on enterprises. Consequently, enterprises have been investing more funds in ecological governance, leading to a short period of regulatory laxity. During this period, the public still actively participates in ecological governance. Therefore, the local government continues to increase its supervision of ecological governance, including over enterprises. Under strict regulation by local governments and active governance by enterprises, the public living environment tends to improve, which increases public trust in local governments and enterprises and encourages a nonparticipation strategy among the public. Some enterprises may divert the government's ecological governance subsidies to other uses instead of using them for the intended purpose, which is tantamount to a waste of government resources. It is critical for the local government to monitor the use of such subsidies to ensure they are being used for ecological governance as intended. Therefore, local governments should focus on strengthening the supervision and assessment of enterprises. They should establish comprehensive mechanisms for subsidy fund supervision and ecological governance assessment and implement full-cycle supervision of fund use and ecological governance.

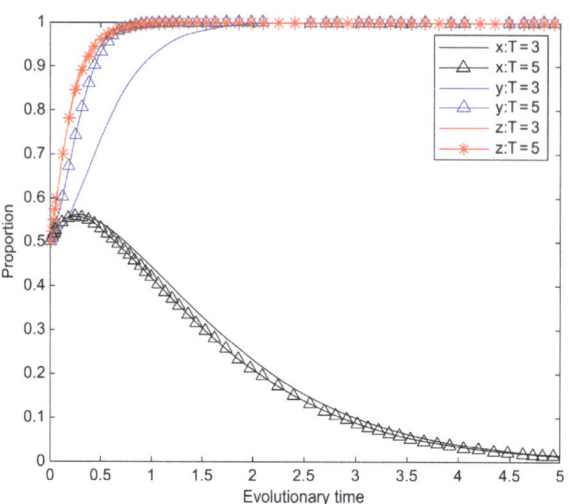

Figure 6. Effect of T on system evolution (prepared by the authors).

3.2.4. Effect of Local Government Parameters on the Stability of Evolutionary Game Systems

This section focuses on the impact of the cost C_5 paid by local governments and penalty B_1 imposed on local governments when the public chooses a participation strategy based on the stability of the evolutionary game system when local governments are loosely regulated. For comparison, the initial state of the evolutionary game was assumed to be (0.5, 0.5, 0.5).

Figure 7 shows the simulation results when $C_5 = 1$ or 5 and the other parameters remain unchanged. The local government's cost of lax regulation influences the evolution of strategies employed by the three parties of interest. In the early stage of system evolution, the public chooses to be involved in environmental governance, as the local government tends to favor a lax regulatory strategy and enterprises tend to favor a negative governance strategy so that the public's living environment cannot be safeguarded. However, in the late stage of system evolution, the public's living environment is continuously optimized under the dual governance of local governments and enterprises, the public's trust in local governments and enterprises is further enhanced, and the public then chooses not to participate in ecological and environmental governance. Local governments that initially implement lax regulatory strategies tend to choose negative governance strategies. However, when the local government implements a strict regulatory strategy, firms shift to an active governance strategy, and the stable time for firms to choose an active governance strategy increases as the local government's strict regulations increase. Local governments will favor strict regulatory strategies as their ecological and environmental governance costs continue to increase. The stabilization time for local governments choosing a strict regulatory strategy at $C_5 = 5$ was much greater than that for local governments choosing a strict regulatory strategy at $C_5 = 1$. The simulation results demonstrate that a good ecological environment is essential for high-quality development. It is vital to maintain a balance between economic growth and environmental protection. Pursuing economic development without strict ecological governance may yield short-term economic benefits for local governments; however, it is not sustainable over the long term. However, ecological destruction can lead to a significant increase in the governance cost for local governments. Additionally, polluted areas are unable to produce economic benefits, which can result in the departure of enterprises and personnel, leading to even greater losses. Therefore, it is critical to consider the long-term consequences of environmental damage. The local government's strategy is crucial for ecological governance and significantly impacts the strategic decisions of the public and enterprises.

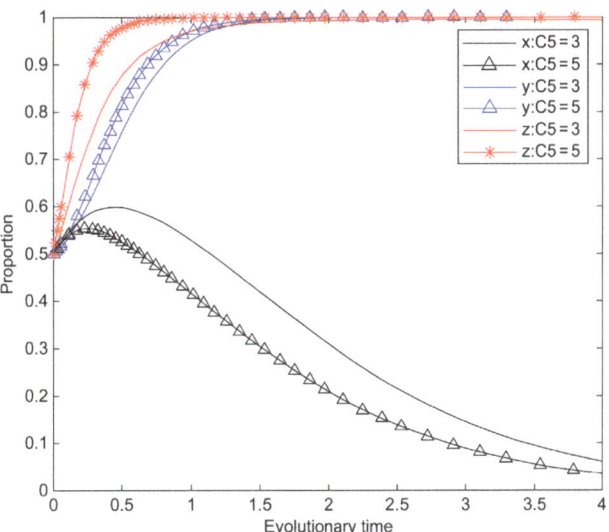

Figure 7. Effect of C_5 on system evolution (prepared by the authors).

Figure 8 shows the simulation results when $B_1 = 1$ or 6 and other parameters remain unchanged. Changes in penalties imposed on local governments affect the system's evolutionary trend, particularly when the public chooses a participation strategy in the context of lax local government regulations. Enterprises do not invest adequately in ecological governance under loose government supervision, resulting in the deterioration of the social and ecological environment. To foster a healthy living environment, the public must participate actively in ecological governance. However, this may have negative consequences for the local government, such as receiving complaints, letters, and public criticism, as well as facing increasing penalties from higher supervisory units. In such a context, local governments will implement stricter regulatory strategies for enterprises, resulting in a rapid increase in government regulation, as shown in Figure 8. Government policies can influence enterprises; however, enterprises could also adopt a more positive attitude toward participation in ecological governance. Notably, the public's focus on the social and ecological environment improves, leading to greater effectiveness in the governance of enterprises and local government. This increases regional satisfaction, and the public's choice of the nonparticipation strategy changes markedly. For the public, the living environment will be effectively improved, the intensity of public complaints and reports will be lower at $B_1 = 6$ than at $B_1 = 1$, and the public will eventually choose not to participate in ecological and environmental governance. The simulation results above demonstrate that lax regulation by local governments has a negative impact on ecological governance, resulting in unfavorable socioeconomic development. Therefore, it is crucial for local governments to maintain strict supervision of ecological governance, avoiding any laxity and adhering strictly to ecological governance regulations to foster a green ecological environment and support socioeconomic development.

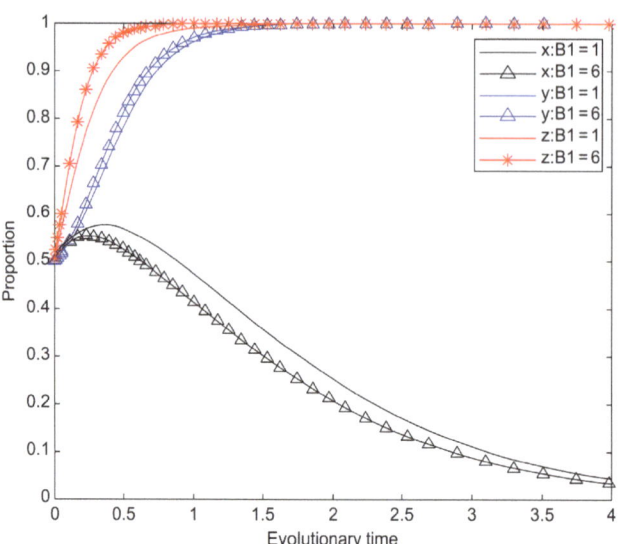

Figure 8. Effect of B_1 on system evolution (prepared by the authors).

4. Discussion

This paper presents a tripartite evolutionary game model of ecological governance in the Yangtze River Delta region. Behavioral strategies and underlying factors were analyzed and are discussed separately below.

(1) Regarding the evolutionary game of ecological governance in the Yangtze River Delta region, studies have been limited to the game between the central government and local governments or between the governments of the three provinces and one city [6,18–24]. Jiang et al. developed a central government–local government evolutionary game model to study the collaborative ecological spatial governance of lakes in the Yangtze River Delta across provincial boundaries [20]. Bo et al. analyzed and explored haze governance behavior in the Yangtze River Delta region [24]. Based on stakeholder theory and evolutionary game theory, this paper presents a tripartite evolutionary game model of public–enterprise–local government from the perspective of multiple stakeholders. The model analyzes the evolutionary game strategy of ecological governance in the Yangtze River Delta region and draws two new conclusions about ecological governance.

Firstly, ecological environment governance in the Yangtze River Delta region is a self-organized process of mutual games among multiple interest groups. This is consistent with the research conclusions of Jiang et al. and Bo et al. [20,24]. The present study further elucidated that the behavioral strategies of participating subjects cannot be analyzed in isolation in ecological governance, which is contrary to the concept of ecological governance. Because ecological governance is a systematic project, one-sided analysis of the game strategy between governments has certain limitations. The behavioral strategy of enterprises and the public also affects the choice of behavioral strategy of a local government. Therefore, in the practice of ecological governance in the Yangtze River Delta region, a closed-loop mechanism should be established for the whole chain and process; additionally, enterprises, the public, and local governments should take responsibility for the corresponding processes and nodes of ecological governance to establish a model area for the collaborative ecological governance in the Yangtze River Delta.

Secondly, local governments provide subsidies to encourage enterprises to engage in active governance, which has positive impacts on their governance strategy. Although the public participates in ecological governance, local governments are penalized for negative public opinion, which encourages the government to adopt a strategy of strict regulation. Fu et al. used quantitative analysis to analyze the data of the main governance indicators of Hangzhou–Shaoxing–Ningbo and another three cities and concluded that there were differences in the regional ecological governance strategies [6]. However, they only analyzed the geographical differences in the governance results of Hangzhou–Shaoxing–Ningbo and another three cities and could not analyze the behavioral strategies of the participating subjects in depth. Therefore, the net increase in profitability for each stakeholder involved in the ecological governance of the Yangtze River Delta region must exceed the cost of implementation for the collaborative ecological governance mechanism to be sustainable and stable.

(2) The study's conclusions on behavioral strategies for ecological governance and their influencing factors in the Yangtze River Delta region align with the findings of Zhao et al. and Cao et al. regarding the variability in the selection of ecological governance strategies by multiple subjects [38,40]. However, the causes of this variability were analyzed from different perspectives, complementing and improving the understanding of the influencing factors. Zhao et al. highlighted variability in the enforcement of government environmental regulations [38]. They also noted that the costs and revenues of enterprises, as well as the costs and public psychology, are key factors influencing eco-governance. Cao et al. noted differences in enterprises' behaviors toward green technological innovation and listed several underlying factors, such as pollution tax and incentive compensation [40].

Researchers have typically examined the variations in ecological governance strategies and their influencing factors from various perspectives, including environmental regulation and green technological innovation [30–32,38–40]. In contrast, the present study focuses on the mechanisms of the whole ecological governance process in the Yangtze River Delta region and considers the public, enterprises, and local governments as game analysis subjects using model simulation. The simulation revealed that the key factors influencing behavioral strategies in the ecological governance of the Yangtze River Delta region are the additional costs paid by the public to participate in ecological governance, the benefits and subsidies received by enterprises when they actively govern the ecological environment, and the costs and penalties paid by local governments when they loosely regulate the ecological environment. Therefore, the establishment of a synergistic governance mechanism of "government–enterprise co-construction of liquid regulatory funds, government–enterprise–public partnerships in low-cost regulatory models, and the sharing of high-quality regulatory outcomes" is proposed as a strategy of motivating and encouraging relevant stakeholders to participate in ecological governance.

5. Conclusions

5.1. Findings

Further research on the behavioral strategies of multi-interested subjects and the mechanisms underlying influencing factors is needed to build a collaborative ecological and environmental governance model in the Yangtze River Delta region. This study applied stakeholder theory to include the public, enterprises, and local governments as interested parties in the collaborative environmental governance mechanism. Through evolutionary game modeling and simulation analysis, this study also explored the behavioral effects and influencing factors of different subjects in environmental governance; simulated the evolutionary trajectories of strategies employed by the three parties under scenarios with governance costs, governance subsidies, and regulatory costs; and analyzed the effectiveness of different policies and measures in promoting the high-quality operation of the collaborative governance of the Yangtze River Delta region. The results of this study enrich the body of theoretical research on the evolutionary game of ecological and environmental governance in the Yangtze River Delta region. The following main conclusions can be drawn from the results of this study:

(1) The ecological governance of the Yangtze River Delta region is a dynamic evolutionary game system involving multiple stakeholders, and the behavioral strategies of each stakeholder will affect the stability of the system. As shown in Figure 2, during the initial phase of system development, local governments continue to strengthen their supervision to encourage enterprises to actively participate in ecological environmental management, and the public will actively participate in ecological environmental management through complaints and reports. The likelihood of enterprises activating their governance strategy increases rapidly at this point. In the final stage of system evolution, local governments and enterprises deliberately implement active governance and strict supervision strategies. Due to reduced publicity and investment in ecological environmental governance by local governments, enthusiasm for public participation decreases. Consequently, local governments transfer some of the regulatory costs to the public, along with incentives or subsidies for enterprises, to encourage public engagement in ecological environment governance. This leads to the stabilization of the three-party game system at the equilibrium point (0, 1, 1).

(2) In the ecological governance of the Yangtze River Delta region, the behavioral strategies of each participant impact the stability of the system, meaning that the strategic choices of one party are influenced by and have a reciprocal effect on the other two parties. Based on the simulation results, under the background of strict regulation by local governments, enterprises are forced to implement active governance strategies regardless of whether the public chooses active participation strategies or not. Local governments pursue economic development without strict regulation of environmental governance, which can achieve certain economic benefits in the short term. However, if the environment is damaged, the environmental management costs paid by the local government increase dramatically, resulting in greater losses. Therefore, the strategic choices of local governments play important roles in the ecological governance of the Yangtze River Delta region and closely influence the strategic choices of the public and enterprises.

(3) The costs and benefits of participating in ecological and environmental governance in the Yangtze River Delta region are the primary factors influencing the behavioral strategies of multiple stakeholders. Specifically, as the cost of public participation in ecological environmental governance increases, the public becomes more inclined towards nonparticipation, enterprises tend to support active governance, and local governments tend to favor strict regulations. The three-party evolutionary game system tends to be stable, and, for enterprises, the increased benefits and subsidies they receive when they actively manage the ecosystem can accelerate the tripartite evolutionary game system to a stable state. For local governments, the cost of lax regulations is a major factor influencing their strategic choices. Additionally, the size of the penalties imposed on local governments plays a crucial role in enterprise decisions.

(4) The ecological governance of the Yangtze River Delta region is a systematic and open project. Adhering to the problem-oriented approach, universal linkage, comprehensive system, development, and change perspectives should be applied to analyze the subjects participating in collaborative ecological governance and their behavioral strategies to maximize the interests of all subjects. In addition, adhering to the system concept, the participating subjects should have clear subject responsibilities and obligations. Based on a people-first principle and considering that the ecological environment is linked the national economy and people's livelihoods, all participating subjects should adhere to the "green mountain is the golden silver mountain" concept, innovate, and work collaboratively to establish a demonstration zone for coordinated ecological governance in the Yangtze River Delta, which could facilitate high-quality economic and social development.

5.2. Recommendations

Based on the analysis of the three-party evolutionary game model and conclusions drawn from previous research, this study proposes the following actions.

(1) Government–enterprise co-development of liquidity supervision funds: The local governments of the three provinces and one city should collaboratively establish a capital supervision pool based on a certain proportion of revenue. Regulatory funds can provide subsidies to the relevant stakeholders when they actively participate in ecological and environmental governance and achieve success. In contrast, the subject of negative ecological environmental governance will be punished, the amount of the punishment will be paid to the regulatory pool, and all information on the flow of funds, rewards, and punishments will be open and transparent.

(2) A low-cost regulatory model in which the government, enterprises, and the public work synergistically. Using a big data information platform, the Yangtze River Delta region has implemented a one-click disclosure of ecological and environmental pollution sources. The local government carries out follow-up verification to determine the real source of pollution, penalizes the relevant interested parties, and establishes a sound mechanism for penalties and rectification, while the government, enterprises, and the private sector are monitored in real time by the tripartite body.

(3) The government, enterprises, and the public share the results of high-quality regulation. With regard to the additional benefits gained from regional ecological and environmental governance, the benefits will be shared by all of the people, enhancing the living conditions for residents, the operational environment for enterprises, and the regulatory circumstances for the government to encourage sustainable development of the ecological environment in the Yangtze River Delta region.

5.3. Limitations and Reflections

(1) Existing studies have only examined the ecological environment of the Yangtze River Delta region as a whole. Future investigations should focus on transboundary water and atmospheric pollution in this area for more specific information and analyses. Using the Yangtze River Delta region as a case study for water pollution governance, future research can be based on evolutionary game theory, with the aims of establishing a government–enterprise–public ecological governance evolutionary game model, exploring the behavioral strategies and the underlying factors for collaborative water pollution management in the Yangtze River Delta region and analysis of the pollution management effects of collaborative water pollution management mechanisms in the Yangtze River Delta region.

(2) The results of the simulation analysis of the tripartite evolutionary game model in this paper are based on the overview of the actual situation of ecological environmental governance in the Yangtze River Delta region, on the basis of which the relevant variables are derived in an ideal situation. Follow-up research can reinforce the cooperation among the public, enterprises, and the government and evaluate the cost of synergistic governance and synergistic governance effect of multi-interested parties to further explore the research questions put forth in this paper.

(3) This study focused on building a scientifically tested dynamic reward and punishment model. Local governments can increase regulatory efforts, penalties for negative corporate governance, and public incentives to report complaints. However, from the perspective of sustainable development, local governments must avoid providing excessive subsidies or incentives to enterprises or the public. Alternatively, they can employ methods such as policy encouragement and technical assistance to encourage eco-friendly practices among both businesses and the public.

Author Contributions: Conceptualization, Q.W.; data curation, Q.W. and C.M.; formal analysis, Q.W.; funding acquisition, C.M.; methodology, Q.W.; resources, C.M.; software, Q.W.; supervision,

C.M.; writing—original draft, Q.W.; writing—review and editing, Q.W. All authors have read and agreed to the published version of the manuscript.

Funding: This work was supported by the National Key Social Science Foundation Program of China "Study on Benefit-Sharing and Compensation Mechanisms for Ecological Synergistic Development in the Yangtze River Delta Region" (Grant No. 20AGL036).

Institutional Review Board Statement: Not applicable.

Informed Consent Statement: Not applicable.

Data Availability Statement: The original contributions presented in the study are included in the article, further inquiries can be directed to the corresponding author.

Conflicts of Interest: The authors declare no conflicts of interest.

References

1. Turiel, J.; Ding, I.; Liu, J.C.E. Environmental governance in China: State, society, and market. *Brill Res. Perspect. Gov. Public Policy China* **2017**, *1*, 1–67. [CrossRef]
2. Mao, C.M.; Cao, X.F. Research on cross-domain synergistic management of atmospheric pollution—Taking the Yangtze River Delta region as an example. *J. Hohai Univ. (Philos. Soc. Sci. Ed.)* **2016**, *18*, 46–51+91.
3. Suo, L.M.; Li, M.Y.; Kan, Y.Q. Collective action, network structure and its model evolution of environmental multi-objects co-governance-an observation based on Hangzhou and Hefei metropolitan area. *J. Gansu Admin. Coll.* **2021**, *1*, 60–71+126.
4. Yuan, C.M.; Gabor, H. Research on the dynamic correlation network of haze pollution and synergistic governance of the Yangtze River Delta city cluster under the perspective of spatiotemporal interaction. *Soft. Sci.* **2019**, *33*, 114–120.
5. Jing, T.; Zhe, J. Role mechanism and empirical test of synergistic environmental governance in the Yangtze River Delta region. *Ecol. Econ.* **2022**, *38*, 192–199+224.
6. Fu, D.Y.; Liu, Y.S. Research on the ecological integration governance strategy of the Yangtze River Delta region-an empirical measurement based on the urban agglomerations of Hangzhou-Shauning-Ningbo, Suzhou-Wuxi-Changzhou and Guangzhou-Foshan-Zhaoqing. *J. Yunnan Univ. Financ. Econ.* **2021**, *37*, 81–90.
7. Ansell, C.; Gash, A. Collaborative governance in theory and practice. *J. Public Admin. Res. Theor.* **2008**, *18*, 543–571. [CrossRef]
8. Emerson, K.; Nabatchi, T.; Balogh, S. An integrative framework for collaborative governance. *J. Public Admin. Res. Theor.* **2012**, *22*, 1–29. [CrossRef]
9. Wu, J.N.; Liu, C.H.; Chen, Z.T. How does a coordinated regional air pollution control mechanism work in China? Experience from the Yangtze River Delta. *China Adm.* **2020**, *5*, 32–39.
10. Si, L.B.; Zhang, J.C. Dynamic mechanism, governance mode and practical situation of cross-administrative region ecological environment cooperative governance: A comparative analysis based on typical cases of national ecological governance key regions. *Qinghai Soc. Sci.* **2021**, *4*, 46–59.
11. Di, Q.B.; Chen, X.L.; Hou, Z.W. Identification of regional differences and critical paths for coordinated governance of pollution reduction and carbon reduction in three major urban agglomerations in China under the "dual-carbon" goal. *Resour. Sci.* **2022**, *44*, 1155–1167.
12. Johari, M.; Hosseeini-Motlagh, S.M.; Basti-Barzoki, M. An Evolutionary Game Theoretic Model for Analyzing Pricing Strategy and Socially Concerned Behavior of Manufacturers. *Transport. Res. E-Log. Transp. Rev.* **2019**, *18*, 506–525. [CrossRef]
13. Traulsen, A.; Glynatsi, N.E. The future of theoretical evolutionary game theory. *Philo. Tran. R. Soc. Lond. B* **2023**, *378*, 20210508. [CrossRef]
14. Sigmund, K.; Nowak, M.A. Evolutionary game theory. *Curr. Biol.* **1999**, *9*, R503–R505. [CrossRef] [PubMed]
15. Sandholm, W.H. Evolutionary game theory. In *Complex Social and Behavioral Systems: Game Theory and Agent-Based Models*; MIT Press: Cambridge, MA, USA, 2020; pp. 573–608.
16. Huberman, B.A.; Glance, N.S. Evolutionary games and computer simulations. *Proc. Natl. Acad. Sci. USA* **1993**, *90*, 7716–7718. [CrossRef] [PubMed]
17. Wang, M.; Li, Y.; Cheng, Z.; Zhong, C.; Ma, W. Evolution and equilibrium of a green technological innovation system: Simulation of a tripartite game model. *J. Clean. Prod.* **2021**, *278*, 123944. [CrossRef]
18. Raman, G.V. Environmental Governance in China. *Theor. Econ. Lett.* **2016**, *6*, 583–595. [CrossRef]
19. Kennny, P.W. Equilibrium pollution taxes in open economies with imperfect competition. *J. Environ. Econ. Manag.* **1994**, *27*, 49–63. [CrossRef]
20. Jiang, X.Y.; Yu, T. A study on synergistic ecological spatial governance of lakes across provinces in the Yangtze River Delta under the perspective of inter-governmental game. *Planner* **2022**, *38*, 67–73.
21. Chou, Z.P.; Liu, C.X.; Zhu, J. Evolutionary game analysis of Beijing-Tianjin-Hebei haze cooperative governance based on collective action logic. *China Popul. Resour. Environ.* **2017**, *27*, 56–65.
22. Wang, H.M.; Xie, Y.L.; Sun, J. Research on the "action" game and synergistic factors of Beijing-Tianjin-Hebei air pollution control in different contexts. *China Popul. Resour. Environ.* **2019**, *29*, 20–30.

23. Li, Y.Y.; Dai, J.; Sheng, Q. Analysis of regional air pollution joint prevention and control eco-compensation mechanism based on evolutionary game—Taking Beijing-Tianjin-Hebei region as an example. *Sci. Technol. Manag. Res.* **2022**, *42*, 202–210.
24. Bo, M.G.; Shi, Z.S.; He, Z. Simulation study of haze synergistic management in the Yangtze River Delta region. *Syst. Sci.* **2020**, *28*, 58–63.
25. Song, Y.; Yang, T.T.; Zhang, M. Research on impact of environmental regulation on enterprise technology innovation-an empirical analysis based on Chinese provincial panel data. *Environ Sci. Pollut. Res. Int.* **2019**, *26*, 835–848. [CrossRef]
26. Nielsen, I.E.; Majumder, S.; Sana, S.S.; Saha, S. Comparative analysis of government incentives and game structures on single and two-period green supply chain. *J. Clean. Prod.* **2019**, *235*, 1371–1398. [CrossRef]
27. Li, G.P.; Yan, B.Q.; Wang, Y.Q. A game study on the evolution of environmental regulatory strategies for pollution management in the Yellow River Basin. *J. Beijing Inst. Technol. (Soc. Sci. Ed.)* **2022**, *22*, 74–85.
28. Wang, X.P.; Zhang, Z.M.; Guo, Z.H.; Su, C.; Sun, L.H. Energy structure transformation in the context of carbon neutralization: Evolutionary game analysis based on inclusive development of coal and clean energy. *J. Clean. Prod.* **2023**, *398*, 136626. [CrossRef]
29. Li, M.; Gao, X. Implementation of enterprises' green technology innovation under market-based environmental regulation: An evolutionary game approach. *J. Environ. Manag.* **2022**, *308*, 114570. [CrossRef]
30. Pan, F.; Liu, Y.; Wang, L. Game analysis of central-land environmental regulation under the perspective of public participation. *Oper. Manag.* **2020**, *29*, 113–123.
31. Chu, Z.P.; Bian, C.; Yang, J. How can public participation improve environmental governance in China? A policy simulation approach with multi-player evolutionary game. *Environ. Impact Assess. Rev.* **2022**, *95*, 106782. [CrossRef]
32. Li, C.F.; Qiu, R.; Wang, Y.H. A tripartite evolutionary game and stability control strategy for monitoring and supervision of green transformation of resource-based enterprises. *Soft Sci.* **2022**, *36*, 99–109.
33. Zhu, X.X.; Mu, Q.R.; Liang, W.Z. An innovative strategic choice for stakeholders in the Chinese traditional commercial street renewal using evolutionary game theory. *J. Innov. Knowl.* **2022**, *7*, 100225. [CrossRef]
34. Wang, Q.; Wang, N.; Wang, H.; Xiu, Y. Study on influencing factors and simulation of watershed ecological compensation based on evolutionary game. *Sustainability* **2022**, *14*, 3374. [CrossRef]
35. Yang, X.Y.; Liang, S.S. Research on governance mechanism of Internet information ecosystem based on evolutionary game. *Contemp. Econ. Sci.* **2023**, *45*, 29–45.
36. Le, X.U.; Ma, Y.G.; Wang, X.F. Research on environmental policy choice of green technology innovation based on evolutionary game: Government behavior vs. public participation. *China Manag. Sci.* **2022**, *30*, 30–42.
37. Cui, M. A three-party evolutionary game analysis of environmental credit regulation in the context of collaborative governance. *Syst. Eng. Theor. Pract.* **2021**, *41*, 713–726.
38. Zhao, L.M.; Chen, Y.Q. Environmental regulation, public participation and corporate environmental behavior-an empirical analysis based on evolutionary game and provincial panel data. *Syst. Eng.* **2018**, *36*, 55–65.
39. You, D.M.; Yang, J.H. Evolutionary game analysis of government environmental regulation and corporate eco-technology innovation behavior under public participation. *Sci. Technol. Manag. Res.* **2017**, *37*, 1–8.
40. Cao, X.; Zhang, L.P. Evolutionary game analysis of the diffusion of corporate green technology innovation. *China Popul. Resour. Environ.* **2015**, *25*, 68–76.
41. Yang, G.M.; Shi, Y.J. Study on ecological compensation of the Three Gorges Basin of the Yangtze River based on evolutionary game theory. *J. Syst. Simul.* **2019**, *31*, 2058–2068.
42. Wang, Y.B.; Lu, Y. A game study on the evolution of China's environmental governance system under the background of fiscal decentralization. *China Popul. Resour. Environ.* **2019**, *29*, 107–117.
43. Liu, Z.; Qian, Q.; Hu, B.; Shang, W.L.; Li, L.; Zhao, Y.; Zhao, Z.; Han, C. Government regulation to promote coordinated emission reduction among enterprises in the green supply chain based on evolutionary game analysis. *Resour. Conserv. Recy.* **2022**, *182*, 106290. [CrossRef]
44. Wang, G.; Chao, Y.; Cao, Y.; Jiang, T.; Han, W.; Chen, Z. A comprehensive review of research works based on evolutionary game theory for sustainable energy development. *Energy Rep.* **2022**, *8*, 114–136. [CrossRef]
45. Zhang, Z.; Zhang, G.; Hu, Y.; Jiang, Y.; Zhou, C.; Ma, J. The evolutionary mechanism of haze collaborative governance: Novel evidence from a tripartite evolutionary game model and a case study in China. *Hum. Soc. Sci. Commun.* **2023**, *10*, 1–14. [CrossRef]
46. Agle, B.R.; Donaldson, T.; Freeman, R.E.; Jensen, M.C.; Mitchell, R.K.; Wood, D.J. Dialogue: Toward superior stakeholder theory. *Bus. Ethics Q.* **2008**, *18*, 153–190. [CrossRef]
47. Tan, Y.; Fang, K. Environmental governance in China. *J. Chin. Gov.* **2016**, *1*, 191–194. [CrossRef]
48. Li, W.X.; Liu, J.Y.; Li, D.D. Getting their voices heard: Three cases of public participation in environmental protection in China. *J. Environ. Manag.* **2012**, *98*, 65–72. [CrossRef]

49. Park, S.H.; Koo, J.C.; Youn, Y.C. Villagers' participation in conservation of village woodlands-Two cases of Namwon City, Korea. *J. Korean For. Soc.* **2013**, *102*, 15–23. [CrossRef]
50. Zhang, T.B.; Zhang, Q.; Fan, Q.Q. Research on corporate governance motivation and public participation externality under government environmental regulation. *China Popul. Resour. Environ.* **2017**, *27*, 36–43.

Disclaimer/Publisher's Note: The statements, opinions and data contained in all publications are solely those of the individual author(s) and contributor(s) and not of MDPI and/or the editor(s). MDPI and/or the editor(s) disclaim responsibility for any injury to people or property resulting from any ideas, methods, instructions or products referred to in the content.

Article

Spatial–Temporal Differentiation and Trend Prediction of Coupling Coordination Degree of Port Environmental Efficiency and Urban Economy: A Case Study of the Yangtze River Delta

Min Wang [1,2,3], Yu Lan [1,2], Huayu Li [1,2,*], Xiaodong Jing [1,4,*], Sitong Lu [5] and Kexin Deng [6]

1. Business School, Hohai University, Nanjing 210098, China; wangm@hhu.edu.cn (M.W.); 221313020021@hhu.edu.cn (Y.L.)
2. School of Economics and Finance, Hohai University, Changzhou 213200, China
3. Low Carbon Economy Research Institute, Hohai University, Nanjing 210098, China
4. Asia Institute, The University of Melbourne, Melbourne, VIC 3010, Australia
5. School of Management and Economics, The Chinese University of Hong Kong, Shenzhen, Shenzhen 518172, China; sitonglu@link.cuhk.edu.cn
6. Business School, New York University Shanghai, Shanghai 200122, China; kd2474@nyu.edu
* Correspondence: 201313060004@hhu.edu.cn (H.L.); jingxiaodong@hhu.edu.cn (X.J.)

Abstract: Green development is a primary path for ports and cities to achieve a low-carbon transition under the Sustainable Development Goals and a powerful driving force to elevate regional port–city relations to a high level of coordination. In this paper, twenty port cities in the Yangtze River Delta (YRD) were selected and port environmental efficiency (PEE) was calculated through the window SBM model, while the EW-TOPSIS model was used to evaluate high-quality urban economic development (HED). The coupling coordination degree (CCD) model, the kernel density model, GIS spatial analysis, and the grey prediction model were used to further explore the spatial–temporal dynamic evolution and prediction of the CCD between PEE and HED. The results suggested that: (1) PEE fluctuation in the YRD is increasing, with a trend of seaports achieving higher PEE than river ports; (2) HED in the YRD shows upward trends, and the polarization of individual cities is obvious; (3) Temporally, the CCD in the YRD has risen from 0.438 to 0.518. Shanghai consistently maintains intermediate coordination, and Jiangsu has experienced the most significant increase in CCD. Spatially, CCD is led by Lianyungang, Suzhou, Shanghai, and Ningbo-Zhoushan, displaying a decreasing distribution pattern from east to west. The projection for 2026 suggests that all port cities within the YRD will have transitioned to a phase of orderly development. To enhance the coordination level in the YRD, policymakers should consider the YRD as a whole to position the ports functionally and manage them hierarchically, utilize the ports to break down resource boundaries to promote the synergistic division of labor among cities, and then tilt the resources towards Anhui.

Keywords: green development; port environmental efficiency; urban economy; coupling coordination degree; spatial–temporal evolution; China

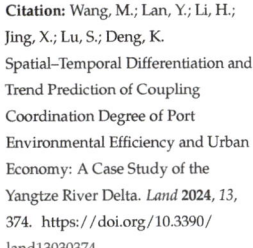

Citation: Wang, M.; Lan, Y.; Li, H.; Jing, X.; Lu, S.; Deng, K. Spatial–Temporal Differentiation and Trend Prediction of Coupling Coordination Degree of Port Environmental Efficiency and Urban Economy: A Case Study of the Yangtze River Delta. *Land* 2024, *13*, 374. https://doi.org/10.3390/land13030374

Academic Editors: Wei Sun, Zhaoyuan Yu, Kun Yu, Weiyang Zhang and Jiawei Wu

Received: 6 February 2024
Revised: 3 March 2024
Accepted: 13 March 2024
Published: 16 March 2024

Copyright: © 2024 by the authors. Licensee MDPI, Basel, Switzerland. This article is an open access article distributed under the terms and conditions of the Creative Commons Attribution (CC BY) license (https://creativecommons.org/licenses/by/4.0/).

1. Introduction

The in-depth implementation of the Coordinated Regional Development Strategy [1] is an inevitable support of China's high-quality development as well as an essential part of promoting the modernization of construction in China. In the construction of coastal cities, the coordination between the port environment and the urban economy has gradually become a focal point. As an essential way to construct ecological civilization, green development has become the guiding principle for the low-carbon transformation of ports and cities [2] and for regional coordination development [3].

The YRD is located on the eastern coast of China and south of the Qinling–Huaihe Line, boasting a mild climate and abundant rainfall. The Yangtze River has endowed the region with vast alluvial plains, providing the necessary foundation for population concentration and industrial and agricultural development in the YRD. Meanwhile, the favorable geographical conditions of the lower reaches of the Yangtze River, such as stable water levels, wide navigation channels, gentle currents, and complex river networks, have catalyzed the region's shipping industry growth. Consequently, the YRD has emerged as the largest port cluster and city cluster in China. The port cluster, relying on the Yangtze River and the ocean, passes through the core area of the YRD and surrounds the left side of the urban agglomeration, forming a close geographical connection with the city cluster. This geographical synergy renders the YRD an ideal case study for exploring port–city coordination dynamics in China.

The YRD port cluster is one of the world's maritime trade centers [4]. It has the broadest hinterland, with the Yangtze River linking 11 provinces in China. Meanwhile, there are numerous harbors and river ports in the YRD, with significant advantages in sea–river intermodal transport. Data from Lloyd's List shows that 7 ports in the YRD entered the world's top 100 container ports in 2022, with Shanghai Port and Ningbo-Zhoushan Port ranking prominently among the top 10. The Ministry of Transport of the People's Republic of China claimed that the annual cargo throughput and container throughput of major ports in the YRD represented 37.56% and 37.39% of the national total in 2022, providing tax revenues and employment opportunities for the YRD city cluster. Furthermore, the YRD port cluster mitigates geographical barriers for goods and resources, enabling coastal industry transfers to inland cities and fostering a synergistic labor division within the cluster [5]. As the center of China's economic construction, the YRD city cluster leads in high-quality development [6]. According to data from the China National Bureau of Statistics, the regional GDP of the YRD accounted for 24% of China's GDP in 2022, providing strong financial support for the development of the port cluster. The advanced transport network in the YRD, with a total operating mileage of high-speed railways exceeding 6600 km, a highway network of 16,700 km, and a network of high-grade waterways exceeding 4270 km, also supports multimodal transport in the port cluster. The port and city clusters are interdependent, supporting the overall prosperity of the YRD [7].

While the port and city clusters of the YRD are synergistically improving in terms of economic benefits, escalating ecological conflict poses a significant challenge to high-quality economic growth. Serving as China's largest port cluster, port operations in the YRD consume plenty of fossil energy [8] and emit large amounts of greenhouse gases and particulate matter [9]. The pollution emissions from ships account for 47.84% of China's three major port clusters [10], affecting urban air quality and the respiratory health of residents of the YRD [11]. Furthermore, the rapid industrialization and urbanization of the YRD city cluster has also exacerbated energy consumption and ecological damage [12]. This contradicts the goal of high-quality and high-efficiency construction of the urban economy and hinders the port cluster's low-carbon transition. In response to these pressing challenges, the Outline of the Yangtze River Delta Regional Integration Development Plan explicitly set out the principles of synergistic environmental management and adherence to green development in December 2019. Consequently, green development has become a critical requirement for the low-carbon transformation of ports, the high-quality development of port cities, and coordinated development between ports and cities.

In summary, green development introduces a new perspective for examining the coordination relationship between the port and the city. It will provide data support for further coordinated development policies of the YRD through reconstructing the PEE and HED evaluation indicators, considering environmental performance, and systematically evaluating the spatial–temporal evolution characteristics of CCD between ports and cities.

2. Literature Review
2.1. Port Environmental Efficiency

Ports serve as the land and water transport intersection, equipped with multifaceted facilities for ship access, berthing, and passenger and cargo transport [13]. Previous papers have explored the technical [14] and operational [15] efficiency of ports in depth. With the popularization of the green development concept, the goal of port upgrading and transformation has become to pursue low energy consumption, low carbon emissions, efficient use of resources, and environmental friendliness [16]. Subsequently, essays on pollution emission measurement [17], emission reduction pathways [18], and green port evaluation [19] have been developed. Aligned with the guiding principles espoused by influential entities such as the Organization for Economic Co-operation and Development (OECD) [20], environmental efficiency is conceptualized as the ratio of the value of products and services obtained by an economy in its production and operation process to the input of production and operation and environmental pollution factors. It measures both economic and environmental efficiency and has become a standard for measuring the efficient allocation of resources. Therefore, PEE reflects the economic value generated per unit of environmental load during the production and operation process of ports, which is a characterization of a port's overall competitiveness [21].

SFA and DEA models are the most commonly used methods in efficiency analysis [22]. Since the DEA model does not require setting a specific functional form, it is more suitable for analyzing multiple inputs and outputs [23]. Wang et al. [24] measured the efficiency of 18 ports in China using a DEA model based on a "game and cooperation" development framework. With China entering the stage of green development, environmental efficiency has become an essential part of port efficiency research. Chin and Low [25] were the first to consider environmental factors in measuring port efficiency. The SBM model, which considers the undesired outputs, has gradually become the mainstream method for studying PEE. Hsu and Huynh [26] explored the PEE evaluation of 12 container terminals in Vietnam by applying SBM models, considering CO_2 as undesirable output. Li et al. [27] measured the PEE in China's Bohai Sea Rim using SBM models. The window SBM model was further employed to analyze the dynamic evolution of environmental efficiency. Nodin et al. [28] examined the efficiency of rice self-sufficiency in Malaysia based on a windowed SBM model. Mamghaderi et al. [29] measured the dynamic environmental performance of 27 OECD countries by using the window SBM model.

2.2. High-Quality Urban Economic Development

HED is constructed under the concept of green development, serving as a comprehensive indicator to measure the sustainable development of various aspects of the urban economy, society, and environment [30]. The economic volume, green technology spillover, and innovative talent cultivation of port cities are the fundamental stones of low-carbon port development.

The existing literature has elaborated HED in multiple dimensions [31]. Most scholars agree that urbanization development, economic growth, and environmental quality are the cornerstones of urban development [32]. With the gradual completion of urban functions, the concept of HED has been gradually extended to three aspects: economic development, quality of life, and ecological environment [33]. Subsequent studies in the literature have successively incorporated social governance, resource utilization, ecological environment, cultural construction, livelihood, and innovation efficiency into the concept of HED [34]. Research methods have focused on principal component analysis [35], the expert survey method [36], and the entropy power method [37]. The New Development Concept encompasses five aspects: innovation, greenness, coordination, openness, and sharing, which represent urban development dynamics, harmony between humanity and nature, balanced development, internal and external linkages, and social justice. It covers economic growth points, which is more in line with the current situation of China's green and high-quality development [38]. Zha et al. [39] used EW-TOPSIS to investigate the development characteristics of HED of the significant city clusters in the Yangtze River Economic Belt. Guo and Sun [40] added economic development indexes to measure the HED of 30 provinces in China.

2.3. Synergistic Effect of Port and Urban Economics

Ports and port cities are an organic whole linked by geographic location. Functioning as pivotal logistics hubs, ports play a dual role by supplying global raw materials [41] and promoting cities' integration into the global industrial chain [42], thus realizing urban commercial service value [43]. Simultaneously, benefiting from the absorptive capacity of ports for productivity factors such as technology, information, and capital, industries like shipping services [44] and bonding [45] are gathering at the borders of ports. This eases urban employment pressure [46] and also strengthens the connection between cities and ports, promoting regional port–city integration. Port cities have a feedback mechanism for ports as well. The cities provide the operational spaces of ports, and their outputs are also related to the economic volume and openness of cities. With the rapid development of the tertiary industry, the support of the advanced management system, green technology overflow, and talent cultivation of the cities are more obvious in the construction of ports. However, it is not sustainable for ports and cities to develop in the traditional way characterized by high pollution and high energy consumption. Negative environmental impacts from ports' production activities can cause serious damage to the ecological environment of port cities; the deterioration of the ecological environment of port cities is detrimental to the survival and development of human beings and ultimately limits the future development of ports. Therefore, modern, high-quality port–city relations require adherence to the direction of green development and the reduction in negative environmental externalities caused by the production of economic agents.

The existing literature has explored the synergistic effect of port and urban development in a quantitative manner. Most documents have measured port–city interactions by constructing evaluation systems. Wang et al. [47] took Lianyungang as an example and built a land–sea coordinated development evaluation system from three aspects of port cities, marine resources, and ports. Guo et al. [48] used a dynamic centralization index to estimate port–city relationships and identify port–city interaction patterns in major coastal port cities in China. Ma et al. [49] measured the integration degree of Chinese ports by the port integration index and conducted a comparative analysis between the YRD and the Bohai Sea Rim. Some of the literature draws on case studies for research. Vroomans et al. [50] took the ports of Hamburg, Antwerp, and Rotterdam as examples and showed that shared values enhance the integration of ports and cities. Russo and Musolino [51] explored port–city relationships in Antwerp, Trieste, Santander, and Algeciras in a transportation system model context. Only a few scholars have used the CCD model to discuss port–city synergistic effects. Liu et al. [52] measured port–city interactions in Liaoning. Guo and Qin [53] constructed a port–city network system to evaluate the theoretical framework and measurement methods of port–city synergistic relationships from the perspective of mobility space.

In summary, the existing literature has researched PEE and HED, but more discussion is needed still. Firstly, the green development concept puts higher demands on the port–city coordination relationship in the YRD, and the related literature should be supplemented. Secondly, the traditional DEA model cannot accurately assess the dynamic changes in PEE in the YRD, and the green dimension should be added to the HED measures to comprehensively assess the level of the urban economy. Ports and cities are interdependent and should be integrated into a system to explore their interaction. Therefore, this article further analyzes the spatial–temporal dynamic evolution of CCD between PEE and HED of the YRD from the green development concept and then makes reasonable predictions to provide data support for the formulation of further coordinated development policies in the YRD. The marginal contribution lies in the following: (1) In terms of research perspective, the existing literature mainly considers the economic performance of port efficiency and urban economic development. This paper explores PEE and HED based on the green development perspective, making up for the existing literature's having ignored environmental efficiency. (2) The traditional PEE and HED measurements have ignored the environmental impact indicators in constructing indicators. This paper adds carbon emission indicators to PEE evaluation, and green dimensions are constructed in the HED evaluation system to compensate for the shortcomings of ignoring environmental factors. (3) Regarding model selection, the window SBM model is used instead of the traditional SBM model to reflect the dynamic development of PEE. The CCD model is also more capable of analyzing the dynamic association of CCD between PEE and HED from a whole perspective than linear regression. The grey prediction model is used to scientifically predict the CCD of the YRD, which assists the port–city coordinated development in the YRD by putting forward reasonable suggestions.

3. Materials and Methods

3.1. Model

3.1.1. Window SBM Model

The SBM model is a non-parametric technical efficiency analysis model for comparing multiple decision-making units (DMUs) based on considering slack variables and non-expected outputs [54]. The window DEA is based on the principle of the moving average; it treats the same assessed object in different periods as separate DMUs and then selects different reference sets for efficiency calculations by moving the window [55]. Therefore, this paper adopted the SBM model of the window DEA to better reflect the time variation in DMUs. Additionally, the method enhances the identification of differences in the efficiency values of decision-making units by expanding the DMUs.

Referring to the extant literature [56], the window width d = 3 was selected.

The model is defined as follows:

$$Min\rho = \frac{1 - \frac{1}{m}\sum_{i=1}^{m}\frac{s_i^-}{x_{i0}}}{1 + \frac{1}{s_1+s_2}(\sum_{r=1}^{s_1}\frac{s_r^g}{y_{r0}^g} + \sum_{r=1}^{s_2}\frac{s_r^b}{y_{r0}^b})}$$

$$s.t. x_0 = X\lambda + s^-$$
$$y_0^g = Y^g\lambda - s^g$$
$$y_0^b = Y^b\lambda + s^b$$
$$s^- \geq 0, s^g \geq 0, s^b \geq 0, \lambda \geq 0$$

(1)

3.1.2. Entropy Weight and TOPSIS (EW-TOPSIS)

The entropy weight method determines the weight of each indicator through the amount of information reflected in the degree of data variability, which reduces the interference of subjective factors caused by artificial assignment. The TOPSIS method ranks quantitatively by comparing the relative distance of each measurement object with the optimal and the worst solutions, with the advantages of simple calculation and reasonable results. In this paper, the EW-TOPSIS model was selected to measure HED.

Firstly, determine the positive and negative ideal solutions of the evaluation indicators.

$$V^+ = (v_j^+)_{j \in J} = \{(\max v_{ij}|j \in J)|i = 1, 2, \cdots, m\}$$

(2)

$$V^- = (v_j^-)_{j \in J} = \{(\min v_{ij}|j \in J)|i = 1, 2, \cdots, m\}$$

(3)

Then, calculate the distance between the indicator to its positive and negative ideal solutions in different years, d_i^+ and d_i^-.

$$\begin{cases} d_i^+ = \sum_{j=1}^{n} (v_{ij} - v_j^+)^2 \\ d_i^- = \sum_{j=1}^{n} (v_{ij} - v_j^+)^2 \end{cases} \quad (4)$$

Finally, calculate the relative fitness of the evaluation object and the optimal scheme C_i.

$$C_i = \frac{d_i^-}{d_i^+ + d_i^-} \quad (5)$$

The value C_i represents the HED, and the calculation results can be used for ranking different provinces.

3.1.3. Coupling Coordination Degree Model

Coupling reflects two systems' degree of interdependence and constraints, the degree of interaction between two systems; coordination measures the virtuous cooperation between two systems [57]. Using the CCD model to explore the coordination level is more consistent with the synergistic relationship between PEE and HED.

$$C = 2 \left[\frac{U_1 U_2}{(U_1 + U_2)^2} \right]^{\frac{1}{2}} \quad (6)$$

$$D = (C \times T)^{1/2} \quad (7)$$

$$T = \alpha U_1 + \beta U_2 \quad (8)$$

where U_1 and U_2 respectively represent the PEE and HED; C is coupling degree, D is CCD; T is a composite evaluation index between the two systems; and α and β are the weight coefficient. This study assumes that PEE and HED are equally important [58], so $\alpha = \beta = 0.5$. The classification of CCD is shown in Table 1.

Table 1. Classification of CCD.

D Value	Level	Stage
$0.000 \leq D \leq 0.090$	Extreme disorder	
$0.090 < D \leq 0.190$	Serious disorder	Disordered development stage
$0.190 < D \leq 0.290$	Moderate disorder	
$0.290 < D \leq 0.390$	Mild disorder	
$0.390 < D \leq 0.490$	Bare coordination	
$0.490 < D \leq 0.590$	Mild coordination	
$0.590 < D \leq 0.690$	Moderate coordination	Orderly development stage
$0.690 < D \leq 0.790$	Intermediate coordination	
$0.790 < D \leq 0.890$	Good coordination	
$0.890 < D \leq 1.000$	Quality coordination	

3.1.4. Grey Prediction Model

The grey system theory indicates that complex system data complexity must also have the overall function and the inherent development of the law. It can effectively predict future data through correlation analysis of the degree of difference in trends among system factors with a small sample size and weak data regularity [59]. This paper used this model to forecast the CCD between PEE and HED in the YRD from 2022 to 2026.

Firstly, assume that the time series $X^{(0)} = \{X^{(0)}(1), X^{(0)}(2), \ldots, X^{(0)}(n)\}$ consists of n observations, and perform a cumulative productive series, $Z^{(1)} = \{Z^{(1)}(2), Z^{(1)}(3), \ldots, Z^{(1)}(n)\}$. The corresponding differential equation for the GM (1.1) model is:

$$X^{(0)}(k) + \alpha Z^{(1)}(k) = \mu \tag{9}$$

where α is the developmental ash number and μ is the endogenous control ash number.

Then, assuming that $\hat{\alpha}$ is the parameter vector to be estimated, $\hat{\alpha} = \begin{pmatrix} \alpha \\ \mu \end{pmatrix}$; then, the least squares method can be used to solve $\hat{\alpha} = (B^T B)^{-1} B^T Y$ and solve the differential equations to obtain the prediction model:

$$\hat{X}^{(1)}(t) = \left[X^{(0)}(0) - \frac{\mu}{\alpha}\right] e^{-\alpha t} + \frac{\mu}{\alpha} \tag{10}$$

3.2. Index System Construction

Following the data requirements of the DEA window model and considering the characteristics of the port industry, the article selected the number of terminals and the length of berths as input variables to reflect the production and service capacity of the port and selected cargo throughput and container throughput as desired output variables to reflect the capacity of the port. In addition, based on the requirements of China's green development and the dual-carbon background, port carbon emissions were taken as the non-desired output. The PEE measurement indicators are presented in Table 2.

Table 2. Indicator system for PEE in ports.

Indicator Category	Specific Indicator
Input indicators	Number of berths (units)
	Length of wharf (meters)
Expected output indicators	Cargo throughput (10,000 tons)
	Container throughput (10,000 TEU)
Unexpected output indicator	Carbon emission (10,000 tons)

The New Development Concept brings green development to the fore of the national development strategy, which is of great guiding significance for China's high-quality construction. Therefore, the evaluation index system of HED was constructed from the New Development Concept of innovation, coordination, greenness, openness, and sharing [60]. It is presented in Table 3.

Table 3. Indicator system for HED in port cities.

Dimension	Rule	Indicator	Attribute
Innovation	Innovative inputs	Science and technological activists (persons)	+
		Internal expenditures on R&D funding (CNY 100 million)	+
		Proportion of science, technology, and innovation in fiscal expenditure (%)	+
		Proportion of education in fiscal expenditure (%)	+
	Innovation outputs	Patent acceptance per 10,000 persons (pieces)	+
Coordination	Urban–rural coordination	Disposable income gap between urban and rural residents (CNY)	-
	Industrial coordination	Proportion of tertiary sector (%)	+
	Demand coordination	Proportion of consumption in GDP (%)	+
		Proportion of consumption in GDP (%)	+
	Economic and social coordination	Inflation rate (%)	-
		Unemployment rate (%)	-

Table 3. *Cont.*

Dimension	Rule	Indicator	Attribute
Green	Resource consumption	Energy consumption per unit of GDP (tons of standard coal/10,000 CNY)	-
	Environmental pollution	Wastewater (tons)	-
		Waste gas (10,000 tons)	-
		Waste solids (tons)	-
	Environmental governance	Share of environmental protection in fiscal expenditure (%)	+
Open	Trade opening	Proportion of imports to GDP (%)	+
		Proportion of exports to GDP (%)	+
	Investment opening	Proportion of FDI to GDP (%)	+
Sharing	Development sharing	GDP per capita (CNY)	+
	Public services	Number of medical beds per 10,000 population (sheets)	+
		Number of students per 10,000 population (persons)	+
		Proportion of general public services in fiscal expenditure (%)	+

3.3. Data Sources

The YRD, located at north latitude 27°12′–35°20′ N and east longitude 114°54′–122°12′ E, covers a total area of 358,000 square kilometers. It is positioned in the lower reaches of the Yangtze River, bordering the central region to the west and the Yellow Sea, the East China Sea, and the Pacific Ocean to the east. The region features flat land, and the terrain slopes from southwest to northeast. According to the Outline of the Plan for the Integrated Development of the Yangtze River Delta Region [61] approved by the State Council of the PRC, the YRD contains four provincial-level administrative units, Shanghai, Jiangsu, Anhui, and Zhejiang, and 41 municipal-level administrative units. Based on data availability, this paper selected the data of 20 major ports and port cities (Figure 1) in the YRD for 2012–2021. The indicator data of PEE were obtained from the China Urban Statistical Yearbook and the China Port Yearbook. The indicator data of HED came from the China Urban Statistical Yearbook, the Statistical Yearbooks of each port city, and the statistical bulletin of national economic and social development of previous years. Individual missing data were filled in by linear interpolation. Data on the administrative divisions of the YRD and the Yangtze River system in Figure 1 were obtained from the Resource and Environmental Science and Data Centre, Chinese Academy of Sciences (https://www.resdc.cn/; accessed on 13 August 2023).

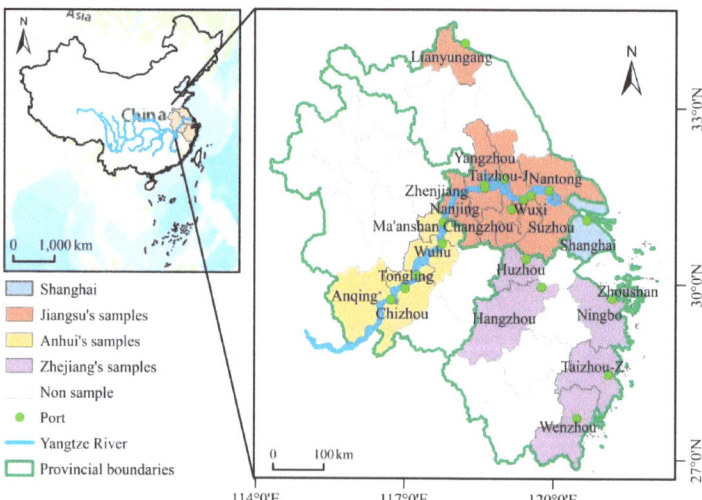

Figure 1. Twenty ports and their hinterlands in the YRD.

4. Results

4.1. Comprehensive Evaluation of PEE and HED

The window SBM and EW-TOPSIS models were applied to measure the PEE and HED of the YRD from 2012 to 2021. The temporal trends at the provincial level are shown in Figure 2. Using 0.4 and 0.2 as the PEE and HED dividing lines, the four-image distribution of the 20 port cities was plotted (Figure 3).

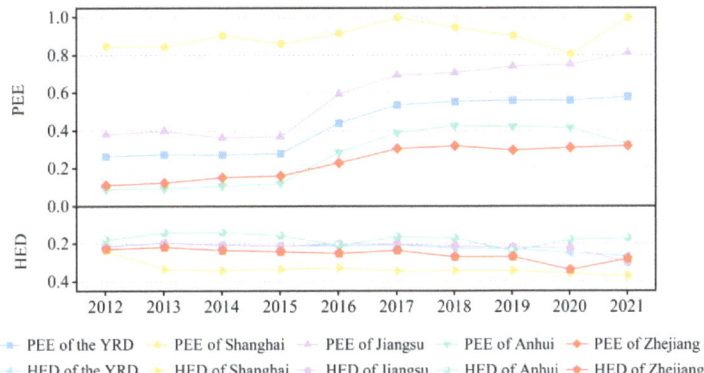

Figure 2. Time evolution of PEE and HED in the YRD from 2012 to 2021.

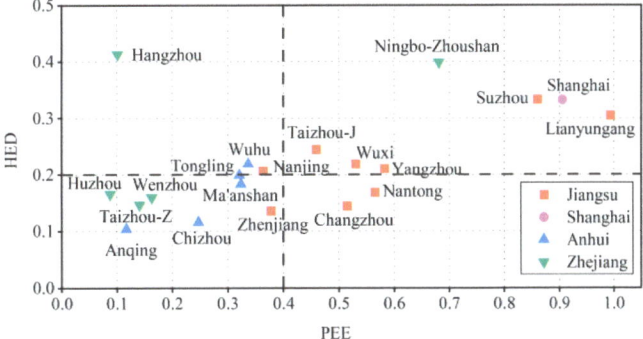

Figure 3. Average PEE and HED of twenty port cities from 2012 to 2021.

The overall PEE of the YRD grew from 0.265 to 0.578 from 2012 to 2021, with an average annual growth rate of 9%. The PEE showed a significant improvement in 2016. In 2015, the Ministry of Transport issued The Opinions on Comprehensively Deepening the Reform of Transportation. The document puts forward a series of requirements, including energy-saving supervision, promotion of emission reduction technologies, and construction of green recycling systems. Since then, the YRD has continued to embark on intensified efforts to bolster vessel emissions management and accelerate the adoption of shore power and clean energy sources. The YRD's PEE has failed to exceed 0.6 each year, which suggests that the ports have significant room for progress in implementing green coordinated development and reducing carbon emissions. Jiangsu carried out a deep-water channel project in 2016, which greatly liberated the capacity of the lower reaches of the Yangtze River, with Jiangsu boasting the fastest PEE growth rate. Among the 20 ports, coastal ports such as Lianyungang Port, Shanghai Port, Suzhou Port, and Ningbo-Zhoushan Port are at the forefront of PEE. Inland river ports, especially Huzhou Port, Hangzhou Port, and Anqing Port, on the other hand, have a weaker performance.

From 2012 to 2021, the HED increased from 0.211 to 0.265, with robust development momentum. Spatially, it shows regional solid imbalances and increasing inter-provincial disparities. Shanghai's HED consistently maintained the highest level in the YRD, with an increase of 55% in the past ten years. Positioned as the national economic center, Shanghai exerts a strong agglomeration effect regarding innovative resources, high-end industries, and high-quality talents. Meanwhile, Shanghai has continuously increased investment in the ecological environment, putting its HED significantly ahead of that of the surrounding provinces and cities. The polarization of HED in Zhejiang is extremely serious. This discrepancy can be attributed to the geographical constraints faced by southern Zhejiang, which is confined to a narrow plain area less conducive to accommodating large-scale industries. Conversely, northern Zhejiang penetrates deep into the core region of the YRD, fostering close economic and industrial ties with Shanghai and southern Jiangsu. This has further widened the developmental gap again. Anhui, being the sole non-coastal entity in the YRD, did not fully integrate into the YRD until 2019. Consequently, the HED of Anhui is relatively low. Except for Wuhu, all the other cities rank in the bottom ten.

4.2. Measurement Results for CCD between PEE and HED

The CCD between PEE and HED in the YRD from 2012 to 2021 is shown in Figure 4. In 2016, the CCD within the YRD experienced a notable 18% increase, marking a transition from bare coordination to mild coordination. This development can be primarily attributed to the implementation of green port construction. The adoption of clean energy and breakthroughs in grid-connected shore power technology have greatly reduced the emission of sulfur dioxide and nitrogen oxides in ports and enhanced the PEE of ports while greatly improving the environment of port cities.

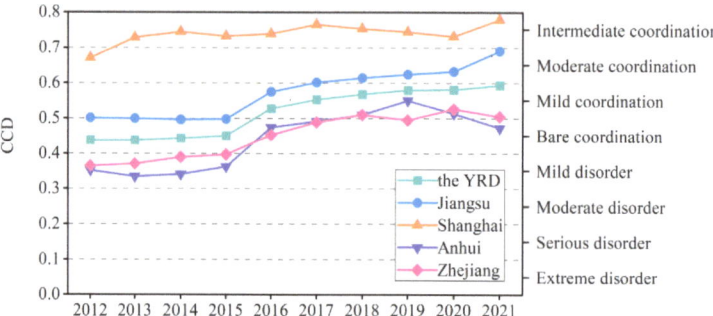

Figure 4. Time evolution of CCD of the YRD from 2012 to 2021.

At the provincial level, Shanghai has consistently maintained an intermediate coordination level. Jiangsu follows closely, with 38% progress in CCD. The coordination level of Zhejiang and Anhui lags behind the overall level of the YRD, with Zhejiang's low PEE and Anhui presenting ample room for improvement in HED. The difference in CCD among the four provinces in the YRD shows a trend of narrowing first and then widening. The outbreak of COVID-19 in 2020 dealt a severe blow to foreign trade and green technological innovation in China's ports and blocked the development of the urban economy, disrupting the YRD's upward trend in harmonization.

Figure 5 displays the CCD for twenty port cities in the YRD from 2012 to 2021. The average coordination level of most port cities is concentrated in bare coordination and mild coordination, showing a spindle shape. The proportion of port cities in the disordered development stage is 45%, and no city reaches good coordination. Lianyungang, Shanghai, Suzhou, and Ningbo-Zhoushan have reached intermediate coordination. The ports in these major cities are closely linked to the cities' economies, with important mutual symbiotics. In contrast, Anqing, Huzhou, Taizhou, and Wenzhou are constrained by their relatively remote locations and weak hinterland economies. These cities struggle to provide adequate cargo volumes and lack robust infrastructure support for the ports. The ports in these locales encounter challenges in attracting industrial elements to congregate within the cities, further exacerbating their economic limitations. Regarding regional distribution, the CCD of port–city systems in Zhejiang and Anhui is below 0.5, while in Jiangsu and Zhejiang it is mostly above 0.5.

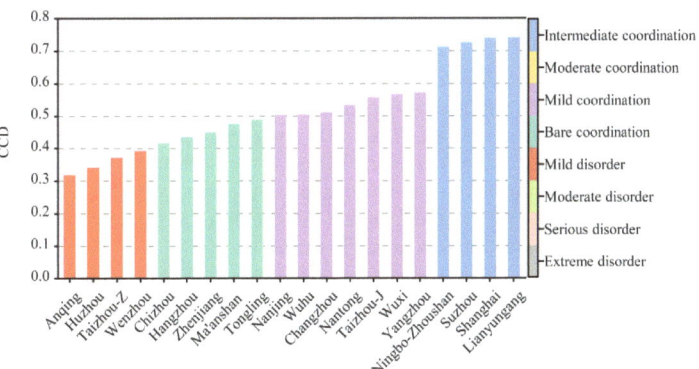

Figure 5. CCD of twenty port cities in the YRD from 2012 to 2021.

4.3. Spatial Dynamic Evolution of CCD between PEE and HED

Kernel density estimation is a non-parametric estimation method for estimating probability density functions. Kernel density curves for the YRD from 2012 to 2021 were plotted using kernel density analysis to further reveal the dynamic evolution of the CCD. Since Shanghai has only one port and does not need kernel density analysis, the YRD, Jiangsu, Anhui, and Zhejiang results are reported (Figure 6).

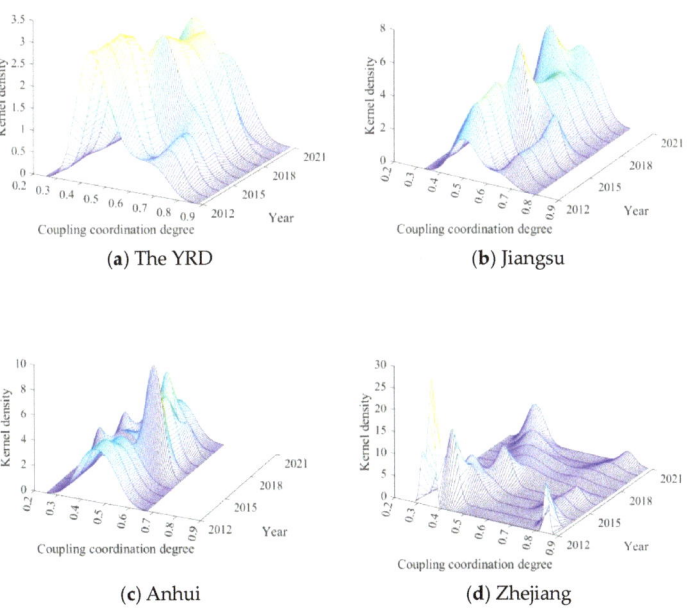

Figure 6. Kernel density map of CCD in the YRD from 2012 to 2021. (**a**) CCD of the YRD; (**b**) CCD of Jiangsu; (**c**) CCD of Anhui; (**d**) CCD of Zhejiang.

As shown in Figure 6a, the center line of the kernel density in the YRD is shifted to the right, which shows an overall increase in the CCD. The number of wave peaks turns from multiple to single, and the wave width of the curve narrows, indicating that the CCD multi-polarization in the YRD is gradually weakening.

As shown in Figure 6b, Jiangsu's kernel density shows a rightward shifting trend. From 2012 to 2016, Suzhou, Lianyungang, and other regions of Jiangsu had significant differences in CCD, with profound polarization and a double-peak pattern of kernel density. The CCD gradually increased after 2016, and the kernel density curve shifted from multiple to single peaks.

Figure 6c shows that Anhui's CCD kernel density curve is highly chaotic. The median line moves from 0.4 to 0.5, and the coordination level is lower overall. From 2016 to 2019, the CCD of Wuhu and Ma'anshan first showed a considerable enhancement, and the curves appeared to have multiple steep peaks. After 2019, the CCD of Tongling, Chizhou, and Anqing gradually enhanced, the differences between the regions decreased, and the wave peaks shifted to a single peak.

The development of CCD in Zhejiang is shown in Figure 6d. The bimodal pattern runs through the sample period, and the polarization is apparent.

4.4. Spatial–Temporal Differentiation of CCD between PEE and HED

Using ArcGIS spatial visualization technology, Figure 7 displays the spatial distribution of the YRD in 2012, 2015, 2018, and 2021 to present the spatial–temporal differentiation of the CCD.

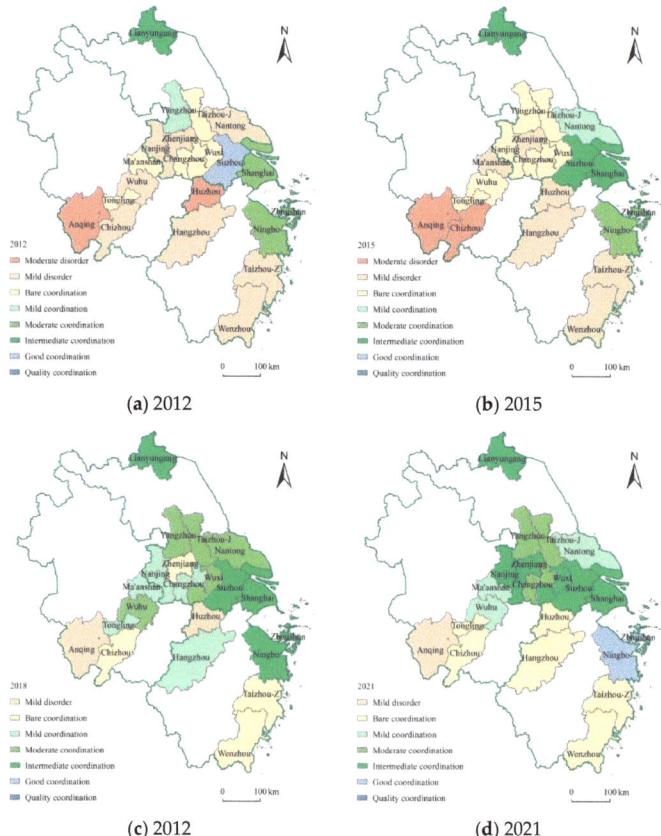

Figure 7. Spatial distribution of CCD in the YRD in 2012, 2015, 2018 and 2021.

From the perspective of the YRD as a whole, the number of cities in moderate disorder changes from 2 to 0, while the number of cities in the orderly development stage changes from 5 to 13, indicating that the YRD's CCD has undergone substantial improvement.

The CCD of the four provinces shows different states of development. Shanghai has maintained an intermediate coordination level since 2013 and is in the leading position in the YRD.

The coordination level in Jiangsu has improved significantly. In 2012, only Yangzhou, Suzhou, and Lianyungang reached the stage of orderly development. After 2016, Jiangsu gradually emphasized the uniqueness of inland ports and conducted meticulous research on inland ports. In 2017, the People's Government of Jiangsu drew up the first Layout Plan for Inland River Ports in Jiangsu Province (2017–2035), which accurately positioned the development of inland ports. Since then, inland river ports such as Nantong Port and Yangzhou Port have accelerated the pace of intensive, large-scale, and green development. By 2021, all ten port cities in Jiangsu had realized the orderly development stage, realizing a leap from 30% to 100%.

Anhui shows a steady and progressive development. In 2012 and 2015, the five cities in Anhui were all in a disorderly development stage. After 2018, Anhui's CCD was subject to an enormous spillover from Jiangsu, which led to a steady increase in Wuhu and Ma'anshan. Therefore, the CCD showed a high development trend in the east and a low development trend in the west.

Zhejiang is seriously polarized. Ningbo-Zhoushan is the only location in the YRD to achieve good coordination in 2021. On the other hand, Huzhou, Hangzhou, Taizhou, and Wenzhou are in a disorderly development stage.

4.5. Trend Prediction of CCD between PEE and HED

Based on the grey prediction model, the CCD between PEE and HED in 20 port cities in the YRD from 2022 to 2026 was predicted by Matlab. The accuracy test of the grey prediction model is described in Table 4. The posterior difference ratio (C) is less than 0.65, and the slight error probability (P) is more significant than 0.7, meaning the model qualifies. The smaller the root mean squared error (RMSE), the higher the approximation between the predicted results of the model and the actual results. According to Table 4, the accuracies of the CCD prediction values all meet the qualified standards, making the prediction results reliable.

Table 4. CCD prediction accuracy test between PEE and HED in the YRD from 2022 to 2026.

Region	C	P	RMSE	Region	C	P	RMSE
Suzhou	0.382	0.8	0.023	Wuhu	0.178	1.0	0.044
Nantong	0.183	0.8	0.038	Anqing	0.345	0.8	0.030
Zhenjiang	0.103	1.0	0.038	Ma'anshan	0.239	0.9	0.046
Nanjing	0.063	1.0	0.028	Tongling	0.567	0.7	0.047
Lianyungang	0.575	0.7	0.017	Chizhou	0.487	0.7	0.091
Wuxi	0.071	1.0	0.033	Ningbo-Zhoushan	0.097	0.9	0.029
Taizhou-J	0.108	1.0	0.032	Wenzhou	0.179	0.9	0.027
Changzhou	0.565	0.7	0.055	Taizhou-Z	0.476	0.7	0.036
Yangzhou	0.357	0.8	0.06	Hangzhou	0.387	0.7	0.053
Shanghai	0.216	0.7	0.013	Huzhou	0.054	1.0	0.012

The prediction results for CCD between the PEE and HED of 20 port cities in the YRD from 2022 to 2026 are reported in Figure 8. In 2024, the coordination level of the YRD will realize the transition from moderate coordination to intermediate coordination. Until 2016, the CCD reached 0.76. Regionally, Jiangsu will surpass Shanghai and step into good coordination. Zhejiang and Anhui will also break through to moderate coordination. All port cities will individually enter an orderly development stage, during which Wuxi and Ningbo-Zhoushan will achieve quality coordination. However, Anqing and Taizhou are just on the edge of bare coordination. In terms of growth rate, Shanghai, Suzhou, and Lianyungang are growing slowly, so it is urgent to provide new power to the breakthrough of port–city coordination.

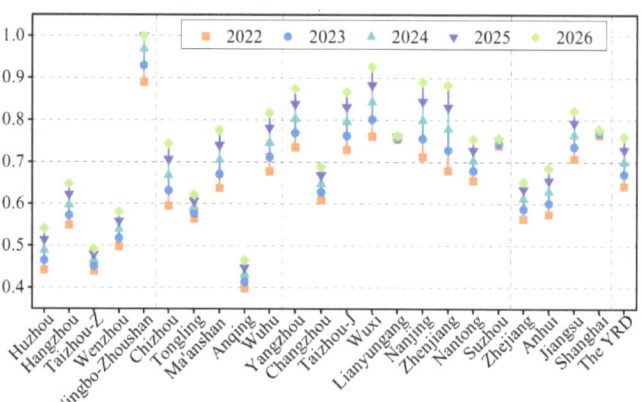

Figure 8. CCD forecast between PEE and HED of 20 port cities in the YRD from 2022 to 2026.

5. Discussion

The proposal of the dual-carbon target means that China takes green development as the keynote of high-quality development. Simultaneously, fostering a coordinated development relationship between ports and cities is fundamental to promoting port cities to realize high-quality development. Therefore, it is of practical significance to explore the characteristics of the CCD of port cities in the YRD under the concept of green development.

In terms of PEE, the majority of the existing studies focus on the environmental efficiency of the world's large ports [62], and a few comparative analyses have been conducted on a port cluster. This article selected the YRD port cluster as its research subject, incorporating both large- and small-scale ports to offer a reference for the green development of global port clusters. According to our research, PEE shows significant improvement (Figure 2) and a spatial difference between seaports and inland river ports, the former having achieved higher PEE (Figure 3). Consistent with the study of Wang et al. [63], seaports are in the absolute leading position in the YRD. As an international shipping center, Shanghai Port has the largest regional business volume, the most complete infrastructure, and the most advanced green technology [64]. Ningbo-Zhoushan Port has seized the historical opportunities of globalization, containerization, and its natural endowments in the depth of its channel and terminal to become an important hub port in China [65]. This paper further extended the study of inland river ports in the YRD. Excessive competition is the main reason for the relatively low PEE in the inner harbor. Due to the market squeeze from large ports such as Shanghai Port, the development scale of these ports is very limited [66]. Excessive and disorderly competition in inland river ports has also led to duplicated construction and resource waste, which reduces port efficiency [67]. Conversely, Lianyungang Port, situated at a far distance from the core of the YRD, avoids the homogeneous competition within the port cluster to a certain extent [68].

Secondly, the article has broken the traditional perspective of focusing on the economic benefits of cities and constructed a comprehensive evaluation index for HED based on the context of green development in China. This innovative approach integrates environmental sustainability with economic growth metrics, offering a more holistic view of urban development. The HED of the YRD shows a rising trend, while the rapid rise of Shanghai and Zhejiang widens the inter-provincial differences in the YRD (Figures 2 and 3). Shanghai has led in the low-carbon economy development, but the growth rate is slow [69]. Yang et al. [70] highlighted that effective technological development and resource planning need to compensate for Shanghai's relatively small area. Consistent with the results of

Du et al. [71], Suzhou's HED performance stands out equally in the YRD. Being adjacent to Shanghai, Suzhou has historically assumed the industrial transfer from Shanghai, forging close ties with its metropolitan neighbor [72]. There exist certain differences between Du, Cardoso, and Rocco [71] and Chen et al. [73], and economically developed cities such as Nanjing and Changzhou are expected to perform better in terms of HED. In fact, due to the divergent functional positioning and development stages of cities, there are differences in the implementation and efficacy of green development initiatives as well as the availability of support mechanisms [74]. Jiangsu, with its developed industry and complex industrial chain, requires further promotion of its carbon emission reduction measures. As Anhui was fully integrated into the YRD in 2019, it was also comparatively analyzed in this paper. Consistent with the results of Wei et al. [75], innovation and green development are the focus of Anhui's HED enhancement.

Thirdly, while prior research such as that of Sun et al. [76] and Wang et al. [77] has predominantly focused on resource coordination within urban systems, this article has highlighted the significant interdependent relationship between port and city, adopting the YRD port and city clusters and creatively putting PEE and HED into the coupled coordination framework as two subsystems to explore their interaction and analyze them comparatively. The results show that the CCD is on the rise (Figure 4), with a regional imbalance (Figures 6 and 7). Consistent with the findings of Qu, Kong, Li, and Zhu [3] and Chen, Zhang, Song, and Wang [58], Shanghai and Ningbo-Zhoushan are well coordinated. As China's premier coastal cities, they boast obvious advantages in terms of both urban economy and port productivity. The substantial economic output of these cities not only ensures ample cargo volume for the ports but also facilitates investment in port infrastructure [78]. Furthermore, the ports transship urban cargo and revitalize urban resources [79], forming an economic cycle. In addition, Kong and Liu [80] further point out that there is overinvestment in ports in Ningbo-Zhoushan, which leads to waste of resources and reduced efficiency. The article also focused on the development of CCD in inner port cities in the YRD. Anhui is the only inland province in the YRD. Due to the limitations of urban space and economic volume, there is no strong mutual support between PEE and HED in Anhui [42].

Based on the background of high-quality development in China, our research constructed HED evaluation indexes around the New Development Concept. As the era progresses, devising a more comprehensive and in-depth index system to examine HED remains an area ripe for further investigation. Moreover, this paper took ports and cities as subsystems in the coupled coordination framework study. Our following research will focus on including more subsystems, such as industries, for a more detailed exploration of the urban economic system.

6. Conclusion and Recommendations

6.1. Conclusions

This paper explores the CCD between PEE and HED in 20 major port cities in YRD from 2012 to 2021 based on the perspective of green development. The conclusions are as follows:

(1) The gradual implementation of green port construction in the YRD has increased PEE to 0.578, but there is still plenty of scope for progress. The inter-provincial gap keeps narrowing, indicating a favorable regional integration development trend. The PEE in Jiangsu and Shanghai is higher than that in Zhejiang and Anhui. The PEE of seaports is higher than that of river ports due to the support of the market, advanced technology, geographic advantages, and robust infrastructure development.

(2) HED fluctuates and rises, widening the gap between provinces. Shanghai performs better in terms of HED. Economic leaders such as Hangzhou and Ningbo-Zhoushan support Zhejiang's HED. Jiangsu underperforms in HED because of its large industrial scale and complex industrial chain. Furthermore, Anhui, being landlocked, has the lowest HED in the YRD.

(3) In terms of temporal evolution, the CCD between PEE and HED in the YRD rose from bare coordination to mild coordination in 2012–2021, exhibiting an overall rising trend. This is mainly because the significant increase in PEE positively affects the overall quality of the city's economy, and the city also guarantees the improvement of green technologies in the port. Shanghai relies on its strong economy and efficient ports to keep the CCD at the highest level in the RD. Jiangsu's initiatives in waterway renovation and green port construction have achieved the most significant rise in CCD. In terms of spatial evolution, Jiangsu's CCD polarization has been weakening. Anhui's overall CCD is lower, with an unstable coordination trend. Zhejiang's polarization is apparent. Across the entire YRD, there exists a spatial distribution characterized by a decreasing trend from the coast to inland. It is expected that by 2026, the average coordination level of the YRD will reach intermediate coordination. Taizhou-Z, Huzhou, Anqing, Tongling, and Changzhou are the critical points for the synergistic, high-quality development of ports and cities in the YRD.

6.2. Recommendations

Based on the findings of our study, we propose the following recommendations:

(1) Promote the construction of green ports based on the existing port integration system and reduce carbon emissions through talent and technology introduction, equipment automation, and pollution emission management. Furthermore, implementing functional positioning and hierarchical management across various seaports and river ports will mitigate excessive competition and maximize the sea–river intermodal transport capabilities of the YRD.

(2) Jiangsu should carry the green concept through industrial development and reduce carbon emissions through clean energy use and optimization of industrial structure. Zhejiang should leverage the influential role of Hangzhou and Ningbo-Zhoushan to stimulate regional development. Anhui should endeavor to absorb industrial and population spillovers from the YRD and focus on ecological benefits alongside economic development.

(3) Recognize and harness the synergistic relationship between ports and cities to foster the development of the green circular economy. Exploit the ports' resource deployment and transportation functions to promote the transfer of coastal industries to Anhui, establishing a balanced division of labor within the YRD city cluster. Strengthen city support for port infrastructure and the application of green technology to enhance port capacity. Accelerate YRD integration efforts to facilitate synergistic regional development.

Author Contributions: Conceptualization, M.W., Y.L., H.L., X.J., S.L. and K.D.; methodology, M.W., Y.L., H.L. and X.J.; software, Y.L. and H.L.; validation, S.L. and K.D.; formal analysis, M.W., Y.L., H.L., X.J., S.L. and K.D.; writing—original draft preparation, M.W., Y.L., H.L., X.J., S.L. and K.D.; writing—review and editing, M.W., Y.L., H.L. and X.J.; visualization, M.W., Y.L. and H.L., supervision, H.L. and X.J.; project administration, M.W. All authors have read and agreed to the published version of the manuscript.

Funding: This research was funded by the National Social Science Foundation of China (22BJL052), the Major Program of the Philosophy and Social Science Foundation of the Jiangsu Educational Committee (2021SJZDA120), and the Fundamental Research Funds for the Central Universities (B220203043).

Data Availability Statement: The original contributions presented in the study are included in the article, further inquiries can be directed to the corresponding author.

Acknowledgments: The authors sincerely thank all the students and staff who provided their input to this study. Our acknowledgments are also extended to the anonymous reviewers for their constructive comments.

Conflicts of Interest: The authors declare no conflicts of interest.

References

1. Central People's Government of China. Report of the 20th National Congress of the Communist Party of China. Available online: https://www.gov.cn/xinwen/2022-10/25/content_5721685.htm (accessed on 17 September 2023).
2. Botana, C.; Fernández, E.; Feijóo, G. Towards a Green Port strategy: The decarbonisation of the Port of Vigo (NW Spain). *Sci. Total Environ.* **2023**, *856*, 159198. [CrossRef]
3. Qu, Y.; Kong, Y.; Li, Z.; Zhu, E. Pursue the coordinated development of port-city economic construction and ecological environment: A case of the eight major ports in China. *Ocean Coast. Manag.* **2023**, *242*, 106694. [CrossRef]
4. Jiang, Z.; Pi, C.; Zhu, H.; Wang, C.; Ye, S. Temporal and spatial evolution and influencing factors of the port system in Yangtze River Delta Region from the perspective of dual circulation: Comparing port domestic trade throughput with port foreign trade throughput. *Transp. Policy* **2022**, *118*, 79–90. [CrossRef]
5. Moretti, B. Technical Land-Sea Spaces. Impacts of the Port Clusterization Phenomenon on Coasts, Cities and Architectures. *J. Contemp. Urban Aff.* **2023**, *7*, 208–223. [CrossRef]
6. Xia, C.; Zhai, G. The spatiotemporal evolution pattern of urban resilience in the Yangtze River Delta urban agglomeration based on TOPSIS-PSO-ELM. *Sustain. Cities Soc.* **2022**, *87*, 104223. [CrossRef]
7. Deng, P.; Song, L.; Xiao, R.; Huang, C. Evaluation of logistics and port connectivity in the Yangtze River Economic Belt of China. *Transp. Policy* **2022**, *126*, 249–267. [CrossRef]
8. Chen, J.; Abbas, J.; Najam, H.; Liu, J.; Abbas, J. Green technological innovation, green finance, and financial development and their role in green total factor productivity: Empirical insights from China. *J. Clean. Prod.* **2023**, *382*, 135131. [CrossRef]
9. Tseng, Y.; Yuan, C.; Wong, K.; Lin, C. Chemical fingerprints and source resolution of atmospheric fine particles in an industrial harbor based on one-year intermittent field sampling data. *Sci. Total Environ.* **2023**, *868*, 161335. [CrossRef]
10. Wan, Z.; Ji, S.; Liu, Y.; Zhang, Q.; Chen, J.; Wang, Q. Shipping emission inventories in China's Bohai Bay, Yangtze River Delta, and Pearl River Delta in 2018. *Mar. Pollut. Bull.* **2020**, *151*, 110882. [CrossRef]
11. Yang, L.; Zhang, Q.; Lv, Z.; Zhang, Y.; Yang, Z.; Fu, J.; Lv, J.; Wu, L.; Mao, H. Efficiency of DECA on ship emission and urban air quality: A case study of China port. *J. Clean. Prod.* **2022**, *362*, 132556. [CrossRef]
12. Liu, C.; Sun, W.; Li, P.X.; Zhang, L.; Li, M. Differential characteristics of carbon emission efficiency and coordinated emission reduction pathways under different stages of economic development: Evidence from the Yangtze River Delta, China. *J. Environ. Manag.* **2023**, *330*, 117018. [CrossRef]
13. Guo, J.; Wang, Z.; Yu, X. Accessibility measurement of China's coastal ports from a land-sea coordination perspective—An empirical study. *J. Transp. Geogr.* **2022**, *105*, 103479. [CrossRef]
14. Kammoun, R.; Abdennadher, C. Seaport efficiency and competitiveness in European seaports. *Transp. Policy* **2022**, *121*, 113–124. [CrossRef]
15. Iris, Ç.; Lam, J.S.L. A review of energy efficiency in ports: Operational strategies, technologies and energy management systems. *Renew. Sustain. Energy Rev.* **2019**, *112*, 170–182. [CrossRef]
16. Sogut, M.Z.; Erdoğan, O. An investigation on a holistic framework of green port transition based on energy and environmental sustainability. *Ocean Eng.* **2022**, *266*, 112671. [CrossRef]
17. Wang, L.; Li, Y. Estimation methods and reduction strategies of port carbon emissions—What literatures say? *Mar. Pollut. Bull.* **2023**, *195*, 115451. [CrossRef]
18. Wang, B.; Liu, Q.; Wang, L.; Chen, Y.; Wang, J. A review of the port carbon emission sources and related emission reduction technical measures. *Environ. Pollut.* **2023**, *320*, 121000. [CrossRef] [PubMed]
19. Hua, C.; Chen, J.; Wan, Z.; Xu, L.; Bai, Y.; Zheng, T.; Fei, Y. Evaluation and governance of green development practice of port: A sea port case of China. *J. Clean. Prod.* **2020**, *249*, 119434. [CrossRef]
20. OECD. Eco-Efficiency. Available online: https://read.oecd-ilibrary.org/environment/eco-efficiency_9789264040304-en (accessed on 23 August 2023).
21. Na, J.; Choi, A.Y.; Ji, J.; Zhang, D. Environmental efficiency analysis of Chinese container ports with CO_2 emissions: An inseparable input-output SBM model. *J. Transp. Geogr.* **2017**, *65*, 13–24. [CrossRef]
22. Lampe, H.; Hilgers, D. Trajectories of efficiency measurement: A bibliometric analysis of DEA and SFA. *Eur. J. Oper. Res.* **2015**, *240*, 1–21. [CrossRef]
23. Roll, Y.; Hayuth, Y. Port performance comparison applying data envelopment analysis (DEA). *Marit. Policy Manag.* **1993**, *20*, 153–161. [CrossRef]
24. Wang, L.; Zhou, Z.; Yang, Y.; Wu, J. Green efficiency evaluation and improvement of Chinese ports: A cross-efficiency model. *Transp. Res. Part D Transp. Environ.* **2020**, *88*, 102590. [CrossRef]
25. Chin, A.; Low, J. Port performance in Asia: Does production efficiency imply environmental efficiency? *Transp. Res. Part D Transp. Environ.* **2010**, *15*, 483–488. [CrossRef]
26. Hsu, W.; Huynh, N. Container terminals' efficiency with the unexpected output: A revised SBM approach. *Environ. Sci. Pollut. Res.* **2023**, *30*, 37845–37858. [CrossRef]
27. Li, J.; Ren, J.; Ma, X.; Xiao, G. Environmental efficiency of ports under the dual carbon goals: Taking China's Bohai-rim ports as an example. *Front. Mar. Sci.* **2023**, *10*, 1129659. [CrossRef]
28. Nodin, M.; Mustafa, Z.; Hussain, S. Eco-efficiency assessment of Malaysian rice self-sufficiency approach. *Socio-Econ. Plan. Sci.* **2023**, *85*, 101436. [CrossRef]

29. Mamghaderi, M.; Mamkhezri, J.; Khezri, M. Assessing the environmental efficiency of OECD countries through the lens of ecological footprint indices. *J. Environ. Manag.* **2023**, *338*, 117796. [CrossRef]
30. Zhu, C.; Fang, C.; Zhang, L. Analysis of the coupling coordinated development of the Population–Water–Ecology–Economy system in urban agglomerations and obstacle factors discrimination: A case study of the Tianshan North Slope Urban Agglomeration, China. *Sustain. Cities Soc.* **2023**, *90*, 104359. [CrossRef]
31. Bell, S.; Morse, S. Sustainability Indicators Past and Present: What Next? *Sustainability* **2018**, *10*, 1688. [CrossRef]
32. Han, Z.; Jiao, S.; Zhang, X.; Xie, F.; Ran, J.; Jin, R.; Xu, S. Seeking sustainable development policies at the municipal level based on the triad of city, economy and environment: Evidence from Hunan province, China. *J. Environ. Manag.* **2021**, *290*, 112554. [CrossRef] [PubMed]
33. Luo, S.; Yu, M.; Dong, Y.; Hao, Y.; Li, C.; Wu, H. Toward urban high-quality development: Evidence from more intelligent Chinese cities. *Technol. Forecast. Soc. Chang.* **2024**, *200*, 123108. [CrossRef]
34. Guo, B.; Wang, Y.; Zhang, H.; Liang, C.; Feng, Y.; Hu, F. Impact of the digital economy on high-quality urban economic development: Evidence from Chinese cities. *Econ. Model.* **2023**, *120*, 106194. [CrossRef]
35. Yang, Z. Comparison and empirical analysis of the urban economic development level in the Yangtze River urban agglomeration based on an analogical ecosystem perspective. *Ecol. Inform.* **2021**, *64*, 101321. [CrossRef]
36. Pan, W.; Wang, J.; Lu, Z.; Liu, Y.; Li, Y. High-quality development in China: Measurement system, spatial pattern, and improvement paths. *Habitat. Int.* **2021**, *118*, 102458. [CrossRef]
37. Zheng, W.; Zhang, L.; Hu, J. Green credit, carbon emission and high quality development of green economy in China. *Energy Rep.* **2022**, *8*, 12215–12226. [CrossRef]
38. Wan, J.; Wang, Z.; Ma, C.; Su, Y.; Zhou, T.; Wang, T.; Zhao, Y.; Sun, H.; Li, Z.; Wang, Y.; et al. Spatial-temporal differentiation pattern and influencing factors of high-quality development in counties: A case of Sichuan, China. *Ecol. Indic.* **2023**, *148*, 110132. [CrossRef]
39. Zha, Q.; Liu, Z.; Song, Z.; Wang, J. A study on dynamic evolution, regional differences and convergence of high-quality economic development in urban agglomerations: A case study of three major urban agglomerations in the Yangtze river economic belt. *Front. Environ. Sci.* **2022**, *10*, 1012304. [CrossRef]
40. Guo, J.; Sun, Z. How does manufacturing agglomeration affect high-quality economic development in China? *Econ. Anal. Policy* **2023**, *78*, 673–691. [CrossRef]
41. Cong, L.; Zhang, D.; Wang, M.; Xu, H.; Li, L. The role of ports in the economic development of port cities: Panel evidence from China. *Transp. Policy* **2020**, *90*, 13–21. [CrossRef]
42. Wu, Z.; Woo, S.; Lai, P.L.; Chen, X. The economic impact of inland ports on regional development: Evidence from the Yangtze River region. *Transp. Policy* **2022**, *127*, 80–91. [CrossRef]
43. Li, Z.; Luan, W.; Zhang, Z.; Su, M. Research on the Interactive Relationship of Spatial Expansion between Estuarine and Coastal Port Cities. *Land* **2023**, *12*, 371. [CrossRef]
44. Zhu, S.; Zheng, S.; Ge, Y.-E.; Fu, X.; Sampaio, B.; Jiang, C. Vertical integration and its implications to port expansion. *Marit. Policy Manag.* **2019**, *46*, 920–938. [CrossRef]
45. Qu, C.; Wang, G.W.Y.; Zeng, Q. Modelling port subsidy policies considering pricing decisions of feeder carriers. *Transp. Res. Part E Logist. Transp. Rev.* **2017**, *99*, 115–133. [CrossRef]
46. Hidalgo-Gallego, S.; NúñezSánchez, R. The effect of port activity on urban employment: An analysis for the Spanish functional urban areas. *J. Transp. Geogr.* **2023**, *108*, 103570. [CrossRef]
47. Wang, C.; Chen, J.; Li, Z.; Abouel Nasr, E.; El-Tamimi, A. An indicator system for evaluating the development of land-sea coordination systems: A case study of Lianyungang port. *Ecol. Indic.* **2019**, *98*, 112–120. [CrossRef]
48. Guo, J.; Qin, Y.; Du, X.; Han, Z. Dynamic measurements and mechanisms of coastal port–city relationships based on the DCI model: Empirical evidence from China. *Cities* **2020**, *96*, 102440. [CrossRef]
49. Ma, Q.; Jia, P.; She, X.; Haralambides, H.; Kuang, H. Port integration and regional economic development: Lessons from China. *Transp. Policy* **2021**, *110*, 430–439. [CrossRef]
50. Vroomans, J.; Geerlings, H.; Kuipers, B. The energetic relationship between ports and cities; how the role of shared values is under pressure. *Case Stud. Transp. Policy* **2022**, *10*, 2358–2368. [CrossRef]
51. Russo, F.; Musolino, G. Port-city interactions: Models and case studies. *Transp. Res. Procedia* **2023**, *69*, 695–702. [CrossRef]
52. Liu, J.; Zhou, J.; Liu, F.; Yue, X.; Kong, Y.; Wang, X. Interaction Analysis and Sustainable Development Strategy between Port and City: The Case of Liaoning. *Sustainability* **2019**, *11*, 5366. [CrossRef]
53. Guo, J.; Qin, Y. Coupling characteristics of coastal ports and urban network systems based on flow space theory: Empirical evidence from China. *Habitat. Int.* **2022**, *126*, 102624. [CrossRef]
54. Tone, K. A slacks-based measure of efficiency in data envelopment analysis. *Eur. J. Oper. Res.* **2001**, *130*, 498–509. [CrossRef]
55. Charnes, A.; Clark, C.; Cooper, W.W.; Golany, B.A.B. A developmental study of data envelopment analysis in measuring the efficiency of maintenance units in the U.S. air forces. *Ann. Oper. Res.* **1984**, *2*, 95–112. [CrossRef]
56. Zarbi, S.; Shin, S.; Shin, Y. An Analysis by Window DEA on the Influence of International Sanction to the Efficiency of Iranian Container Ports. *Asian J. Shipp. Log.* **2019**, *35*, 163–171. [CrossRef]
57. Meehl, G.; Senior, C.; Eyring, V.; Flato, G.; Lamarque, J.; Stouffer, R.; Taylor, K.; Schlund, M. Context for interpreting equilibrium climate sensitivity and transient climate response from the CMIP6 Earth system models. *Sci. Adv.* **2020**, *6*, eaba1981. [CrossRef]

58. Chen, J.; Zhang, W.; Song, L.; Wang, Y. The coupling effect between economic development and the urban ecological environment in Shanghai port. *Sci. Total Environ.* **2022**, *841*, 156734. [CrossRef]
59. Huang, Y.; Huang, C. The integration and application of fuzzy and grey modeling methods. *Fuzzy Sets Syst.* **1996**, *78*, 107–119. [CrossRef]
60. Zhou, C.; Li, X.; Lin, X.; Cheng, M. Influencing factors of the high-quality economic development in China based on LASSO model. *Energy Rep.* **2022**, *8*, 1055–1065. [CrossRef]
61. The State Council of the PRC. Outline of the Plan for the Integrated Development of the Yangtze River Delta Region. Available online: https://www.gov.cn/zhengce/2019-12/01/content_5457442.htm (accessed on 17 September 2023).
62. Erdas, C.; Fokaides, P.A.; Charalambous, C. Ecological footprint analysis based awareness creation for energy efficiency and climate change mitigation measures enhancing the environmental management system of Limassol port. *J. Clean. Prod.* **2015**, *108*, 716–724. [CrossRef]
63. Wang, Z.; Wu, X.; Guo, J.; Wei, G.; Dooling, T. Efficiency evaluation and PM emission reallocation of China ports based on improved DEA models. *Transp. Res. Part D Transp. Environ.* **2020**, *82*, 102317. [CrossRef]
64. Li, D.; Xin, X.; Zhou, S. Integrated governance of the Yangtze River Delta port cluster using niche theory: A case study of Shanghai Port and Ningbo-Zhoushan Port. *Ocean Coast. Manag.* **2023**, *234*, 106474. [CrossRef]
65. Feng, H.; Grifoll, M.; Zheng, P. From a feeder port to a hub port: The evolution pathways, dynamics and perspectives of Ningbo-Zhoushan port (China). *Transp. Policy* **2019**, *76*, 21–35. [CrossRef]
66. Shi, J.; Jiao, Y.; Chen, J.; Ye, J.; Gong, J. A study on the evolution of competition pattern of inland container ports along the Yangtze River in China. *J. Transp. Geogr.* **2023**, *109*, 103591. [CrossRef]
67. Jiang, L.; Wang, X.; Yang, K.; Gao, Y. Bilevel optimization for the reorganization of inland river ports: A niche perspective. *Socio-Econ. Plan. Sci.* **2023**, *86*, 101466. [CrossRef]
68. Wang, M.; Ji, M.; Wu, X.; Deng, K.; Jing, X. Analysis on Evaluation and Spatial-Temporal Evolution of Port Cluster Eco-Efficiency: Case Study from the Yangtze River Delta in China. *Sustainability* **2023**, *15*, 8268. [CrossRef]
69. Li, Z.; Galeano Galván, M.; Ravesteijn, W.; Qi, Z. Towards low carbon based economic development: Shanghai as a C40 city. *Sci. Total Environ.* **2017**, *576*, 538–548. [CrossRef]
70. Yang, H.; Li, X.; Elliott, M. Integrated quantitative evaluation framework of sustainable development—The complex case of the Yangtze River Delta. *Ocean Coast. Manag.* **2023**, *232*, 106426. [CrossRef]
71. Du, Y.; Cardoso, R.V.; Rocco, R. The challenges of high-quality development in Chinese secondary cities: A typological exploration. *Sustain. Cities Soc.* **2024**, *103*, 105266. [CrossRef]
72. Wang, D.; Shi, Y.; Wan, K. Integrated evaluation of the carrying capacities of mineral resource-based cities considering synergy between subsystems. *Ecol. Indic.* **2020**, *108*, 105701. [CrossRef]
73. Chen, X.; Di, Q.; Jia, W.; Hou, Z. Spatial correlation network of pollution and carbon emission reductions coupled with high-quality economic development in three Chinese urban agglomerations. *Sustain. Cities Soc.* **2023**, *94*, 104552. [CrossRef]
74. Wang, D.; Li, Y.; Yang, X.; Zhang, Z.; Gao, S.; Zhou, Q.; Zhuo, Y.; Wen, X.; Guo, Z. Evaluating urban ecological civilization and its obstacle factors based on integrated model of PSR-EVW-TOPSIS: A case study of 13 cities in Jiangsu Province, China. *Ecol. Indic.* **2021**, *133*, 108431. [CrossRef]
75. Wei, X.; Zhao, R.; Xu, J. Spatiotemporal Evolution, Coupling Coordination Degree and Obstacle Factors of Urban High-Quality Development: A Case Study of Anhui Province. *Sustainability* **2023**, *15*, 852. [CrossRef]
76. Sun, J.; Zhai, N.; Mu, H.; Miao, J.; Li, W.; Li, M. Assessment of urban resilience and subsystem coupling coordination in the Beijing-Tianjin-Hebei urban agglomeration. *Sustain. Cities Soc.* **2024**, *100*, 105058. [CrossRef]
77. Wang, L.; Yuan, M.; Li, H.; Chen, X. Exploring the coupling coordination of urban ecological resilience and new-type urbanization: The case of China's Chengdu–Chongqing Economic Circle. *Environ. Technol. Innov.* **2023**, *32*, 103372. [CrossRef]
78. Song, L.; van Geenhuizen, M. Port infrastructure investment and regional economic growth in China: Panel evidence in port regions and provinces. *Transp. Policy* **2014**, *36*, 173–183. [CrossRef]
79. Shan, J.; Yu, M.; Lee, C.-Y. An empirical investigation of the seaport's economic impact: Evidence from major ports in China. *Transp. Res. Part E Logist. Transp. Rev.* **2014**, *69*, 41–53. [CrossRef]
80. Kong, Y.; Liu, J. Sustainable port cities with coupling coordination and environmental efficiency. *Ocean Coast. Manag.* **2021**, *205*, 105534. [CrossRef]

Disclaimer/Publisher's Note: The statements, opinions and data contained in all publications are solely those of the individual author(s) and contributor(s) and not of MDPI and/or the editor(s). MDPI and/or the editor(s) disclaim responsibility for any injury to people or property resulting from any ideas, methods, instructions or products referred to in the content.

 *land*

Article

Assessment of the Implementation Effects of Main Functional Area Planning in the Yangtze River Economic Belt

Ming Wei [1,2], Wen Chen [1,3,*] and Yi Wang [4]

1. Key Laboratory of Lake and Watershed Science for Water Security, Nanjing Institute of Geography and Limnology, Chinese Academy of Sciences, Nanjing 210008, China; weiming17@mails.ucas.ac.cn
2. University of Chinese Academy of Sciences, Beijing 100049, China
3. College of Resources and Environment, University of Chinese Academy of Sciences, Beijing 100049, China
4. School of Water Resources and Hydropower Engineering, North China Electric Power University, Beijing 102206, China; hywy02@foxmail.com
* Correspondence: wchen@niglas.ac.cn

Abstract: The Yangtze River Economic Belt, relying on the golden waterway of the Yangtze River, serves not only as a vital industrial and urban stronghold in China but also bears the significant responsibility of the Yangtze River's major conservation efforts. The implementation of the main functional zones within the economic belt can provide regional synergies for development and protection through the optimization and organization of spatial structures, which is conducive to promoting the green and high-quality development of the Yangtze River Economic Belt in accordance with local conditions. In pursuit of these objectives, this paper utilizes multi-source data and selects corresponding indicators based on the main form of functional zoning to analyze the land protection and development patterns of the Yangtze River Economic Belt and to assess the effectiveness of the main functional zone planning implementation. The findings reveal that the enactment of main functional area planning has incrementally enhanced the level of land development and conservation in terms of certain aspects across the Yangtze River Economic Belt. This is evidenced by the burgeoning expansion of construction land in areas earmarked for optimization and pivotal development, bolstered by robust population and economic concentration capabilities, alongside a surge in per capita output. Moreover, ecological lands within critical ecological function zones exhibited signs of rejuvenation. Nonetheless, the outcomes are not universally aligned with the anticipated goals: the expanse of arable land in primary agricultural production zones has contracted, accompanied by a downturn in the proportion of grain output; the proliferation of construction land within key ecological function zones continues unabated, and ecological lands have experienced reductions over various intervals. The main functional zones have yet to fully embrace and enact protective strategies, highlighting an urgent need for more formidable institutional frameworks to guarantee their rigorous enforcement.

Keywords: Yangtze River Economic Belt; main functional area; land use/cover change; development and protection; ecosystem services

Citation: Wei, M.; Chen, W.; Wang, Y. Assessment of the Implementation Effects of Main Functional Area Planning in the Yangtze River Economic Belt. *Land* **2024**, *13*, 940. https://doi.org/10.3390/land13070940

Academic Editor: Dagmar Haase

Received: 7 May 2024
Revised: 21 June 2024
Accepted: 24 June 2024
Published: 28 June 2024

Copyright: © 2024 by the authors. Licensee MDPI, Basel, Switzerland. This article is an open access article distributed under the terms and conditions of the Creative Commons Attribution (CC BY) license (https://creativecommons.org/licenses/by/4.0/).

1. Introduction

The Yangtze River Economic Belt, a vital geographical axis that stretches from east to west and extends its influence from north to south, stands as one of the pivotal corridors for territorial spatial development in China [1,2]. In 2022, the total population of the Yangtze River Economic Belt reached 608 million, with a GDP of CNY 55.98 trillion, accounting for 43.1% and 46.5% of the nation's total, respectively, bearing nearly half the weight of China's economic and social development. At present, the region is characterized by robust population density and economic concentration, reflecting a consistent elevation in the standards of economic and social advancement [3,4]. Nevertheless, it grapples with substantial constraints due to resource and environmental limitations, and the expansion of construction land has, to a certain degree, encroached upon the sanctity of arable and ecological territories [5,6]. There are considerable disparities between the upper, middle, and lower reaches of the river, which increase the difficulty and pressure of achieving coordinated regional development [7–9]. The quest to refine spatial structuring and enhance regional governance, thereby catalyzing the green and high-caliber growth of the Yangtze River Economic Belt, remains a focal point of societal attention across various sectors [10–12]. In alignment with the guiding ethos of "emphasizing conservation over development", the Yangtze River Economic Belt is suggested to more intimately align with the national land development strategies, carving out innovative and tailored pathways for its protection and advancement [13].

The quest for regional green and high-quality development hinges on the optimization of comprehensive benefits. The main form of functional area planning has sketched a novel territorial spatial development and protection scheme for the Yangtze River Economic Belt, facilitating the construction of differentiated development and assessment pathways [14,15]. This holds significant strategic importance for the region's coordinated and sustainable development, making the implementation of main functional area planning an intrinsic requirement for high-quality development [16].

"National Main Functional Area Planning", promulgated in 2010, takes into account the varying resource and environmental capacities of distinct regions. It integrates current developmental intensities and prospective growth, assigning explicit territorial functions at the county echelon [17,18]. This demarcation engenders a zoning control paradigm imbued with distinctive Chinese features, offering a steadfast trajectory for the spatial organization of regions over the long haul [19–21]. Main functional areas are categorized into four types according to development approaches: optimized development regions, key development regions, restricted development regions, and prohibited development regions. The first two categories encompass urban territories, whereas the restricted development regions are subdivided into major agricultural production areas and key ecological function areas [22].

According to the policy, targeted spatial development and protection optimization strategies for the Yangtze River Economic Belt have been proposed. For instance, Fan Jie and others (2015) have analyzed the strategic position of the Yangtze River Economic Belt under the context of globalization and regional integration, outlining the spatial pattern and innovation-driven development blueprint of the belt [23]. Similarly, Chen Wen and associates (2015) have painted a picture of the developmental and protective landscape of the belt, informed by an analysis of regional disparities and spatial development suitability assessments, and have provided strategic guidance and policy recommendations based on the ecological–economic traits of various regions [24].

Moreover, the main form of functional area planning offers a fresh perspective for the regional classification assessment of high-quality development in the Yangtze River Economic Belt: urbanized areas should promote economic growth and enhance quality and efficiency, while in restricted development areas, the major agricultural production areas should be charged with safeguarding food security, and the key ecological function areas should be entrusted with the preservation of the integrity of natural and cultural resources [25]. Land, as a fundamental element, is intricately linked to high-caliber development and necessitates alignment with the varied functional zone typologies [26–28]. Under the anticipated state of functional zoning implementation, the increase in construction land in optimized development regions needs to be controlled. The arable land in major agricultural production areas and various types of ecological land in key ecological function areas should see stable growth. By constructing diversified local schemes for development and protection according to the zoning guidance of main functional area planning, and thereby enhancing the overall economic–ecological benefits from the perspective of the regional spatial division of labor, we can effectively propel the construction of ecological civilization in the new era to a higher level [29–31].

Existing research has delved into the domain of functional zoning, evaluating regional development from a multitude of perspectives. Academics have devised indicators across economic, social, and ecological dimensions to gauge governmental efficacy within distinct zonal classifications [32–34]. Some have integrated functional zones into the municipal level, combining multiple subsystems to assess the development of the Yangtze River Economic Belt through coupled coordination and scheduling [35]. Others have conducted performance evaluations based on specific provinces or districts [36,37]. In current studies, most scholars have pointed out the shortcomings in the implementation of functional zoning plans [38–40]. For instance, Wu Dan et al. (2018) have noted an uptick in the average vegetation coverage across the Yangtze River Economic Belt, signaling a positive trend [41]. However, they observed that the land use conversion rate in key ecological function areas surpasses that in both optimized and key development regions, a trend at odds with their designated developmental constraints.

Therefore, to address whether the development and protection of the Yangtze River Economic Belt are being effectively implemented, one could consider whether or not the changes in resource allocation factors in various regions are consistent with the requirements of functional zoning. This includes a specific analysis of the following issues: Are the regions adjusting the structure of arable land, ecological land, and construction land in accordance with functional zoning plans? Have they achieved the effect of increasing the proportion of grain production in major agricultural production areas and enhancing the ecosystem service functions in key ecological function areas? Have they achieved the effect of promoting the continuous concentration of population and economic total volume towards urbanized areas within a certain limit, particularly those earmarked for significant development?

To achieve this objective, the present study conducts a county-level assessment of the efficacy of main functional zone implementation, leveraging statistical data to analyze the conservation and development status of the Yangtze River Economic Belt. It also takes into account the expected requirements of the functional zones. Specifically, the main functional zones are analyzed in four categories: zones earmarked for optimized development, areas designated for key development, regions identified as major agricultural producers, and sectors critical for ecological functions. Considering that the "National Main Functional Area Planning" and the "Outline of the Yangtze River Economic Belt Development Plan" were released at the end of 2010 and in September 2016, respectively, this study uses the years 2011, 2016, and 2021 as time points for analysis, with five-year intervals. Based on the core concept of the Yangtze River's major protection, the study analyzes the on-the-ground implementation status and effects of the main form of functional area planning. This includes assessing the impact on food security and ecological safety by examining changes in arable land and ecological land area, as well as the implementation of protection requirements in the main functional zones. It also evaluates the effects of grain production and ecological function changes. The implementation status of development requirements in the main functional zones is observed through the area of construction land, while urbanization and industrialization effects in the Yangtze River Economic Belt are assessed from aspects such as permanent population and economic output (Figure 1). We conduct a relatively comprehensive indicator analysis of the Yangtze River Economic Belt in the Section 3, while in the Section 4, we focus on the main functions of different main functional areas and analyze corresponding indicators to see if the functions were implemented effectively. The overarching goal is to deliver an encompassing appraisal of the symbiotic relationship between development and conservation efforts in the Yangtze River Economic Belt, thus furnishing a scientifically grounded reference to inform and bolster high-caliber, sustainable growth. This study emphasizes the importance of understanding the land use land cover change model through the policies of the main functional areas, which is a possible direction for future sustainable development research.

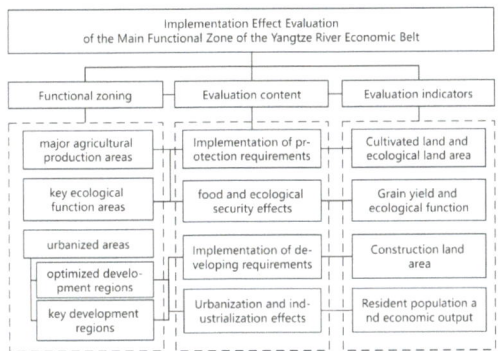

Figure 1. Analytical framework.

2. Research Methods and Data Sources

2.1. Main Functional Zones in the Yangtze River Economic Belt

The existing provincial main functional area plans were independently compiled, resulting in inconsistencies in naming and unit division. To address this, this paper refers to "National Main Functional Area Planning" and uniformly organizes the main functional zoning of the Yangtze River Economic Belt into three categories: urbanized areas, major agricultural production areas, and key ecological function areas. Urbanized areas further include optimized development regions and key development regions. Considering the significant changes in zoning and the differences in statistical units among provinces, for statistical convenience, this study will merge prefecture-level city districts belonging to the same main functional zones. In total, 843 regions were obtained, including 30 optimized development zones, 242 key development zones, 306 main agricultural product production zones, and 265 key ecological function zones (Figure 2). Among these, the urban districts of most prefecture-level cities are designated as key development regions, while the optimized development regions only involve the core development zones of Shanghai, Jiangsu, and Zhejiang, thereby exhibiting a pronounced spatial gradient from east to west.

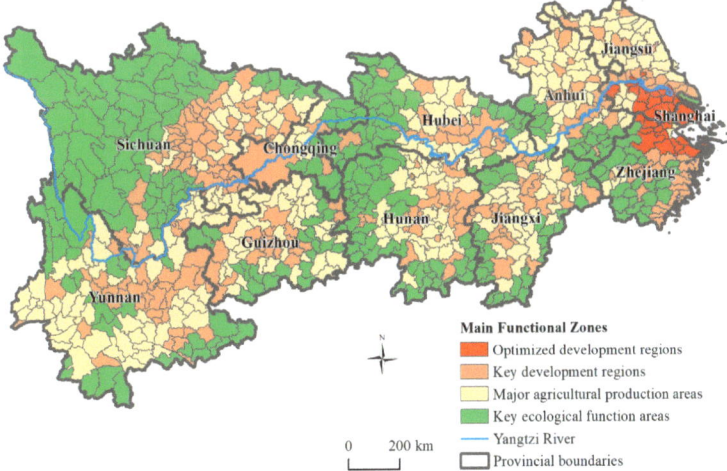

Figure 2. Main functional zones in the Yangtze River Economic Belt.

2.2. Calculation of Relevant Indicators

2.2.1. Single Land Use Dynamics Index and Relative Land Use Change Rate

In this research, the single land use dynamics index (SLDI) is utilized to characterize the evolution within a particular category of land use, while the relative land use change rate (RLCR) is applied to gauge the disparities between localized alterations and the aggregate transformation [42]. The respective equations are as follows:

$$K = \frac{U_b - U_a}{U_a} \cdot \frac{1}{T} \times 100\% \tag{1}$$

$$R = \frac{L_b}{L_a} / \frac{C_b}{C_a} \tag{2}$$

In the formulas, K and R represent the single land use dynamics index and the relative land use change rate, respectively. U_a and U_b denote the initial and final area of a specific land use category, while T stands for time. L_a and L_b refer to the initial and final area of a specific land use type in a local region, and C_a and C_b correspond to the initial and final area of the same land use type in the entire region. If R is greater than 1, it indicates that the change in the local area is greater than that in the entire region.

2.2.2. Ecosystem Service Value

The valuation of ecosystem service value (ESV) can be effectively determined through land use analysis [43]. In this study, drawing upon the methodologies of Xie Gaodi and others (2003), and further refining the parameters set forth by Costanza (1997) in his evaluation of ecosystem service values, we established a valuation table for the per unit area service value of China's terrestrial ecosystems [44,45]. To calculate the ecosystem service value equivalence factor of the Yangtze River Economic Belt, we use one-seventh of the annual per hectare market price of grain. The average grain yield in China from 2011 to 2021 was around 5565 kg/hm², and with the 2021 minimum procurement price for wheat being CNY 2.26 /kg, the ecosystem service value equivalence factor for the Yangtze River Economic Belt is determined to be CNY 1796.7. Due to the limited and fragmented wetland patches, they are incorporated into the water area for treatment. Utilizing this factor, we computed the unit area ecosystem service values for various land use categories (as shown in Table 1), which in turn allows for an analysis of the ecosystem service value fluctuations within the Yangtze River Economic Belt. The formula for computation is as follows:

$$ESV = \sum A_k \times VC_k \tag{3}$$

In the formula, ESV denotes the ecosystem service value, A_k represents the area of the k-th land use category, and VC_k is the value coefficient indicating the unit area service value of the k-th category.

Table 1. Ecological service value per unit area of different land use types (CNY/hm²).

Ecosystem Service Functions	Forest	Grassland	Cropland	Water	Barren	Construction Land
Gas regulation	6288.45	1437.36	898.35	0	0	0
Climate regulation	4851.09	1617.03	1599.06	826.48	0	0
Water conservation	5749.44	1437.36	1078.02	36,616.75	53.90	0
Soil formation and protection	7007.13	3503.57	2623.18	17.97	35.93	0
Waste disposal	2353.68	2353.68	2946.59	32,664.01	17.97	0
Biodiversity conservation	5857.24	1958.40	1275.66	4473.78	610.88	0
Food production	179.67	539.01	1796.70	179.67	17.97	0
Raw material	4671.42	89.84	179.67	17.98	0	0
Entertainment	2299.78	71.87	17.97	7797.68	17.97	0
Total	39,257.90	13,008.11	12,415.20	82,594.30	754.61	0

2.3. Data Sources

The data of the main functional area come from relevant plans released by the Chinese government. The land use data utilized in this study were sourced from the publication "30 m annual land cover and its dynamics in China from 1990 to 2019" in the *Earth System Science Data* journal (DOI: 10.5281/zenodo.8176941). Due to the absence of snowland and the fragmentation and scarcity of shrubs and wetland patches in the research area, this dataset, which encompasses nine land use categories, is integrated into six categories: forest, grassland, cropland, water, barren, and construction land. The socio-economic data referenced in this study were obtained from the "China City Statistical Yearbook", the "China County Statistical Yearbook", and respective annual reports from various provinces and cities. Data gaps were filled using information retrieved from official government websites.

3. Analysis of Protection and Development Indicators for the Yangtze River Economic Belt

3.1. Implementation and Effectiveness of Protection

3.1.1. Status of Indicator Implementation for Land Protection

Our analysis has assessed the land use dynamics and the relative rates of land use change across various functional zones within the Yangtze River Economic Belt for two periods, 2011–2016 and 2016–2021, as detailed in Table 2. It should be noted that the year 2016 will not be counted twice because when we use 2011–2016, we calculate the growth rate rather than the total amount, which is the difference between the 2016 and 2011 data, and the same goes for 2016–2021. Due to the limited and fragmented wetland patches, they will be incorporated into the water area for treatment. Looking at the aggregate data, the past decade has seen a generally stable pattern in the fluctuations of both cultivated and ecological lands. Despite this overall stability, there has been a discernible downward trend in the extent of cultivated and ecological lands within all main functional zones, signaling that the objectives set for land conservation and management have not been entirely met. More specifically, the interval between 2011 and 2016 witnessed a contraction in the areas designated as grasslands, forests, and cultivated lands. In the subsequent period from 2016 to 2021, there was a continued shrinkage in the expanses of water bodies, grasslands, and cultivated lands, although forested areas experienced a marginal recovery. The annual average decrease in cultivated land for the two respective time frames stood at 0.03% and 0.21%.

Table 2. Dynamic degree and relative change rate of land use in various functional areas of the Yangtze River Economic Belt.

Functional Zoning	Periods	Index	Forest	Grassland	Cropland	Water	Barren	Construction Land
Optimized development regions	2011–2016	SLDI	−0.94	−15.60	−0.81	−0.84	−0.20	3.19
		RLCR	0.96	0.22	0.96	0.96	0.85	0.97
	2016–2021	SLDI	−0.20	−13.82	−0.04	−0.89	1.23	1.55
		RLCR	0.98	0.32	1.01	1.01	1.01	0.99
Key development regions	2011–2016	SLDI	−0.07	−1.86	−0.25	0.11	−0.14	4.51
		RLCR	1.00	0.92	0.99	1.01	0.85	1.03
	2016–2021	SLDI	0.33	−2.97	−0.28	−1.22	1.27	2.23
		RLCR	1.01	0.88	1.00	1.00	1.01	1.02
Major agricultural production areas	2011–2016	SLDI	−0.13	−1.78	−0.03	0.26	−6.64	3.34
		RLCR	1.00	0.93	1.00	1.01	0.57	0.98
	2016–2021	SLDI	0.23	−3.30	−0.15	−1.62	−2.91	1.47
		RLCR	1.00	0.86	1.00	0.98	0.81	0.98
Key ecological function areas	2011–2016	SLDI	−0.12	−0.18	0.66	−0.23	3.37	4.38
		RLCR	1.00	1.01	1.03	0.99	1.00	1.02
	2016–2021	SLDI	0.09	−0.24	−0.23	0.36	0.99	2.63
		RLCR	1.00	1.02	1.00	1.08	1.00	1.03
Yangtze River Economic Belt	2011–2016	SLDI	−0.12	−0.37	−0.03	0.01	3.28	3.84
	2016–2021	RLCR	0.17	−0.54	−0.21	−1.17	0.98	1.87

In terms of land type, the major agricultural production areas saw a greater reduction in the area of cultivated land, with the decrease expanding from 0.134% in the period from 2011 to 2016 to 0.744% in the period from 2016 to 2021 (Table 3), accompanied by a reduction in grassland and water body areas. In key ecological function areas, ecological lands such as grasslands, forest lands, and water bodies experienced declines during different periods. All categories of ecological land in optimized development regions showed a tendency to decrease, while in key development areas, the area of land used for construction purposes significantly increased, with corresponding reductions in grasslands, water bodies, and cultivated lands (Table 4).

Table 3. Changes in cropland in the Yangtze River Economic Belt from 2011 to 2021 (10,000 km^2).

Functional Zoning	2011	2016	2021	Yearly Change from 2011 to 2016 (%)	Yearly Change from 2016 to 2021 (%)
Optimized development regions	2.308	2.214	2.210	−0.81	−0.04
Key development regions	24.857	24.553	24.207	−0.25	−0.28
Major agricultural production areas	31.349	31.307	31.074	−0.03	−0.15
Key ecological function areas	10.470	10.816	10.689	0.66	−0.23
Yangtze River Economic Belt	68.985	68.889	68.180	−0.03	−0.21

Table 4. Changes in ecological land in the Yangtze River Economic Belt from 2011 to 2021 (10,000 km^2).

Functional Zoning	2011	2016	2021	Yearly Change from 2011 to 2016 (%)	Yearly Change from 2016 to 2021 (%)
Optimized development regions	1.386	1.323	1.288	−1.13	−0.66
Key development regions	22.323	22.161	22.245	−0.18	0.10
Major agricultural production areas	34.937	34.652	34.647	−0.20	−0.003
Key ecological function areas	76.313	75.791	75.832	−0.17	0.01
Yangtze River Economic Belt	134.959	133.926	134.012	−0.19	0.02

3.1.2. Impact of Food Security and Ecological Security in the Yangtze River Economic Belt

In agricultural production, the Yangtze River Economic Belt has witnessed a contraction in arable land. However, this has been offset by a surge in grain yield per unit, propelling a consistent increase in total grain production. Over the decade spanning 2011 to 2021, the Belt's grain output swelled by 25.76 million tons, with an annual growth rate exceeding 1% (Table 5). Although optimized development regions recorded a marginal dip in grain production, key development areas experienced robust gains, bolstering the overall grain output in urbanized areas and elevating their contribution to the Belt's aggregate production. The decrease in arable land within the principal agricultural zones was modest, at 0.03% and 0.15% for the intervals 2011–2016 and 2016–2021, respectively. Yet, the share of the Belt's total arable land area climbed from 45.44% in 2010 to 45.45% in 2016, and to 45.58% by 2021. The increment in the grain production of major agricultural production areas also saw a significant leap, from 679,700 tons in the initial five-year period to 4,761,500 tons in the subsequent five years. Contributions to grain production growth were observed across all regional types, with the exception of optimized development areas.

Table 5. Changes in grain yield in the Yangtze River Economic Belt from 2011 to 2021 (10,000 tons).

Functional Zoning	2011	2016	2021	Yearly Change from 2011 to 2016 (%)	Yearly Change from 2016 to 2021 (%)
Optimized development regions	676.31	634.74	601.49	−1.23	−1.05
Key development regions	5997.99	6842.23	7174.17	2.82	0.97
Major agricultural production areas	12,070.25	12,138.22	12,614.37	0.11	0.78
Key ecological function areas	2952.38	3513.01	3883.31	3.80	2.11
Yangtze River Economic Belt	21,696.93	23,128.21	24,273.34	1.32	1.24

In terms of ecological security, the ecosystem service value in key ecological function areas saw a modest recovery between 2016 and 2021. Nevertheless, the value of ecosystem services in optimized development regions, key development areas, and primary agricultural production zones all exhibited declining trends (refer to Table 6), with the optimized development regions facing the steepest reduction. This highlights the critical necessity and significance of establishing functional zones for ecological conservation. Moreover, an analysis of the ecosystem service value change map (see Figure 3) reveals that areas with an uptick in ecosystem service values are predominantly located in the main agricultural production zones and key ecological function areas, with a particular concentration in the central and western regions. In the period from 2011 to 2016, the increase in ecosystem service values was more evenly distributed across regions. However, between 2016 and 2021, a distinct pattern emerged, characterized by expansive growth in the central and western areas and a consistent decline in the eastern parts.

Table 6. Ecosystem service value in various functional areas of the Yangtze River Economic Belt (100 million CNY).

Functional Zoning	2011	2016	2021	Yearly Change from 2011 to 2016 (%)	Yearly Change from 2016 to 2021 (%)
Optimized development regions	1112.76	1064.72	1038.59	−1.08	−0.61
Key development regions	12,207.98	12,136.80	12,126.17	−0.15	−0.02
Major agricultural production areas	18,264.80	18,187.78	18,124.24	−0.11	−0.09
Key ecological function areas	26,753.53	26,630.06	26,699.96	−0.12	0.07
Yangtze River Economic Belt	58,339.07	58,019.36	57,988.96	−0.14	−0.01

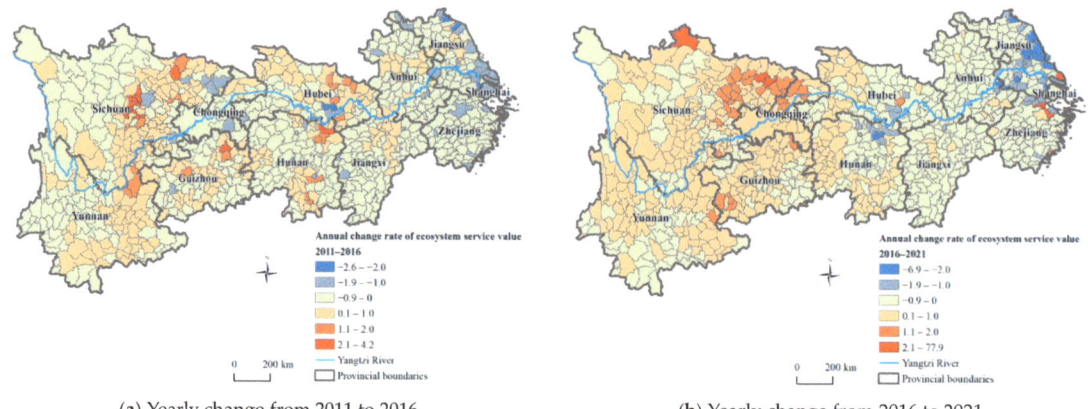

(a) Yearly change from 2011 to 2016   (b) Yearly change from 2016 to 2021

Figure 3. Ecosystem service value change rate of the Yangtze River Economic Belt (%).

We extracted the top five and bottom five cities in terms of ecosystem service value growth rates from 2011 to 2016 and from 2016 to 2021 (Table 7). Between 2011 and 2016, the regions with the fastest growth rates were all key development areas located in Central and Western China, with Sichuan Province being the most important. This was mainly due to the impact of policies such as the pilot program of returning farmland to forests and grasslands in Sichuan, which resulted in a large amount of farmland being converted into ecological land. From 2016 to 2021, although the overall ecosystem service value of the Yangtze River Delta region showed a downward trend, the fastest growing areas still shifted to the coastal areas of Jiangsu and Zhejiang. Some of them are key ecological function areas. For example, the ecosystem service value of Shengsi County increased by 77.93% annually from 2016 to 2021, because of its small original size and significant expansion of forest land over the past five years. Others are the counties and cities in key and optimized development areas where the mudflat resources are continuously increasing.

Table 7. Top 5 and bottom 5 cities with growth rate of ecosystem service value in 2011–2021.

City	Functional Zoning	2011–2016 (%)	Province	City	Functional Zoning	2016–2021 (%)	Province
Dujiangyan	Key development regions	4.20	Sichuan	Shengsi	Key ecological function areas	77.93	Zhejiang
Danling	Key development regions	4.19	Sichuan	Haiyan	Major agricultural production areas	19.11	Zhejiang
Yueyang	Key development regions	2.88	Hunan	Qidong	Key development regions	7.73	Jiangsu
Qingshen	Key development regions	2.74	Sichuan	Yuyao	Optimized development regions	5.01	Zhejiang
Qionglai	Key development regions	2.67	Sichuan	Haining	Optimized development regions	4.99	Zhejiang
Lengshuijiang	Key development regions	−2.06	Hunan	Xinghua	Major agricultural production areas	−3.34	Jiangsu
Xiantao	Key development regions	−2.32	Hubei	Sheyang	Major agricultural production areas	−3.58	Jiangsu
Lianyungang urban districts	Key development regions	−2.37	Jiangsu	Gaochun	Major agricultural production areas	−3.86	Jiangsu
Xiangshui	Major agricultural production areas	−2.39	Jiangsu	Dongtai	Major agricultural production areas	−5.99	Jiangsu
Kunshan	Optimized development regions	−2.59	Jiangsu	Dafeng	Major agricultural production areas	−6.89	Jiangsu

For the last five, they were mainly in key development areas from 2011 to 2016, while from 2016 to 2021, they were main agricultural production zones, all located in Jiangsu. The reason is that in the early stage, due to rapid economic development and urban expansion, ecological land and arable land were squeezed out, and a large amount of barren and ecological land such as mudflat wetlands were converted into agricultural land in the later period.

3.2. Implementation and Effectiveness of Development

3.2.1. Implementation Status of Land Development Indicators

Over the decade from 2011 to 2021, there was a consistent trend of growth in the development of construction land across the various principal functional zones within the Yangtze River Economic Belt (refer to Table 8). The increase in construction land as a percentage of the total land area occurred in the following order: optimized development regions, key development areas, main agricultural production zones, and key ecological function zones. Nevertheless, the expansion rate in each functional zone experienced a deceleration in the period from 2016 to 2021. Optimized development regions sustained their growth in construction land, with the proportion of new development reaching a significant 3.35% of the regional area from 2011 to 2016. Key development areas intensified their consolidation of construction land, accounting for the largest addition in area. The main agricultural production zones exhibited a relatively modest growth rate over the decade, whereas the key ecological function zones experienced an increase in construction land that surpassed the average level. These trends suggest that, despite the implementation of the main functional area strategy, the control over construction land use has been somewhat effective, but the containment of land expansion in optimized development regions and critical ecological function zones has not been adequately enforced.

Table 8. Changes in land use for construction in the Yangtze River Economic Belt from 2011 to 2021 (10,000 km^2).

Functional Zoning	2011	2016	2021	Yearly Change from 2011 to 2016 (%)	Yearly Change from 2016 to 2021 (%)
Optimized development regions	0.984	1.141	1.229	3.19	1.55
Key development regions	2.069	2.536	2.819	4.51	2.22
Major agricultural production areas	1.965	2.293	2.461	3.34	1.47
Key ecological function areas	0.392	0.478	0.541	4.38	2.63
Yangtze River Economic Belt	5.410	6.449	7.050	3.84	1.87

3.2.2. Impact of Urbanization and Industrialization on Population and Industry

From 2011 to 2021, the population density in the Yangtze River Economic Belt increased further, and the population growth rate of each functional area became relatively balanced, with the rise in permanent residents predominantly occurring in urbanized areas (refer to Table 9). Optimized development regions have consistently been focal points for surges in population density. In the period from 2011 to 2016, population growth rates were generally higher in urbanized areas, especially in key development zones of Central and Western China. This trend decelerated between 2016 and 2021, with the most rapid population increases localized in Southern Jiangsu, Zhejiang, and the central sections of the Yangtze River Economic Belt (Figure 4). The trajectory of population density changes mirrored this pattern, with a more equitable increase in population density from 2011 to 2016, whereas from 2016 to 2021, the growth in population density became more pronounced in optimized and key development regions. This shift indicates a deceleration in the large-scale urbanization of the population, with a new trend emerging of population concentration in areas with more advanced development.

Table 9. Population Changes in the Yangtze River Economic Belt from 2011 to 2021 (ten thousand people).

Functional Zoning	2011	2016	2021	Yearly Change from 2011 to 2016 (%)	Yearly Change from 2016 to 2021 (%)
Optimized development regions	7770	8665	9149	2.302	1.118
Key development regions	24,140	25,982	28,278	1.526	1.767
Major agricultural production areas	18,288	18,139	16,551	−0.163	−1.751
Key ecological function areas	8054	8087	7689	0.083	−0.986
Yangtze River Economic Belt	58,252	60,873	61,666	0.900	0.261

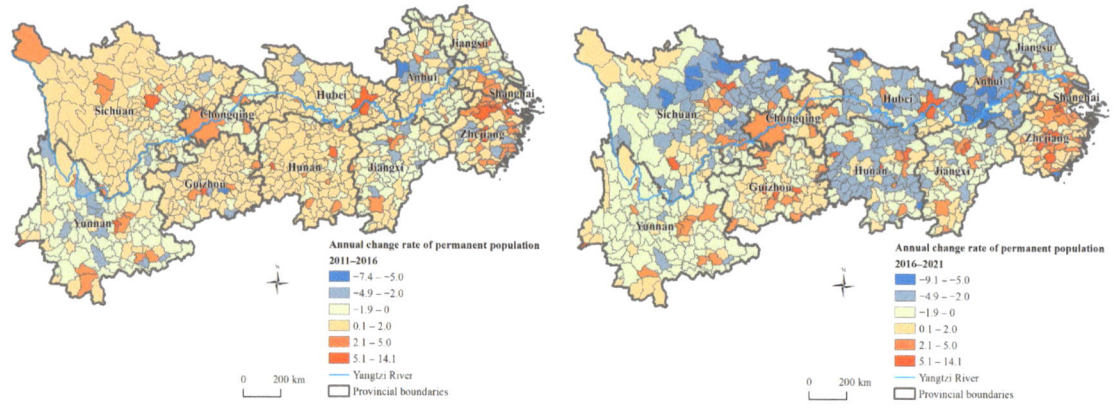

(a) Yearly change from 2011 to 2016   (b) Yearly change from 2016 to 2021

Figure 4. Permanent population change rate of the Yangtze River Economic Belt (%).

In terms of economic output, from 2011 to 2021, all functional areas within the Yangtze River Economic Belt advanced in economic development, with an annual average GDP growth rate exceeding 10% (refer to Table 10). The GDP growth rate slowed down from 2016 to 2021, but there was an increase in the absolute growth amount to varying degrees. Urbanized areas accounted for approximately 75% of the GDP increment in the Yangtze River Economic Belt, making a significant contribution to the economic growth of the region. Major agricultural production areas have the fastest annual growth rate, and the proportion of GDP in key development areas to the Yangtze River Economic Belt has increased the most. From 2016 to 2021, the areas with high growth rates shifted from the central part of the Yangtze River Economic Belt to the western part, with some areas in Anhui and Jiangxi also exhibiting high economic growth trends (Figure 5).

Table 10. Changes in GDP of the Yangtze River Economic Belt from 2011 to 2021 (trillion CNY).

Functional Zoning	2011	2016	2021	Yearly Change from 2011 to 2016 (%)	Yearly Change from 2016 to 2021 (%)
Optimized development regions	6.74	9.95	15.33	9.525	10.818
Key development regions	10.19	16.63	25.19	12.632	10.302
Major agricultural production areas	3.54	6.14	9.60	14.710	11.265
Key ecological function areas	1.43	2.44	3.69	14.162	10.295
Yangtze River Economic Belt	21.90	35.16	53.82	12.111	10.616

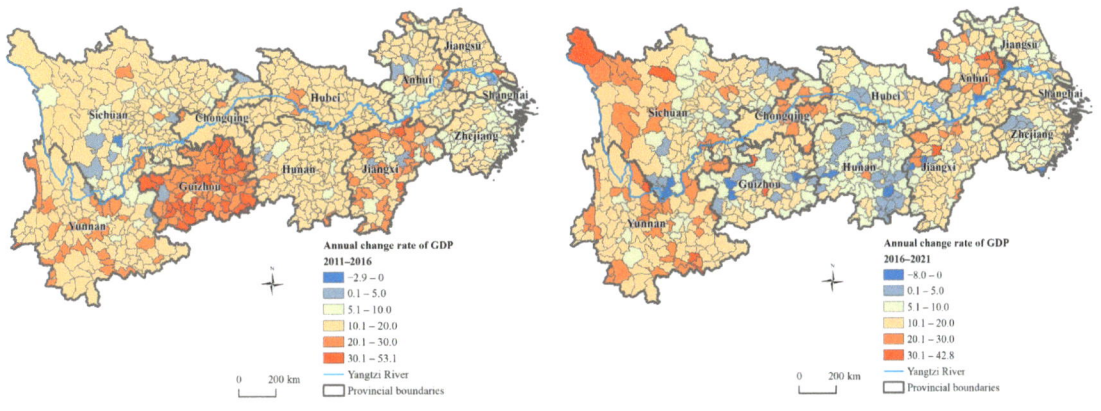

(a) Yearly change from 2011 to 2016 (b) Yearly change from 2016 to 2021

Figure 5. GDP growth rate of the Yangtze River Economic Belt (%).

4. Analysis of the Implementation Effects of the Main Functional Zones in the Yangtze River Economic Belt

4.1. Main Agricultural Production Zones

In the years 2011, 2016, and 2021, the total grain output of the main agricultural production zones in the Yangtze River Economic Belt was 120.7 million tons, 121.4 million tons, and 126.1 million tons, respectively, showing a stable growth trend. However, the proportion of grain output from these main agricultural production zones in both the Yangtze River Economic Belt and the entire country declined from 55.63% and 21.13% to 51.95% and 18.47%, respectively (refer to Table 11). This indicates that the main agricultural production zones have not fully played their role in accordance with their primary function. The function of grain production has, to some extent, shifted to other types of regions, and there is a trend of spreading beyond the Yangtze River Economic Belt. The role of the main agricultural production zones in ensuring grain security still requires further attention.

Table 11. Changes in the proportion of grain yield in the main agricultural production areas.

Year	Grain Yield (100 million tons)	Proportion in the Economic Belt (%)	Proportion in China (%)
2011	1.207	55.63	21.13
2016	1.214	52.49	19.70
2021	1.261	51.95	18.47

Specifically, the distribution of grain output share and growth in the main agricultural production zones is uneven, with the central and western regions outperforming the eastern region. In the 306 statistical units of major agricultural production areas, the majority have seen a decline in their share of the national grain output. Compared to 2011, only 105 units experienced an increase in their proportion of the national grain production in 2016. Between 2016 and 2021, merely 66 units saw an uptick in their share. Concurrently, only 16 units (5.23%) managed to increase their share during both the 2011–2016 and 2016–2021 periods, while 89 regions were unable to maintain their grain supply levels from 2016.

Figure 6 depicts the shifts in the proportion of grain production from the primary agricultural zones relative to the national output (measured in %). From 2011 to 2016, regions including Yunnan, Guizhou, and some areas in Northern Jiangsu and Eastern Sichuan witnessed an uptick in their grain production shares, whereas the remaining provinces saw a decline, with Northern Anhui experiencing the most pronounced drop. This phase demonstrated a pattern of grain production shares increasing from the east to the west. In the subsequent period from 2016 to 2021, Northern Anhui and certain areas in Hubei, Guizhou, and Western Yunnan observed growth in their grain output shares, while Jiangsu, Jiangxi, Hunan, and Western Yunnan experienced reductions in their respective shares.

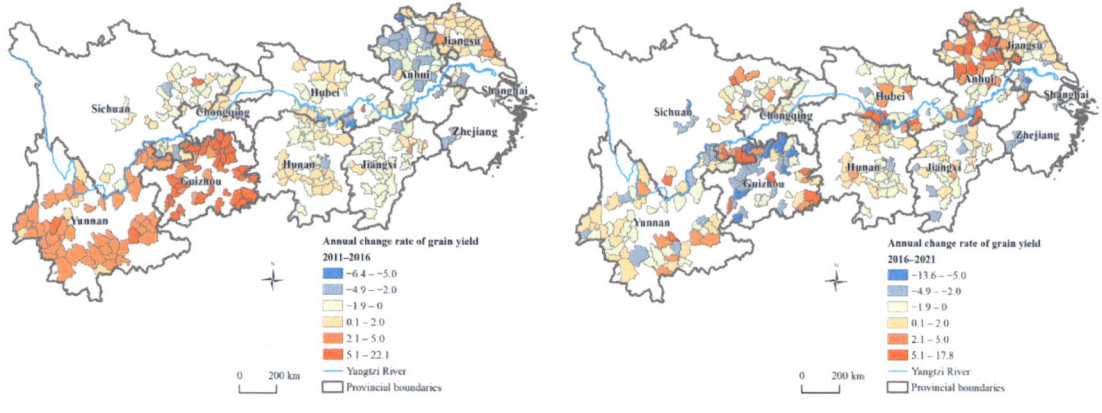

(a) Yearly change from 2011 to 2016   (b) Yearly change from 2016 to 2021

Figure 6. Grain yield change rate in the main agricultural production areas (%).

4.2. Key Ecological Functional Areas

The ecological land area in key ecological functional areas has shown a trend of decreasing to rebounding. In recent years, after the introduction of the Yangtze River Economic Belt planning scheme, the effect of ecological function protection has gradually become prominent. According to Tables 3 and 5, the ecological land area of the key ecological functional areas in the Yangtze River Economic Belt decreased by 5222 square kilometers from 2011 to 2016, with an average annual decrease of 0.17%. However, there was a slight recovery from 2016 to 2021, with an additional area of 408 square kilometers. At the same time, the growth rate of cultivated land and the relative change rate of water body growth in key ecological functional areas between 2011 and 2016 are leading the entire region.

In terms of the value of ecosystem services, among the 264 key ecological function zones, the number of areas showing an upward trend increased from 80 (30.3%) during 2011–2016 to 130 (49.2%) in the period of 2016–2021. Moreover, the decline in ecosystem service values across various regions has moderated, which to some extent highlights the effective implementation of ecological function conservation (Figure 7). From 2011 to 2016, the areas with growth in ecosystem service value were concentrated in a few regions, such as northern Hubei, southern Hunan, and southern Sichuan. Between 2016 and 2021, these areas expanded to include western Hubei, western Hunan, northern Sichuan, and northern Yunnan, showing a trend of contiguous development. The eastern districts, however, still generally exhibited a downward trend, indicating that there is still a need to strengthen ecological protection in key ecological function zones to safeguard ecological resources effectively.

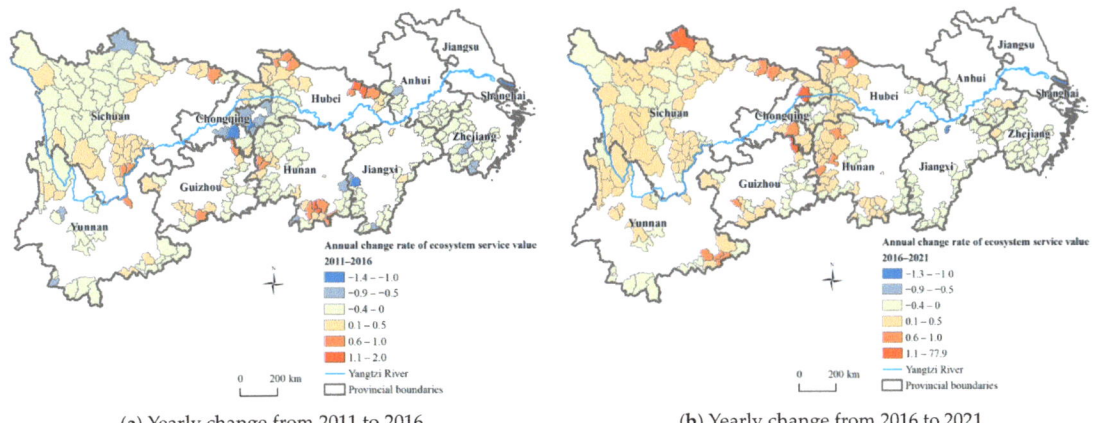

(a) Yearly change from 2011 to 2016 (b) Yearly change from 2016 to 2021

Figure 7. Ecosystem service value change rate in key ecological functional areas (%).

4.3. Urbanized Areas

Urbanized areas take the critical responsibility of economic development and serve as the key engine driving high-quality growth along the Yangtze Economic Belt, with a relatively stable share of GDP. Among the 272 optimized and key development zones, 192 and 136 zones saw an increase in their GDP share of the national total during the periods of 2011–2016 and 2016–2021, respectively, demonstrating the strong economic agglomeration capabilities of these optimized and key development areas. In terms of GDP share, the optimized development zones have experienced steady growth, maintaining a stable yet important share of the national total; the economic contributions of key development zones have seen a significant increase in 2011–2016, with a slight decline observed in the period from 2016 to 2021 (Table 12).

Table 12. Changes in the proportion of GDP in urbanized areas.

Year	Optimized Development Regions		Key Development Regions	
	Proportion in the Economic Belt (%)	Proportion in China (%)	Proportion in the Economic Belt (%)	Proportion in China (%)
2011	30.78	13.81	46.54	20.89
2016	28.30	13.33	47.30	22.28
2021	28.49	13.34	46.81	21.92

Over the decade, land use efficiency in the urbanized regions of the Yangtze Economic Belt has seen an enhancement. The period from 2016 to 2021 marked a more prevalent and concentrated increase in GDP per unit construction land across the provinces within the Belt (Figure 8). The fastest growth in GDP per unit area from 2011 to 2016 was in Lincang urban district, with an average annual increase of 25.3%; the slowest was in Bazhong urban district, with a growth rate of −6.7%. In the subsequent period from 2016 to 2021, Hekou Yao Autonomous County recorded the highest average annual growth rate at 29.9%, in stark contrast to Pukou District, which experienced a decline of 9.35%. Part of the reason for this may be due to a mismatch between urban expansion and the level of economic development, leading to a period of low land use efficiency. For example, in the 2016–2021 period, the expansion of construction land in Bazhong urban district was considerably less pronounced, and with the uptick in urban economic development, the land use efficiency saw a recovery. Furthermore, areas with less favorable locational conditions and a weaker developmental base have also managed to make significant strides in GDP per unit area, propelled by tourism and urban development initiatives.

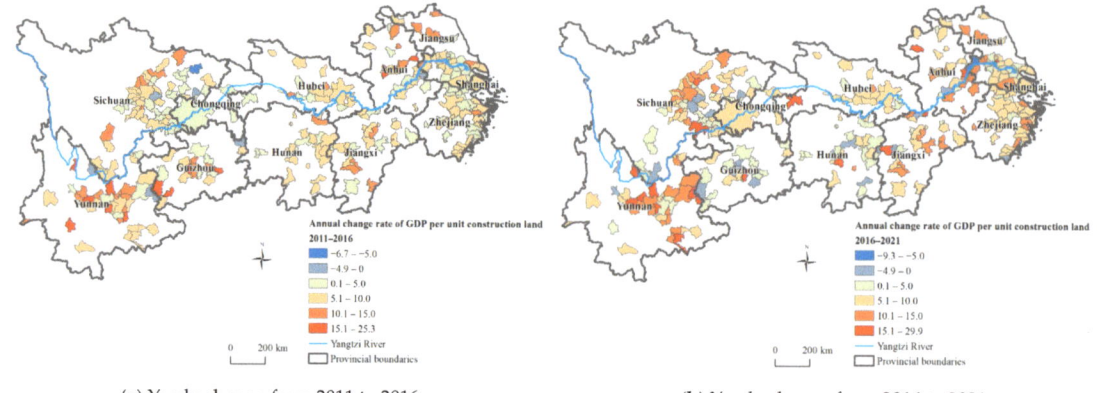

(a) Yearly change from 2011 to 2016   (b) Yearly change from 2016 to 2021

Figure 8. GDP per unit construction land change rate in urbanized areas (%).

5. Conclusions and Discussion

5.1. Conclusions

This paper focuses on the green and high-quality development of the Yangtze River Economic Belt, adopting a multi-index analysis at the county level to examine the patterns of land space protection and development over the past decade. It evaluates the effectiveness of the implementation of various principal functional areas. This study will contribute to breaking through the monolithic standards of regional development assessment, better coordinating population, resources, and the environment, and maximizing the comprehensive benefits of the economy, agriculture, and ecology. This will lead to more regionally accurate assessment results. The conclusions of this study are as follows:

(1) The trajectory of territorial space conservation and development within the Yangtze River Economic Belt, guided by functional zoning strategies, has maintained a consistent and stable course, with marked effectiveness, particularly post-2016, following the introduction of strategic planning. This period has been characterized by a significant uptick in results. Urban regions have exhibited robust capabilities for population and economic concentration, coupled with a progressive enhancement in land utilization efficiency. The economic belt as a whole has seen a steady climb in grain output, alongside an appreciable increase in the ecological worth of pivotal ecological function zones.

(2) The current state of protection and development does not fully align with the anticipated outcomes of the implementation of main functional area planning. The main functional indicators of some areas tend to weaken. In optimized development regions, urban space continues to expand, while ecological land and arable land in key development areas have decreased. The scale of arable land and the proportion of grain production in major agricultural production areas have declined nationally. Construction land in key ecological function areas has rapidly increased, with ecological land experiencing declines in different periods. Furthermore, the annual valuation of ecosystem services across the various functional areas has been on a downward trajectory.

(3) The prospects for the implementation of main functional area planning in the Yangtze River Economic Belt are promising. Moving forward, it is essential to advance protection and development in tandem, strategically coordinate efforts, and leverage the strengths of different areas. By refining the functional division of labor, the region can further capitalize on its comparative advantages and establish a cooperative mechanism based on shared costs and mutual benefits. This approach will help to alleviate the tensions between population dynamics and resource–environmental constraints.

5.2. Discussion

This paper analyzes the status of protection and development indicators in the Yangtze River Economic Belt from the perspective of principal functional areas, assessing the effectiveness of its green development. This approach helps to overcome the limitations imposed by traditional administrative boundaries, taking into account China's population and economic patterns, as well as land use configurations. It explores differentiated high-quality development paths from a multi-functional regional perspective, aiming to further optimize policy layouts based on the current principal functional area schemes and development plans. Depending on the resource endowments and development trajectories of different regions, urbanized areas should pursue high-level development while balancing food security and ecological protection; restricted development areas should prioritize protection and engage in reasonable development within the limits of resources and the environment.

However, the Yangtze River Economic Belt is a large-scale system with strong stability, and fluctuations in its development process can be easily overlooked, making it challenging to delve into the complex mechanisms behind observed phenomena. The division of principal functional areas at the county level may face difficulties due to the extensive size of counties, leading to development constraints on planning in some regions and making it hard to fully implement the requirements of main functional area planning. Furthermore, land use changes stand as one of the primary indicators that directly reflect governmental policy decisions. The government's steering of protection and development activities, as guided by main functional area planning, is predominantly manifested through land policy. However, the actual impact of these policies is not exclusively contingent upon governmental conduct; market dynamics and societal actors may also play a role, and these interactions merit further scholarly exploration. Furthermore, the indicators and analysis methods we use are not complicated, as it is difficult to collect county-level data at a large scale. We look forward to breakthroughs in data and methods in future research. In addition, with the advent of the "dual circulation" paradigm, the strategic significance and spatial organizational framework of the Yangtze River Economic Belt are poised for potential shifts, which calls for more in-depth research.

Author Contributions: Conceptualization, M.W. and W.C.; methodology, W.C.; data curation, M.W.; writing—original draft preparation, M.W. and Y.W.; writing—review and editing, W.C. and Y.W.; supervision, W.C.; funding acquisition, W.C. All authors have read and agreed to the published version of the manuscript.

Funding: This research was funded by the Natural Resources Development Special Fund (Marine Science and Technology Innovation) Project of Jiangsu Province, grant number JSZRHYKJ202205; Independently Deployed Scientific Research Project of Nanjing Institute of Geography and Limnology, Chinese Academy of Sciences, grant number NIGLAS2022GS06.

Data Availability Statement: Data are contained within the article.

Acknowledgments: We would like to thank the Regional Economic Transformation and Management Reform Collaborative Innovation Center of Nanjing University for supporting this paper.

Conflicts of Interest: The authors declare no conflicts of interest.

References

1. Liu, L.; Yang, Y.; Liu, S.; Gong, X.; Zhao, Y.; Jin, R.; Duan, H.; Jiang, P. A comparative study of green growth efficiency in yangtze river economic belt and yellow river basin between 2010 and 2020. *Ecol. Indic.* **2023**, *150*, 110214. [CrossRef]
2. Yang, X.; Feng, Z.; Chen, Y.; Xu, X. A comparative analysis of the levels and drivers of regional coordinated development in the yangtze river economic belt and yellow river basin, China. *Heliyon* **2024**, *10*, e26513. [CrossRef]
3. Chen, C. Changes in the spatial distribution of the employed population in the yangtze river delta region since the 21st century: An analysis and discussion based on census data. *Land* **2023**, *12*, 1249. [CrossRef]
4. Feng, Y.; Sun, M.; Pan, Y.; Zhang, C. Fostering inclusive green growth in china: Identifying the impact of the regional integration strategy of yangtze river economic belt. *J. Environ. Manag.* **2024**, *358*, 120952. [CrossRef] [PubMed]
5. Ma, X.; Wu, H.; Qin, B.; Wang, L.C. Spatiotemporal change of landscape pattern and its eco-environmental response in the yangtze river economic belt. *Sci. Geogr. Sin* **2022**, *42*, 1706–1716.
6. Wang, J.; Zhang, X. Land-based urbanization in China: Mismatched land development in the post-financial crisis era. *Habitat Int.* **2022**, *125*, 102598. [CrossRef]
7. Guan, X.; Wei, H.; Lu, S.; Dai, Q.; Su, H. Assessment on the urbanization strategy in China: Achievements, challenges and reflections. *Habitat Int.* **2018**, *71*, 97–109. [CrossRef]
8. Xu, Y.; Tang, Q.; Fan, J.; Bennett, S.J.; Li, Y. Assessing construction land potential and its spatial pattern in China. *Landsc. Urban Plan.* **2011**, *103*, 207–216. [CrossRef]
9. Wang, N.; Li, S.; Kang, Q.; Wang, Y. Exploring the land ecological security and its spatio-temporal changes in the yangtze river economic belt of China, 2000–2020. *Ecol. Indic.* **2023**, *154*, 110645. [CrossRef]
10. Liu, Y.; Zhou, Y. Territory spatial planning and national governance system in China. *Land Use Policy* **2021**, *102*, 105288. [CrossRef]
11. Chen, M.; Liang, L.; Wang, Z.; Zhang, W.; Yu, J.; Liang, Y. Geographical thoughts on the relationship between 'beautiful China' and land spatial planning. *J. Geogr. Sci.* **2020**, *30*, 705–723. [CrossRef]
12. Wang, L.; Shen, J. Spatial planning and its implementation in provincial China: A case study of the Jiangsu region along the yangtze river plan. *J. Contemp. China* **2016**, *25*, 669–685. [CrossRef]
13. Fu, H.; Liu, J.; Dong, X.; Chen, Z.; He, M. Evaluating the sustainable development goals within spatial planning for decision-making: A major function-oriented zone planning strategy in China. *Land* **2024**, *13*, 390. [CrossRef]
14. Chen, W.; Duan, X.J.; Chen, J.L.; Xu, G. The methods of spatial development function regionalization. *Acta Geogr. Sin.* **2004**, *59*, 53–58.
15. Fan, J. Spatial organization pathway for territorial function-structure: Discussion on implementation of major function zoning strategy in territorial spatial planning. *Geogr. Res.* **2019**, *38*, 2373–2387.
16. Fan, J.; Li, P. The scientific foundation of major function oriented zoning in China. *J. Geogr. Sci.* **2009**, *19*, 515–531. [CrossRef]
17. Fan, J.; Wang, Y.; Wang, C.; Chen, T.; Jin, F.; Zhang, W.; Li, L.; Xu, Y.; Dai, E.; Tao, A. Reshaping the sustainable geographical pattern: A major function zoning model and its applications in China. *Earth's Future* **2019**, *7*, 25–42. [CrossRef]
18. Wang, Y.; Fan, J. Multi-scale analysis of the spatial structure of China's major function zoning. *J. Geogr. Sci.* **2020**, *30*, 197–211. [CrossRef]
19. Bao, S.; Lu, L.; Zhi, J.; Li, J. An optimization strategy for provincial "production–living–ecological" spaces under the guidance of major function-oriented zoning in China. *Sustainability* **2024**, *16*, 2248. [CrossRef]
20. Fan, J.; Sun, W.; Zhou, K.; Chen, D. Major function oriented zone: New method of spatial regulation for reshaping regional development pattern in China. *Chin. Geogr. Sci.* **2012**, *22*, 196–209. [CrossRef]
21. Qi, F.; Qiu, S.; Zhao, C.; Chen, J.; Shao, S.; Liu, B. Improve policy system of main functional zones in national spatial planning. *Bull. Chin. Acad. Sci. Chin. Version* **2024**, *39*, 714–725.
22. Fan, J. Draft of major function oriented zoning of China. *Acta Geogr. Sin.* **2015**, *70*, 186–201.
23. Fan, J.; Wang, Y.F.; Chen, D.; Zhou, C.H. Analysis on the spatial development structure of the yangtze river economic belt. *Prog. Geogr.* **2015**, *34*, 1336–1344.
24. Chen, W.; Sun, W.; Wu, J.W.; Chen, C.; Yan, D.S. Constructing a spatial pattern of development and protection in the yangtze river economic belt and its analysis. *Prog. Geogr.* **2015**, *34*, 1388–1397.
25. Wang, C.S.; Zhu, S.S.; Fan, J.; Liu, H. Key indicators and their data requirements for supervision and evaluation of mfoz planning. *Prog. Geogr.* **2012**, *31*, 1678–1684.

26. Chen, S.; Yao, S. Evaluation of the protection effectiveness of national key ecological functional area based on land use and ecosystem service value. *Environ. Dev. Sustain.* **2024**, *26*, 12467–12487. [CrossRef]
27. Wang, J.; Lin, Y.; Glendinning, A.; Xu, Y. Land-use changes and land policies evolution in China's urbanization processes. *Land Use Policy* **2018**, *75*, 375–387. [CrossRef]
28. Zhu, C.; Dong, B.; Li, S.; Lin, Y.; Shahtahmassebi, A.; You, S.; Zhang, J.; Gan, M.; Yang, L.; Wang, K. Identifying the trade-offs and synergies among land use functions and their influencing factors from a geospatial perspective: A case study in Hangzhou, China. *J. Clean. Prod.* **2021**, *314*, 128026. [CrossRef]
29. Wen, C. *Economic Analysis of Spatial Equilibrium*; The Commercial Press: Beijing, China, 2008.
30. Xu, X.; Na, R.; Shen, Z.; Deng, X. Impact of major function-oriented zone planning on spatial and temporal evolution of "three zone space" in China. *Sustainability* **2023**, *15*, 8312. [CrossRef]
31. Liu, J.Y.; Liu, W.C.; Kuang, W.H.; Ning, J. Remote sensing-based analysis of the spatiotemporal characteristics of built-up area across China based on the plan for major function-oriented zones. *Acta Geogr. Sin* **2016**, *71*, 355–369.
32. Xia, H.; Zhang, W.; He, L.; Ma, M.; Peng, H.; Li, L.; Ke, Q.; Hang, P.; Wang, X. Assessment on China's urbanization after the implementation of main functional areas planning. *J. Environ. Manag.* **2020**, *264*, 110381. [CrossRef] [PubMed]
33. Guo, Y.; Tong, Z.; Chen, H.; Wang, Z.; Yao, Y. Heterogeneity study on mechanisms influencing carbon emission intensity at the county level in the yangtze river delta urban agglomeration: A perspective on main functional areas. *Ecol. Indic.* **2024**, *159*, 111597. [CrossRef]
34. Tang, C.C.; Liu, H.D. Construction of government performance appraisal system of the major function oriented zones in the yangtze river basin. *Econ. Geogr.* **2015**, *35*, 36–44.
35. Ma, Z.Y.; Duan, X.J.; Wang, L.; Wang, Y.Z. Study on spatial coupling characteristics of regional development and resource-environment carrying capacity and path of high-quality development in yangtze river economic belt. *Resour. Environ. Yangtze Basin* **2022**, *31*, 1873–1883.
36. Li, Z.; Hu, B.; Ren, Y. The supply–demand budgets of ecosystem service response to urbanization: Insights from urban–rural gradient and major function-oriented areas. *Remote Sens.* **2022**, *14*, 5670. [CrossRef]
37. Lu, Y.; Wu, P.; Xu, K. Multi-time scale analysis of urbanization in urban thermal environment in major function-oriented zones at landsat-scale: A case study of Hefei city, China. *Land* **2022**, *11*, 711. [CrossRef]
38. Geng, L.; Zhang, Y.; Hui, H.; Wang, Y.; Xue, Y. Response of urban ecosystem carbon storage to land use/cover change and its vulnerability based on major function-oriented zone planning. *Land* **2023**, *12*, 1563. [CrossRef]
39. Huang, X.; Pan, B. Progress, problems and suggestions on the implementation of major function area planning. *Nat. Res. Econ. Chin* **2020**, *33*, 4–9.
40. Wang, J.F.; Li, L.F.; Li, Q.; Hu, S.X.; Wang, S. Monitoring spatio-temporal dynamics and causes of habitat quality in yellow river basin from the perspective of major function-oriented zone planning. *Contemp. Probl. Ecol.* **2022**, *15*, 418–431. [CrossRef]
41. Wu, D.; Zou, C.X.; Lin, N.F.; Xu, M.J. Characteristic analysis of ecological status in the yangtze river economic belt based on the plan for major function-oriented zones. *Resour. Env. Yangtze Basin* **2018**, *27*, 1676–1682.
42. Liu, J.Y.; Liu, M.L.; Zhuang, D.F.; Deng, X.Z.; Zhang, Z.X. Spatial pattern analysis of recent land use changes in China. *Sci. China Ser. D* **2002**, *32*, 1031–1040.
43. Chen, W.X.; Zhao, H.B.; Li, J.F.; Zhu, L.J.; Wang, Z.Y.; Zeng, J. Land use transitions and the associated impacts on ecosystem services in the middle reaches of the yangtze river economic belt in China based on the geo-informatic tupu method. *Sci. Total Environ.* **2020**, *701*, 134690. [CrossRef] [PubMed]
44. Xie, G.D.; Lu, C.X.; Leng, Y.F.; Zheng, D.; Li, S.C. Ecological assets valuation of the tibetan plateau. *J. Nat. Resour.* **2003**, *18*, 189–196.
45. Costanza, R.; D'Arge, R.; de Groot, R.; Farber, S.; Grasso, M.; Hannon, B.; Limburg, K.; Naeem, S.; O'Neill, R.V.; Paruelo, J.; et al. The value of the world's ecosystem services and natural capital. *Nature* **1997**, *387*, 253–260. [CrossRef]

Disclaimer/Publisher's Note: The statements, opinions and data contained in all publications are solely those of the individual author(s) and contributor(s) and not of MDPI and/or the editor(s). MDPI and/or the editor(s) disclaim responsibility for any injury to people or property resulting from any ideas, methods, instructions or products referred to in the content.

Article

Carbon Balance Zoning and Spatially Synergistic Carbon Reduction Pathways—A Case Study in the Yangtze River Delta in China

Hui Guo [1,2] and Wei Sun [1,3,*]

[1] Key Laboratory of Lake and Watershed Science for Water Security, Nanjing Institute of Geography and Limnology, Chinese Academy of Sciences, Nanjing 210008, China; huiguo@niglas.ac.cn
[2] University of Chinese Academy of Sciences, Beijing 100049, China
[3] Institute for Carbon Neutral Development, Nanjing 210096, China
* Correspondence: wsun@niglas.ac.cn

Abstract: The concept of major function-oriented zones is highly compatible with the idea of spatially synergistic carbon reduction. In this study, 2005–2020 is taken as the research period, and 305 counties in the Yangtze River Delta (YRD) region are taken as the research unit. The S0M-K-means clustering model and GeoDetector are adopted on the basis of carbon emission/absorption accounting to analyse the spatial and temporal variations in the carbon balance in the YRD region. Furthermore, carbon balance zoning and influencing factors are analysed. Then, a regional spatially synergistic carbon reduction pathway is proposed. The results show that carbon absorption in the YRD region struggles to offset carbon emissions; the regional carbon imbalance is gradually becoming worse; and each county's carbon emission/absorption shows a significant spatial imbalance. Optimised development zones and key development zones are high-value agglomerations of carbon emissions, while the main sources of carbon sinks in the YRD region are the key ecological functional zones. The YRD region has 87 high carbon control zones, 167 carbon emission optimisation zones, and 51 carbon sink functional zones, which are further subdivided into 9 types of carbon balance zones in accordance with the major function-oriented zones (MFOZs). Based on the driving factors of carbon balance changes in the YRD region, this study proposes differentiated spatially synergistic carbon reduction paths for each zone in accordance with the carbon balance zones. As the Yangtze River Delta is an essential engine for China's economic development, the study of its carbon balance is highly relevant in formulating differentiated low-carbon development pathways for each functional zone and promoting regional spatially synergistic carbon reduction to realise the target of "dual-carbon" development.

Keywords: major function-oriented zoning; Yangtze River Delta; carbon budget; carbon balance zoning; spatially synergistic carbon reduction pathways

Citation: Guo, H.; Sun, W. Carbon Balance Zoning and Spatially Synergistic Carbon Reduction Pathways—A Case Study in the Yangtze River Delta in China. *Land* **2024**, *13*, 943. https://doi.org/10.3390/land13070943

Academic Editor: Marko Scholze

Received: 29 May 2024
Revised: 22 June 2024
Accepted: 26 June 2024
Published: 28 June 2024

Copyright: © 2024 by the authors. Licensee MDPI, Basel, Switzerland. This article is an open access article distributed under the terms and conditions of the Creative Commons Attribution (CC BY) license (https://creativecommons.org/licenses/by/4.0/).

1. Introduction

As the most important factor causing global climate change, anthropogenic greenhouse gas emissions have caused ecological and environmental problems that have severely constrained the sustainable development of human civilization [1]; furthermore, contradictions between economic growth and ecological conservation have become more and more prominent. Controlling and reducing carbon emissions has become important for countries worldwide [2]. Since the beginning of this century, driven by industrialisation and urbanisation, China's economy has made great progress, but increased fossil energy consumption and the destruction of surface vegetation often go hand in hand with rapid economic growth, leading to a rapid increase in the regional carbon emissions, a corresponding shift in the spatial pattern of carbon sources and sinks, and a growing imbalance between regional carbon revenues and expenditures [3,4]. As the world's economic and demographic power, China's CO_2 emissions account for nearly 30% of global CO_2 emissions [5]. Against this background, the Chinese government has proposed the goal of "carbon peak and carbon neutrality" by 2030. This measure is not only important for implementing an ecological civilisation and promoting the green transformation of economic and social development, it is also important for committing to addressing and responding to global climate change. The imbalance between carbon emissions and carbon absorption is a major obstacle to achieving regional carbon reduction and carbon neutrality [6,7]. China's "dual-carbon" goal involves both achieving effective control of the overall amount of carbon emissions and realising synergies and equity in carbon emission reduction between regions [8,9].

With the intensification of global warming, studies related to carbon emissions have become increasingly abundant, and numerous studies have been conducted from different perspectives on carbon emission/absorption accounting [10,11], carbon balance/compensation zoning [8,12], and carbon efficiency [13]. These studies reveal the features of the spatial and temporal evolution of carbon emission/absorption at different scales and regions, as well as their driving factors [14,15]. Carbon emission/absorption accounting provides the basis for these relevant studies. The emission factor method based on the carbon emission factors provided in the Guidelines for National Greenhouse Gas Inventories compiled by the Intergovernmental Panel on Climate Change (IPCC) is considered the basic reference for carbon emission accounting [16]; methods of accounting for carbon emissions based on remotely sensed interpreted data, such as land use data and night-time lighting data, have also been widely applied [17,18]. In terms of accounting for carbon absorption, most of the studies that have been conducted have used inventory, modelling, or atmospheric inversion methods to assess and account for ecosystem carbon absorption [19,20]. In addition, some scholars and research institutes have constructed and released carbon emission and absorption datasets [21,22], which provide data support for macroscale studies. Chinese scholars have carried out a wealth of studies on the characteristics of the temporal and spatial evolution of carbon revenues and expenditures on the basis of carbon accounting at different geographical scales, such as the provincial, prefectural, and watershed scales; the results of this research show that China's carbon emissions have significant spatial differentiation and agglomeration effects and that the imbalance between carbon sources and sinks in the national territory is considerable [23,24]. The differences among regional carbon budgets stem mainly from complex interactions among economic, social, policy, technological, and ecological systems [25]. Some scholars have explored the roles of different socioeconomic factors in driving changes in regional carbon emissions via the logarithmic mean divisia index model (LMDI) [26], the stochastic impacts by regression on population, affluence, and technology model (STIRPAT) [27], and the extended Kaya's constant [28]. Furthermore, the population size, economic model, energy consumption structure, and emission reduction measures are the main factors driving regional carbon budget changes [29].

To alleviate the imbalance between carbon sources and sinks in China, much research has been conducted on carbon compensation; a consensus exists on the importance of interregional carbon compensation systems in promoting regional low-carbon development and achieving carbon neutrality [30,31]. Furthermore, the division of carbon balance zones or carbon offset zones based on the scales of regional carbon budgets and spatial differences is the key to establishing these systems and promoting spatially synergistic carbon reduction [32,33].

Major function-oriented zoning (MFOZ) is a comprehensive geographic zoning method with applicability, innovation, and foresight that was introduced by the Chinese government in 2010 [9,33]. Based on the differences in regional resource and environment carrying capacity, existing development intensity, and future development potential, the plan groups state territories into four categories: optimised development zones (ODZs), key development zones (KDZs), restricted development zones (RDZs), and prohibited development zones (PDZs). Among them, RDZs are subdivided into agricultural production zones (APZs) and key ecological functional zones (EFZs). Defining the main function of each geographical unit and determining the development or protection methods for it according to local conditions are highly important for building a spatial development pattern of the national territory that is in harmony with the socioeconomic and ecological environment [31,34]. Research on carbon balance involves characterising the balance between sources and sinks of carbon in a specific region within a certain period [35,36]. Regional differences in natural ecosystems and socioeconomic activities result in different patterns of territorial space development, which directly or indirectly cause changes in regional carbon emission/absorption [37–39]. Objectively, differences in carbon revenues and expenditures among regions occur and are manifested mainly in the strengths or weaknesses of regional carbon sources/sinks and the imbalance of the spatial distribution of carbon emissions/absorption at different geographic scales; therefore, exploring spatially synergistic carbon reduction pathways can promote the dual-carbon goal [34,40]. Moreover, major function-oriented zoning is highly compatible with spatially synergistic carbon reduction. Specifically, the differentiated spatial planning policies of different functional zones can guide and control human activities, thus affecting regional carbon emissions and carbon absorption [41,42]. Therefore, an analysis of regional carbon balances from a MFOZ perspective is highly significant for realising regional carbon equity and synergistic carbon reduction [8].

Generally, research on carbon balance has received extensive attention from academics and achieved rich research results, but some deficiencies still exist. First, most city- or county-level research focuses only on carbon emissions, though some do estimate regional carbon absorption via the coefficient estimation method; however, due to differences in the climate and environment of each region, estimation data based on the same coefficient have large errors, which affects the accuracy of carbon balance research. Second, the resource endowment and development paths of county units are somewhat different; the county is the fundamental unit of China's MFOZs, and the existing studies are still relatively limited and urgently need to be enriched. In addition, most research on the influencing factors related to carbon balance has utilized the STIRPAT, the LMDI, and other models to portray the extent of each factor's influence; research on the analysis of influencing factors combined with spatial heterogeneity needs to be strengthened.

The Yangtze River Delta (YRD) region is the most economically active and integrated region in China and has the closest regional cooperation. The levels of economic development, industrialisation, urbanisation, and population concentration in the region are all at high levels, and imbalances in the spatial distribution of carbon sources and sinks within the region are increasing; therefore, how to contribute to low-carbon development through spatial synergies is an important topic that needs to be addressed urgently in the region. We take 305 county units in the YRD region of China as the study area and 2005–2020 as the study period. On the basis of carbon emission/absorption accounting, we explore the spatial and temporal evolution characteristics of the carbon balance in the YRD region; propose a carbon balance zoning scheme based on the perspective of MFOZs; quantitatively analyse the factors influencing the changes in the carbon balance; and propose a path of spatially synergistic carbon reduction on this basis. The findings provide practical ideas and policy inspiration for the green and low-carbon transition and the realisation of the dual-carbon target in the YRD region.

2. Study Area and Methodologies

2.1. Study Area

The YRD region is located in the region where the Yangtze River and seas meet, is an impact plain formed by the sedimentary action of rivers, and has a flat topography and a dense river network, making it among the most economically vibrant regions in China. The geographical area includes the Shanghai, Jiangsu, Zhejiang, and Anhui provinces, totalling 305 county study units (see Figure 1). With a long history of development, a densely populated area, and a relatively high level of development, the urbanisation rate of the YRD region increased from 50.92% to 70.85%, and the gross domestic product (GDP) increased by a factor of 5.31, from 2005 to 2020. As of 2020, the region accounts for 16.67% of China's population and 24.12% of its GDP output, but less than 4% of China's territory. Regional resources are abundant and environmental pressures are high, and the contradictions of ecological conservation and economic growth are very prominent. The 305 counties in the YRD region vary substantially in their resource endowment and development level, and there is significant spatial heterogeneity in the scale of carbon emissions and carbon absorption. Therefore, analysing the differences in the carbon balance characteristics of the counties in the YRD region based on major function-oriented zoning and exploring spatially synergistic carbon reduction pathways can help in building a new territorial spatial pattern in line with the dual-carbon goal.

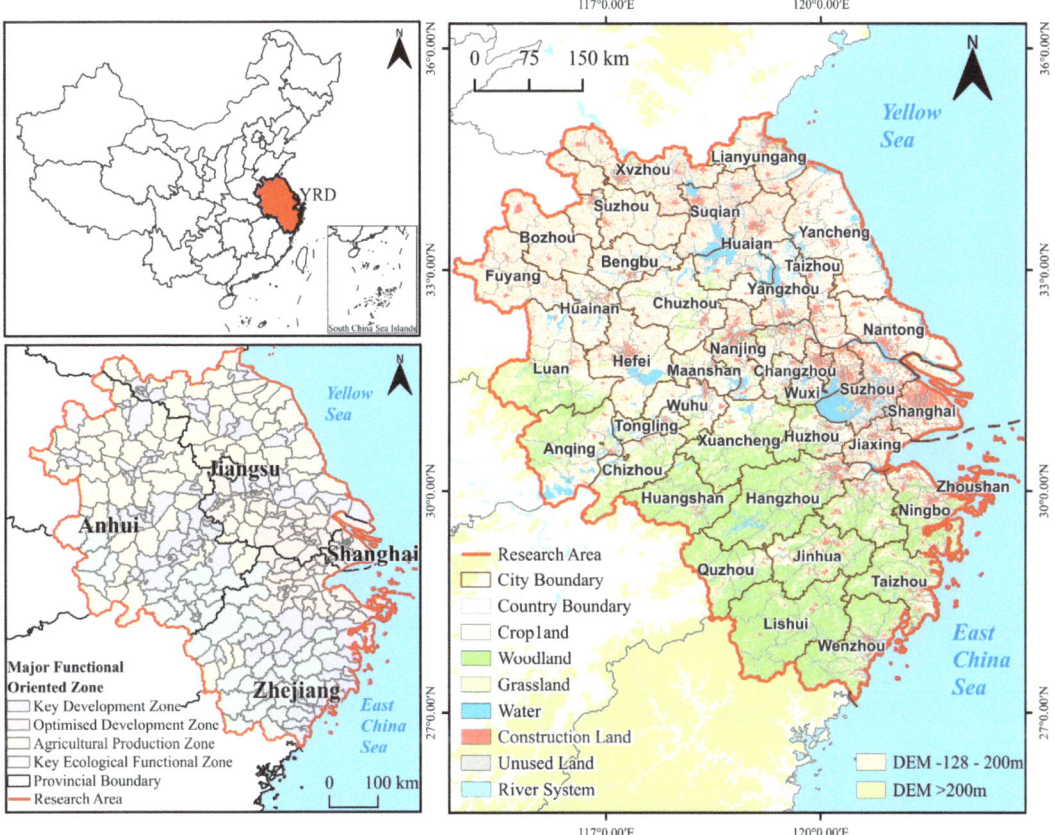

Figure 1. Location Map of YRD Region.

2.2. Data Sources

The administrative division data required for this paper were obtained from the National Platform for Common GeoSpatial Information Services (https://www.tianditu.gov.cn, accessed on 29 April 2024). The land use data were obtained from the Resource and Environment Science Data Platform [43]. The net primary productivity (NPP) of vegetation was obtained by extracting the MOD17A3GV6 product from the Google Earth Engine. The soil-related data were obtained from HWSD v2.0 [44]. Climate data such as precipitation and temperature data were obtained from the National Tibetan Plateau Scientific Data Centre [45,46]. The number of medium and large industrial enterprises were obtained from the website of Qichacha (https://www.qcc.com, accessed on 4 February 2024). The economic and social development data required for this paper were obtained mainly from county and city statistical yearbooks or statistical bulletins, with some missing data made up using linear interpolation. The energy consumption data were obtained from the China

Energy Statistical Yearbook. The MFOZs were obtained from the draft of provincial major function-oriented zoning for Shanghai, Jiangsu, Zhejiang, and Anhui.

2.3. Methods

2.3.1. Carbon Emission Accounting

As the current statistical unit of China's energy consumption data is province-based, it is difficult to directly account for carbon emissions in counties; furthermore, some scholars believe that the difference in energy use efficiency within a province is larger among cities and smaller among counties within a city [47]. Therefore, drawing on the methodology of existing studies, under the premise of ensuring that the total carbon emissions of the counties in the province are equal to the provincial carbon emissions and that the total subsectoral carbon emissions of the counties in the province are equal to the provincial subsectoral carbon emissions, we introduce city-level indicators to indirectly account for the carbon emissions of the counties; the calculation steps are as follows [47,48]:

With reference to the accounting method used by relevant scholars to decompose regional carbon emissions into four sectors [48,49], this paper groups the seven categories of economic sectors in the energy end-consumption section of the China Energy Statistical Yearbook into four categories (see Table 1). The accounting is carried out based on the statistical data of various types of energy consumption in different economic sectors of provincial-level administrative regions in the YRD region with the help of the reference methodology provided by the IPCC. The formula is as follows [16]:

$$rCE = \sum_i \sum_j EC_{i,j} \times K_i + \sum_i EC_{i,p} \times K_p \qquad (1)$$

$$K_i = NCV_i \times CC_i \times 10^{-3} \times COF_i \qquad (2)$$

where rCE denotes the provincial carbon emissions; i denotes the energy type; j denotes the three industries; p denotes the residential sector; $EC_{i,j}$ and $EC_{i,p}$ denote the consumption by the corresponding energy i for industrial sector j and the residential sector p, respectively; K_i and K_p denote the carbon emission factor for energy i and the residential sector p; NCV_i denotes the net caloric value of energy i; CC_i denotes the CO_2 emissions per net caloric value produced by energy i; and COF_i denotes the oxidation ratio during energy i combustion. In this study, we use the carbon oxidation rate of each type of energy published by the IPCC [16], and the net caloric value and the CO_2 emissions per net calorific value of each type of energy based on China's actual measurement [49] (see Table 2).

Table 1. Economic sectoral division.

Economic Sectors in the China Energy Statistics Yearbook	Economic Sectors Delineated in This Paper
Farming, forestry, animal husbandry and fishery	Primary industry
Industry and construction	Secondary industry
Transportation, logistics and postal	
Wholesale and retail trade services	Tertiary industry
Accommodation and catering services	
Consumption for living	Residential sector

Carbon emissions per unit of output value and per capita living carbon emissions are calculated for the three industries at the provincial level according to the provincial three-industry output value and population. Then, they are multiplied by the city three-industry output value and population to get the hypothetical carbon emissions of each city. The theoretical carbon emissions of each city are accounted for with the help of city-level data on energy consumption per unit GDP; the ratio of the two types of carbon emissions is utilised to correct the carbon emission intensity of each sector of the economy at the city scale. The formula is as follows [47]:

$$COCE = \frac{sTCE}{sHCE} = \frac{EI \times tGDP \times ACC}{\sum_j \frac{rCE_j}{rGDP_j} \times sGDP_j + \frac{rCE_p}{rGDP_p} \times sP} \quad (3)$$

$$ACC = \frac{\sum_j EC_j \times K_i}{\sum_k EI_k \times GDP_k} \quad (4)$$

where $COCE$ denotes the city carbon emission correction coefficient; $sTCE$ denotes the city theoretical carbon emission; $sHCE$ denotes the city hypothetical carbon emission; EI denotes the energy consumption per unit of GDP; $tGDP$ denotes the total regional GDP; ACC denotes the average carbon emission coefficient of each type of energy; rCE_j denotes the provincial carbon emissions from sector j; $rGDP_j$ is the GDP of industry j at the provincial level; $sGDP_j$ denotes the GDP of industry j at the city level; rCE_p denotes the per capita carbon emission at the provincial level; $rGDP_p$ denotes the provincial GDP per capita; sP denotes the city population size; EC_j denotes the energy consumption of industry j; k_i denotes cities in the province; EI_k is the energy consumption per unit of GDP in city k; and GDP_k is the GDP in city k.

Table 2. Energy types and emission factors.

Fuels in This Study	NCV (PJ/10⁴ t, 10⁸ m³)	CC (tC/TJ)	COF	Carbon Emission Coefficient
Raw coal	0.21	26.32	0.94	1.91
Cleaned coal	0.26	26.32	0.93	2.33
Other washed coal	0.15	26.32	0.93	1.35
Briquettes	0.18	26.32	0.90	1.56
Coke	0.28	31.38	0.93	2.30
Coke oven gas	1.61	21.49	0.98	1.24
Other gas	0.83	21.49	0.98	0.64
Other coking products	0.28	27.45	0.98	2.76
Crude Oil	0.43	20.08	0.98	3.10
Gasoline	0.44	18.90	0.98	2.99
Kerosene	0.44	19.60	0.98	3.10
Diesel oil	0.43	20.20	0.98	3.12
Fuel oil	0.43	21.10	0.98	3.26
Other petroleum products	0.51	17.20	0.98	3.15
Liquefied petroleum gas	0.47	20.00	0.98	3.38
Refinery gas	0.43	20.20	0.98	3.12
Nature gas	3.89	15.32	0.99	2.16

Note: Briquettes in the table include briquettes and gangue from the China Energy Statistical Yearbook; other gasses in the table include blast furnace gas, converter gas, and other gas from the China Energy Statistical Yearbook; other petroleum products in the table include naphtha, lubricants, paraffin, white spirit; bitumen asphalt, petroleum coke and other petroleum products from the China Energy Statistical Yearbook.

Accounting for carbon emissions in counties is conducted with the help of the corrected carbon intensity for city-level economic sectors. The calculation formula is as follows [47]:

$$xCE = \sum_j COCE \times \frac{rCE_j}{rGDP_j} \times xGDP_j + COCE \times \frac{rCE_p}{rGDP_p} \times xP \quad (5)$$

where xCE denotes county carbon emissions, $xGDP_j$ denotes county j-industry GDP, and xP denotes the county population size.

2.3.2. Carbon Absorption Accounting

Net ecosystem productivity (NEP), a key indicator of carbon absorption, is used to characterise regional carbon absorption. The formula is as follows [50]:

$$CS = NPP - RH \quad (6)$$

$$RH = 0.6163RS^{0.7918} \qquad (7)$$

$$RS = 1.55e^{0.031T} \times \frac{P}{P+0.68} \times \frac{SOC}{SOC+2.23} \qquad (8)$$

$$SOC = SC \times \gamma \times H \times (1-\delta_{2mm}) \times 10^{-3} \qquad (9)$$

where CS denotes the regional annual carbon absorption; NPP denotes the regional annual net primary productivity of vegetation; RH denotes the regional annual soil heterotrophic respiration; RS denotes the annual soil respiration; T denotes the annual mean temperature; P denotes the annual precipitation; SOC denotes the density of soil organic carbon within a depth of 20 cm; SC denotes the organic carbon content of the top soil layer from 0 to 20 cm; γ denotes the soil bulk weight; H denotes the thickness of the soil; and δ_{2mm} represents the percentage of gravel in the soil smaller than 2 mm.

2.3.3. Carbon Balance Analysis and Zoning Methods

The economic contribution coefficient (ECC) of carbon emissions is the ratio of the economic contribution rate to the regional carbon emission rate; it reflects the size of the regional carbon production capacity from the economic perspective. The ecological support coefficient (ESC) of carbon emissions is the ratio of the regional carbon absorption rate to the regional carbon emission rate; it reflects the ecological carbon capacities of the area from the ecological perspective [8]. The formula for the calculations is as follows:

$$ECC = \frac{GDP_i}{GDP_s} \Big/ \frac{CE_i}{CE_s}; \quad ESC = \frac{CA_i}{CA_s} \Big/ \frac{CE_i}{CE_s} \qquad (10)$$

where GDP_i and GDP_s denote the GDP of county i and the YRD region, respectively; CE_i and CE_s denote the carbon emissions of county i and the YRD region, respectively; and CA_i and CA_s denote the carbon absorption of county i and the YRD region, respectively. An $ECC > 1$ means that the county unit's economic contribution is higher than its carbon emission contribution, and the reverse is true for an $ECC < 1$; $ESC > 1$ indicates that a county unit has a strong offsetting capacity for regional carbon emissions, and the reverse is true for $ESC < 1$.

Referring to the literature, we adopt the SOM-K-means cluster analysis method to carry out zoning studies and select indicators from four dimensions—scale of total carbon emissions, socioeconomic development level, resource-environmental carrying capacity, and territorial exploitation degree—as the basis for zoning [8,51] (see Table 3). The calculation of these indicators is based on data from 2020. The regional main functional areas were overlaid on top of the zoning results to further subdivide the carbon balance zones. The calculation formula is as follows:

Table 3. Economic sectoral division.

Dimensions	Indicators
Carbon emission	Total carbon emissions
Socioeconomic development level	Economic contribution coefficient of carbon emission (ECC)
Resource-environment carrying capacity	Ecological support coefficient of carbon emission (ESC)
Territorial exploitation degree (TED)	Share of built-up land area in the area of national territory space

Note: The calculations of relevant indicators are based on the 2020 dataset.

The normalised revealed comparative advantage index (NRCA) is useful for measuring and identifying the dominant attributes of carbon balance zones [52]. The formula is as follows:

$$NRCA_{ij} = \frac{X_{ij}}{X} - \frac{X^i \times X_j}{X^2} \qquad (11)$$

where $NRCA_{ij}$ denotes the NRCA of attribute j in county i; X_{ij} denotes the indicator value of attribute j in county i; X_i denotes the sum of the indicator values of all attributes in county i; X_j denotes the sum of the indicator values of attribute j in each county unit of the YRD region; and X denotes the total indicator of all county units and attributes.

The self-organizing map (SOM) is an unsupervised artificial neural network [53]. K-means is a cluster analysis algorithm with an iterative solution using the sum-of-squares-of-errors criterion function as the clustering criterion function. The SOM-K-means is a hybrid cluster analysis method for spatial partitioning; it combines the advantages of the SOM algorithm in terms of self-organisation and adaptivity and the advantages of K-means in terms of easy interpretation, fast convergence, and high efficiency.

2.3.4. Analysis of Factors Affecting Carbon Compensation Rates

The carbon compensation rate refers to the proportion of carbon absorption to carbon emission, which can more accurately reflect the current status of the regional carbon balance. Therefore, the dependent variable is the carbon compensation rate, and 10 indicator variables in five dimensions, including urbanisation, industrialisation, ecological foundation, agricultural production, and governance-technological level, are selected to detect the driving factors that affect the regional carbon balance. The selected indicators are shown in Table 4.

Table 4. Regional carbon balance influencing factors: indicator system and interpretation.

Dimensions	Indicators	Interpretation of Indicators
Urbanisation	Urbanisation rate (X_1)	Urban population/Total population
	Territorial exploiting degree (X_2)	Construction land area/Total territorial area
Industrialisation	Secondary industry share of GDP (X_3)	Added value of secondary industry/Gross domestic product (GDP)
	Number of medium and large industrial enterprises (X_4)	Number of medium and large industrial enterprises in the region
Ecological foundation	Vegetation richness (X_5)	Normalised difference vegetation index
	Vegetation cover (X_6)	Area of vegetation cover/Total territorial area
Agricultural production	Primary industry share of GDP (X_7)	Added value of primary industry/Gross domestic product (GDP)
	Share of cropland (X_8)	Cropland area/Total territorial area
Governance-technological level	General public budget expenditure (X_9)	General public budget expenditure in the region
	Number of patents received for inventions (X_{10})	Number of patents received for inventions in the region

GeoDetector (GD) can be used to effectively measure spatial heterogeneity; it is a set of statistical methods for detecting the causes of spatial heterogeneity and revealing the driving forces of different influencing factors, and is widely used at different research scales, such as the national, provincial, city, and township scales, as well as in various research fields, like the eco-environment, land use, regional planning, and public health fields [54]. The factor detector measures the degree to which different influences explain spatial heterogeneity; its expression is as follows:

$$q = 1 - \frac{\sum_i n_i \theta_i^2}{n \theta^2} \qquad (12)$$

where q denotes the explanatory power of the influencing factors; it is a number between 0 and 1, and larger values have more explanatory power. i denotes the number of strata of the independent variables; n and n_i denote the sample size and the sample size of the ith stratum, respectively; and θ and θ_i denote the variance in the dependent variable in the research zone and the variance in the dependent variable in stratum i, respectively.

3. Results

3.1. Characteristics of the Spatial-Temporal Evolution of the Carbon Balance

Carbon emissions in the YRD region have shown a significant growth trend over time, but with the implementation of green and low-carbon development in recent years, the growth of regional carbon emissions has gradually levelled off. At the county scale (see Figure 2), the YRD region's carbon emissions show a significant spatial imbalance, with overall high values in the northeast and low values in the southwest. During 2005–2020, the high-value zone expanded outwards along the "Shanghai–Nanjing–Hangzhou–Ningbo" line, while the northern part of the YRD was also an area of rapid growth in carbon emissions. The "Shanghai–Nanjing–Hangzhou–Ningbo" line, with its high population concentration, high level of economic development, and highly developed manufacturing industry, is an important economic engine of the YRD region, and its energy consumption is higher than that of other areas in the YRD region; thus, it is a high carbon emission area in the YRD region. Northern Jiangsu and northern Anhui have rich mineral resources and a good base for industrial development, so carbon emissions in the region are growing rapidly. The spatial structure of carbon emissions in the YRD region has gradually evolved from a "core-periphery" structure with Shanghai, Nanjing, and Hangzhou as the high-value areas in 2005 to a polycentric structure with Shanghai, Nanjing, Suzhou, Wuxi, Xuzhou, Hefei, Hangzhou, and Ningbo as the core cities.

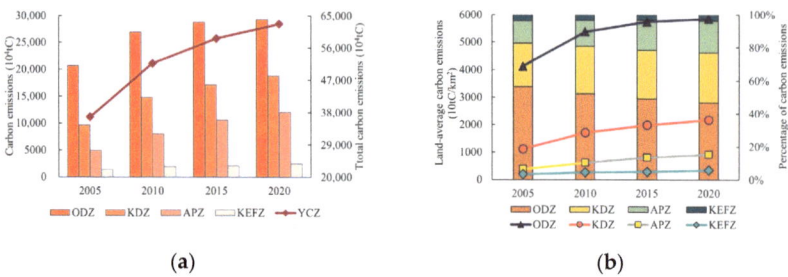

Figure 2. Changes in carbon emissions in the YRD region. (**a**) Changes in carbon emissions from the MFOZs of the YRD region; (**b**) carbon emissions share and land-average carbon emissions of each MFOZ.

From the perspective of major function-oriented zoning (see Figure 3), the YRD region's carbon emissions are mainly concentrated in optimised development zones (ODZs) and key development zones (KDZs), which are important high-population and economically dense areas. They are responsible for absorbing industrial agglomeration and driving regional development. Due to the optimisation and escalation of the industrial structure, the share of carbon emissions in the optimised development zones shows a decreasing trend but still accounts for nearly half of the carbon emissions in the region as a whole; the land-average carbon emissions of this zone are much greater than those in other regions. Furthermore, as a key development area for current and future industrialisation and urbanisation, carbon emissions in this area are increasing as a proportion of the YRD region. Key development zones, which are the focus of current and future industrialisation and urbanisation, have seen their shares of carbon emissions in the region increase. The carbon emissions from the agricultural production zones and key ecological functional zones have remained at a low level; among them, with the advancement of mechanised cultivation and the strengthening of regional development, carbon emissions from the agricultural production zones have increased, while carbon emissions from the restricted development zones, such as key ecological functional zones, have remained basically the same as in previous years.

Between 2005 and 2020, changes in carbon absorption in the YRD region continued to grow over time, but the growth trend was relatively flat. After 2010, the construction of an ecological civilisation gradually became a key issue of concern in China, with effective improvement in the eco-environment and significant growth in carbon absorption. At the county scale (see Figure 4), carbon absorption in the YRD region as a whole shows a spatial feature of being higher in the southwest and lower in the northeast. The high carbon absorption zones in the region are consistently clustered in southwestern Zhejiang and southern Anhui, which have a wide distribution of hilly and mountainous terrain, high vegetation cover, and a relatively low level of industrialisation, making them important ecological spaces in the YRD region; therefore, the carbon absorption in this region is greater than what it is in other regions of the YRD region. In addition, the coastal area of Jiangsu is a high-value area for carbon absorption due to its rich wetland resources. The low-value areas are concentrated primarily along the "Shanghai–Nanjing–Hangzhou–Ningbo" line, which has a high level of urbanisation and industrialisation; furthermore, the space for carbon sinks has been reduced and encroached upon by construction land. In addition, the centres of city administrative regions are low carbon-absorption areas, which increase outwards from the core. The regions with the greatest changes in carbon absorption from 2005 to 2020 are concentrated mainly in the northern region of Anhui, where carbon absorption has been effectively enhanced as a result of the steady progress of ecological restoration.

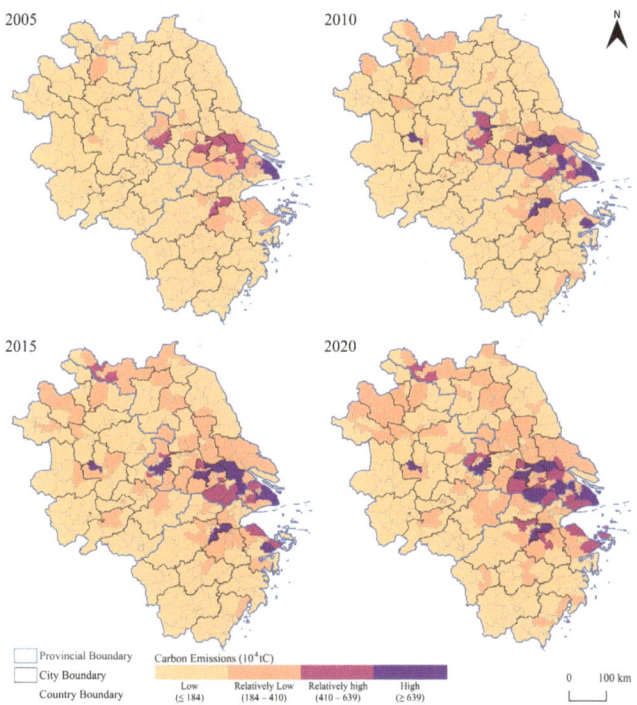

Figure 3. The YRD region's carbon emissions, 2005–2020.

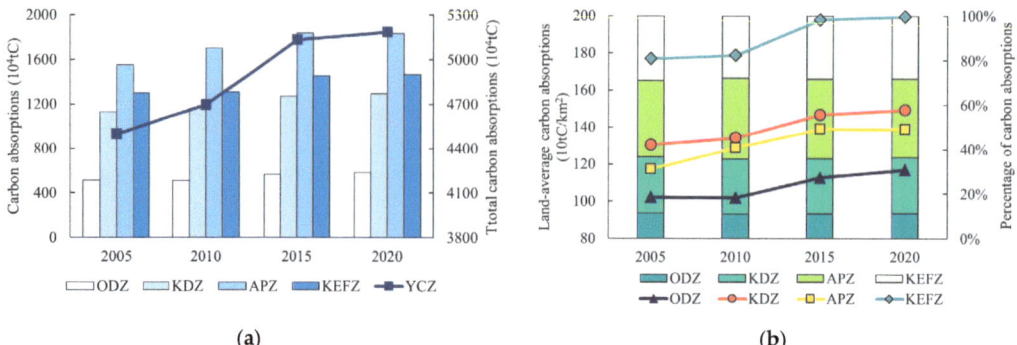

Figure 4. Changes in carbon absorption in the YRD region. (**a**) Changes in carbon absorption from the MFOZs of the YRD region; (**b**) carbon absorption share and land-average carbon absorption of each MFOZ.

From the perspective of MFOZs (see Figure 5), the carbon absorption in the YRD region is concentrated mainly in agricultural production zones and key ecological functional zones, which are responsible for guaranteeing food security and ecological safety; high-intensity industrialisation and urbanisation are restricted in these zones. Therefore, the ecological environment of these zones is relatively good, and the space for carbon sinks is relatively rich, as these zones account for more than 60% of the YRD region's carbon absorption; furthermore, they are the main spatial carriers of carbon absorption in the YRD region. In addition, because of the relatively weak carbon sink function of cropland, the land-average carbon absorption in the main agricultural production areas where cropland is widely distributed is relatively low. The key development zones in the YRD region likewise contribute a certain extent of carbon absorption due to their high resource and environmental carrying capacity, but due to the small land area of the region, the land-average carbon absorption is relatively high.

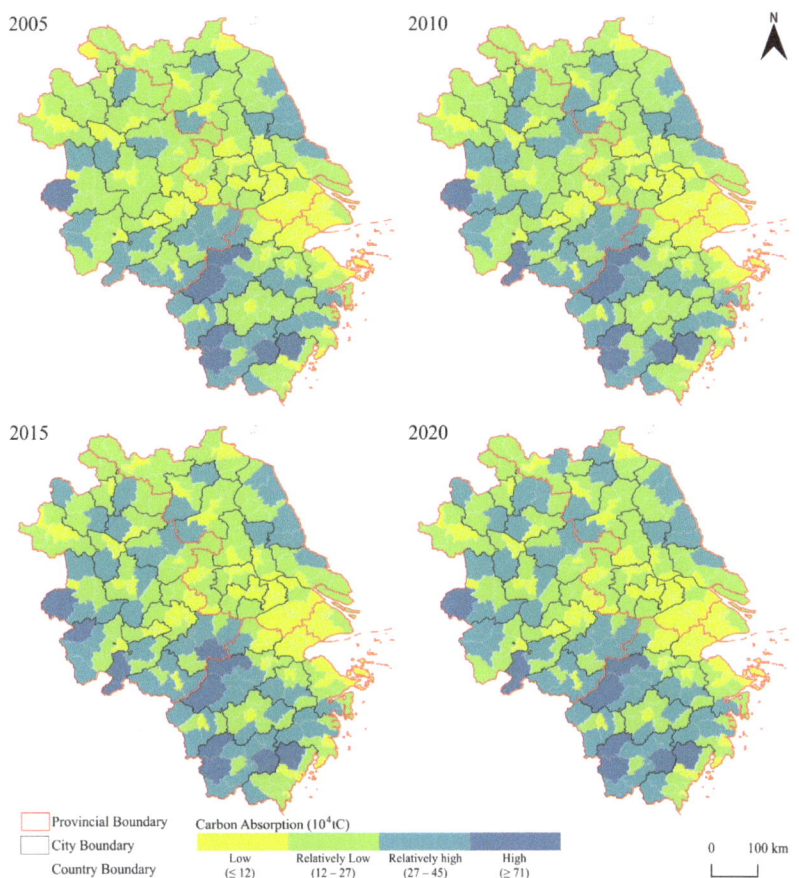

Figure 5. The YRD region's carbon absorption, 2005–2020.

3.2. Carbon Balance Zoning Results

On the basis of the NRCA index, with the help of the SOM-K-means algorithm, the 305 county units in the YRD region were classified into three types of carbon balance zones—high carbon control zones (I), carbon emission optimisation zones (II), and carbon sink functional zones (III), which were overlaid with major function-oriented zoning and further classified into nine types of zones (see Table 5 and Figure 6).

Table 5. Main indicators of the different carbon balance zones.

Carbon Balance Zoning (Number of Units)		Share of GDP/%	Share of Carbon Emissions/%	Share of Territory /%	ECC	ESC	TED /%
High carbon control zone	Optimised development zone (32)	30.41	30.14	9.77	1.01	0.27	25.46
	Key development zone (34)	13.56	16.22	12.43	0.84	0.76	14.06
	Agricultural production zone (21)	5.03	7.79	9.23	0.65	1.09	11.20
Carbon emissions optimisation zone	Optimised development zone (49)	22.95	16.15	4.65	1.42	0.21	34.59
	Key development zone (70)	13.95	14.19	12.90	0.98	0.89	17.30
	Agricultural production zone (4)	8.96	9.91	21.13	0.90	1.89	17.90
	Key ecological function zone (44)	0.49	0.90	1.33	0.54	1.79	3.91
Carbon sink functional zone	Agricultural production zone (14)	1.79	1.68	8.44	1.06	4.67	8.15
	Key ecological function zone (37)	2.87	3.00	20.12	0.96	8.87	2.15

1. **High Carbon Control Zones.** The high carbon control zones consist of 87 county units, accounting for 31.43% of the YRD's land area, generating 48.99% of the YRD's GDP, and 54.15% of the YRD's carbon emissions. These zones are located mainly in the eastern part of Shanghai, the southern part of Jiangsu, the Ningbo-Hangzhou line in Zhejiang, and surrounding the central urban areas of some cities in Anhui, which are the most important area of the YRD region in terms of carbon emissions. Among them, the high carbon control zone-optimised development zones include 32 county units, which are the concentrated distribution areas of the manufacturing industry, and account for 30.14% of the YRD region's carbon emissions. The ECC (1.01) of this kind of zone is less than that of other optimised development zones, as is the ESC (0.27), indicating that the economic benefits of carbon emissions and the ecological support capacity of these zones are at relatively low levels. The high carbon control zone-key development zones include 34 county units, which are the current key areas for industrialisation and urbanisation development in the YRD region. The ECC (0.84) and ESC (0.76) of this type of zone are lower than those of the YRD region, which shows that the high carbon emissions in these zones do not bring about a better economic benefit and have a low ecological support capacity for carbon emissions. The high carbon control zone-agricultural production zone includes 21 county units, which are more disturbed by human activities, and the economic efficiency and ecological support capacity of carbon emissions of these zones are lower than those of other agricultural production zones.
2. **Carbon Emission Optimisation Zones.** The carbon emission optimisation zones, with a total of 167 county units, are the most numerous carbon balance zones in the YRD region, accounting for 40.01% of the YRD region's land area and 46.34% of the YRD's GDP, and generating 41.16% of the YRD's carbon emissions. These zones are located mainly in the northern part of Jiangsu, the northern and central parts of Anhui, the Taihu Lake basin, and the central urban areas of some cities. Among them, the carbon emission optimisation zone-optimisation development zones include 49 county units. These zones have a long history of development; some high carbon emission industries have been transferred to other regions; and their industrial structures have been optimised and upgraded. The ECC (1.42) of this type of zone is the highest in the YRD region, and the ESC (0.21) is the lowest in the YRD region, which indicates that the economic efficiency of carbon emission in these zones is high, but the carbon ecological support capacity is lower than that of other zones due to the high degree of territorial spatial development. The carbon emission optimisation zone-key development zones comprise 70 county units, which have an ECC of 0.98, an ESC of 0.89, and a low level of territorial spatial exploitation, indicating that the economic benefits of carbon emissions and the ecological function of carbon sinks are in a relatively matched state in these zones. The carbon emission optimisation zone-agricultural production zones and the carbon emission optimisation zone-key ecological functional zones comprise 48 county units, which account for 10.82% of the YRD's carbon emissions. The carbon ecological support capacities of these two types of zones are relatively high, but the economic efficiency of carbon emissions is low, especially in the key ecological functional zones, where the ECC is only 0.54. Regional carbon emissions should be optimised in the future development process; furthermore, the regional economy needs to be improved through the development of green industries in the future.
3. **Carbon Sink Functional Zones.** There are 51 county-level units in the carbon sink functional zones, which account for 28.56% of the territorial space of the YRD region but produce only 4.6% of the YRD's carbon emissions. These zones are primarily situated in the southwestern part of Zhejiang and the southern part of Anhui and have the strongest carbon sink functions in the YRD region. Among them, the carbon sink functional zone-agricultural production zones include 14 county units, and have an ECC and ESC greater than 1; the economic efficiency of carbon emission and carbon

ecological support capacity are at a high level. The carbon sink functional zone-key ecological functional zones comprise 37 county units, primarily situated in hilly and mountainous regions, with high vegetation cover and ESCs as high as 8.87. These zones have the highest carbon ecological support capacity in the YRD region and are important ecological barriers in the YRD region.

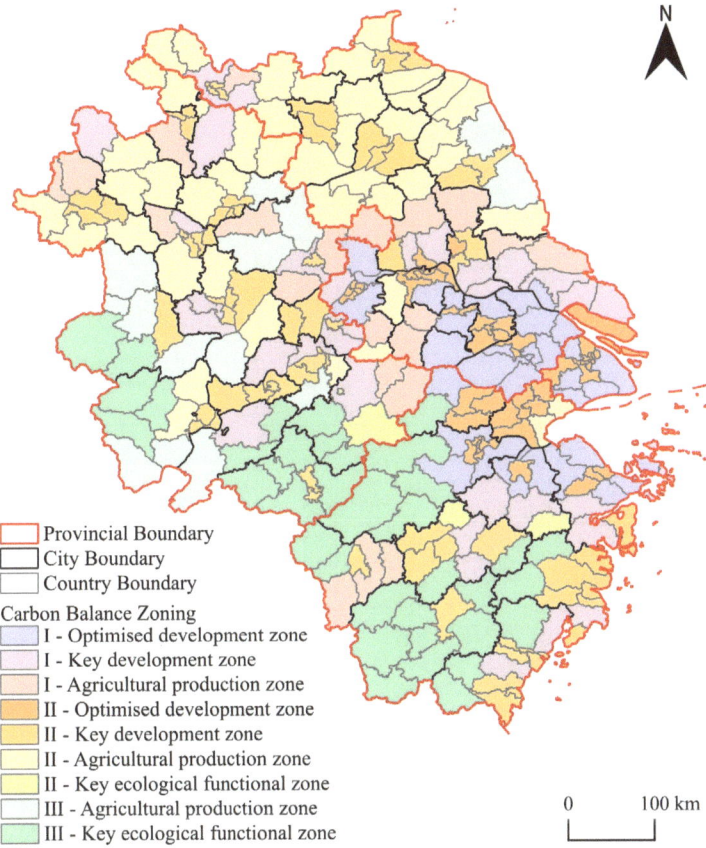

Figure 6. Carbon balance zoning in the YRD region.

3.3. Results of Geodetector Analyses

Using the GeoDetector algorithm package in R [55], the carbon compensation rate of each county unit is taken as the dependent variable, and 10 indicators selected from the urbanisation, industrialisation, ecological foundation, agricultural production, and governance-technological level of each county unit are taken as independent variables, which are inputted into the GeoDetector model together; the capacity value of each driver is calculated to influence the regional carbon compensation rate. The results of the analysis are displayed in Table 6.

Table 6. Detection of carbon compensation rate drivers in the YRD region and its major functional zones.

Carbon Balance Zoning	Year	q-Value of the Driving Indicators									
		X_1	X_2	X_3	X_4	X_5	X_6	X_7	X_8	X_9	X_{10}
Yangtze River Delta	2005	0.22 ***	0.44 ***	0.15 ***	0.16 ***	0.29 ***	0.61 ***	0.20 ***	0.24 ***	0.17 ***	0.08 ***
	2010	0.25 ***	0.46 ***	0.13 ***	0.19 ***	0.43 ***	0.63 ***	0.16 ***	0.23 ***	0.37 ***	0.15 ***
	2015	0.19 ***	0.47 ***	0.10 ***	0.18 ***	0.62 ***	0.69 ***	0.16 ***	0.21 ***	0.45 ***	0.16 ***
	2020	0.18 ***	0.65 ***	0.11 ***	0.23 ***	0.58 ***	0.65 ***	0.19 ***	0.22 ***	0.38 ***	0.21 ***
Optimised development zone	2005	0.41 ***	0.50 ***	0.16 **	0.15 **	0.67 ***	0.30 ***	0.61 ***	0.24 ***	0.35 ***	0.21 ***
	2010	0.45 ***	0.60 ***	0.22 ***	0.12 *	0.62 ***	0.29 ***	0.57 ***	0.24 ***	0.11	0.22 ***
	2015	0.53 ***	0.74 ***	0.16 ***	0.17 ***	0.68 ***	0.36 **	0.59 ***	0.24 ***	0.05	0.12
	2020	0.71 ***	0.52 ***	0.12 **	0.09	0.51 ***	0.26 **	0.83 ***	0.26 **	0.09	0.10
Key development zone	2005	0.32 ***	0.31 ***	0.20 ***	0.06	0.39 ***	0.23 ***	0.38 ***	0.16 ***	0.10 *	0.07
	2010	0.32 ***	0.45 ***	0.15 ***	0.13 **	0.44 ***	0.37 ***	0.39 ***	0.14	0.12	0.08
	2015	0.26 ***	0.53 ***	0.06	0.14 **	0.34 ***	0.47 ***	0.17 ***	0.16 *	0.11	0.06 *
	2020	0.31 ***	0.54 ***	0.11 *	0.10	0.42 ***	0.31 ***	0.41 ***	0.08	0.21 ***	0.05
Agricultural production zone	2005	0.21 ***	0.14	0.33 ***	0.20 **	0.04	0.20 ***	0.39 ***	0.11	0.24 ***	0.08
	2010	0.17 **	0.21 **	0.41 ***	0.29 ***	0.07	0.17 *	0.40 ***	0.08	0.34 ***	0.16 **
	2015	0.21 **	0.35 ***	0.21 *	0.28 ***	0.07	0.22 **	0.38 ***	0.11	0.34 ***	0.07
	2020	0.24 ***	0.24 **	0.35 ***	0.28 ***	0.11	0.22 *	0.48 ***	0.06	0.32 **	0.18 ***
Key ecological function zone	2005	0.28 *	0.42 **	0.38 ***	0.30	0.28 *	0.47 ***	0.27	0.45 ***	0.64 ***	0.19
	2010	0.34 **	0.39 **	0.41 **	0.31 **	0.35 **	0.49 ***	0.24	0.40 **	0.70 ***	0.32 **
	2015	0.21	0.50 ***	0.44 **	0.28 **	0.54 ***	0.57 ***	0.34 **	0.40 *	0.55 ***	0.19
	2020	0.13	0.39 **	0.48 ***	0.29 *	0.46 ***	0.53 ***	0.24	0.49 ***	0.74 ***	0.29 *

Note: ***, **, and * denote 1%, 5%, and 10% significance levels, respectively. X_1 denotes the urbanisation rate; X_2 denotes the territorial exploiting degree; X_3 denotes the secondary industry share of GDP; X_4 denotes the number of medium and large industrial enterprises; X_5 denotes the vegetation richness; X_6 denotes the vegetation cover; X_7 denotes the primary industry share of GDP; X_8 denotes the share of cropland; X_9 denotes the general public budget expenditure; X_{10} denotes the number of patents received for inventions.

At the YRD regional level, the degrees of territorial exploitation (X_2), vegetation richness (X_5), and vegetation cover (X_6) are the main drivers influencing the regional carbon compensation rate; among them, the effect of the degree of territorial exploitation (X_2) increases over time, while the effect of vegetation cover (X_6) is always at a high level.

In terms of optimised development zones, the urbanisation rate (X_1), territorial exploitation degree (X_2), vegetation richness (X_5), and primary industry share of GDP (X_7) are the main factors influencing changes in the regional carbon compensation rate, while the influence of factors related to industrialisation, such as the secondary industry share of GDP (X_3) and the number of medium and large industrial enterprises (X_4), gradually decreases or is not significant over time. In terms of key development zones, the influence of each driver of the regional carbon compensation rate varies less, and the main driver is similar to that of the optimised development zones; however, the influence of the urbanisation rate (X_1) is much lower than that of the optimised development zones, and the influence of the degree of territorial exploitation (X_2) gradually strengthens over time, becoming the most dominant driver. In terms of agricultural production zones, the primary industry share of GDP (X_7), the secondary industry share of GDP (X_3), the general public budget expenditure (X_9), and the number of medium and large industrial enterprises

(X_4) are the most important driving factors. The influence of the primary industry share of GDP (X_7) gradually strengthens with time, and the influence of factors in the area of industrialisation, such as the secondary industry share of GDP (X_3) and the number of medium and large industrial enterprises (X_4), is relatively stable, which is in line with the major functional characteristics of the region. In terms of key ecological function zones, the general public budget expenditure (X_9), vegetation cover (X_6), and share of cropland (X_8) are the main driving factors, and the secondary industry share of GDP (X_3), degree of territorial exploitation (X_2), and vegetation richness (X_5) are also at high levels, while the influence of the other factors has low significance.

4. Discussion

4.1. Characteristics of Carbon Balance Evolution and Carbon Balance Zoning in the YRD Region

This study, as an empirical study, reflects the evolution process and current situation of the carbon balance in the YRD region, and provides a reference case for similar regions to carry out carbon reduction and emission reduction work.

This study shows that the spatial structure of carbon emissions in the YRD region has gradually evolved from a "core-periphery" structure with Shanghai, Nanjing, and Hangzhou as the high-value areas in 2005 to a polycentric structure with the central city of the region as the core, while carbon absorption has always shown a spatial characteristic of being high in the southwest and low in the northeast, which is basically consistent with previous studies [13,37]. In addition, this study used the SOM-K-means algorithm to divide the carbon balance zones under the perspective of the main functional areas. This study finds that optimised development zones and key development zones are the key areas in the YRD region in which to promote carbon reduction, while key ecological functional zones are the dominant areas for carbon sink functions in the YRD region, which indicates that the concept of MFOZs is highly compatible with spatially synergistic carbon reduction.

4.2. Analysis of the Drivers of Changes in the Regional Carbon Balance

Previous studies have generally concluded that regional carbon emission/absorption changes are affected by the integrated impacts of natural ecosystems and human economic activities [38,56], and that the total and spatial-temporal differences between carbon emissions and carbon absorption directly affect the regional carbon balance status, which is further confirmed by this study. The results of the GeoDetector analysis are shown in Table 6; the YRD region's carbon compensation rate was influenced mainly by factors such as the territorial exploiting degree (X_2) and the vegetation cover (X_6), which indicates that territorial spatial development and utilisation in the YRD region have profound impacts on the regional carbon emissions and carbon absorption. As a major engine of China's economic growth, construction land expansion in the YRD region has resulted in increases in carbon emissions while fragmenting the original land cover and weakening the region's carbon sequestration capacity; thus, the incremental increase in carbon absorption is much lower than the incremental increase in carbon emissions, and the imbalance of the regional carbon balance has become more and more prominent.

This study also further supports some of the previous studies, which suggested that different MFOZs have differentiated drivers of carbon balance changes [31,37], which is verified in this study through empirical analyses. As the results in Table 6 show, the main driving factors affecting the change in the carbon compensation rate in the optimized development zone and the key development zone are mainly urbanisation (X_1, X_2) and ecological foundation (X_5, X_6); this indicates that this type of region, as the core area for the development of the YRD, has a high degree of population concentration driven by urbanisation, with a large amount of natural cover and cropland converted to construction land, seriously squeezing the space for agricultural production and the space for carbon sink. This leads to the fragmentation and low-quality of land for carbon sinks, and difficulty for the carbon absorption capacity to offset the continuous growth of carbon emissions, resulting in the continuous decline in the regional carbon compensation rate. Variations in the carbon compensation rate of the agricultural production zones are mainly affected by the secondary industry share of GDP (X_3), the primary industry share of GDP (X_7), and the general public budget expenditure (X_9), that is, the improvement in the infrastructure and industrialisation aggravate the carbon imbalance of the region, which is in line with the characteristics of some of the agricultural production zones of the YRD region that combine the important tasks of industrialisation, urbanisation, and agricultural modernisation. Due to the special functional position and ecological vulnerability of key ecological function zones, ecological improvement, over-exploitation, industrialisation, and urbanisation will all affect the carbon balance status of the region [11,17]. This study shows that the general public budget expenditure (X_9), vegetation cover (X_6), share of cropland (X_8), and secondary industry share of GDP (X_3) are the main driving factors for the change in carbon balance, which, to some extent, validates the relevant viewpoints; however, they still need to be further analysed and explored.

4.3. Spatially Synergy Carbon Reduction Pathways in the YRD Region

Through the above analysis, we can see that there are significant carbon balance differences among different functional zones in the YRD region, and the regional carbon imbalance phenomenon is prominent. Therefore, how to explore the differentiated economic development mode, spatial development mode, and carbon neutralisation path of each region with the help of carbon balance zoning in the YRD region is a key question that should be resolved urgently. To this end, we propose the following differentiated spatial synergistic carbon reduction pathways for different functional zones.

Optimised Development Zones. These zones represent growth poles for economic development in the YRD region, with sound urban systems and strong regional competitiveness. These zones are classified in carbon balance zoning as high carbon control zones or carbon emission optimisation zones. Among them, high carbon control zones should place the transformation of the development mode as their highest priority; actively absorb the high concentration of innovative elements; create an advanced manufacturing base; and improve the region's economic efficiency in terms of carbon emissions. Carbon emission optimisation zones are mostly central urban areas, which are at the forefront of economic transformation and development by virtue of their first-mover advantages in terms of market environment, science and education, and infrastructure. The future of this type of area should focus on the development of modern service and high-tech industries and explore the construction of innovative and ecological cities.

Key Development Zones. These zones have good foundations for economic development, with greater development potential and greater resource and environmental carrying capacities. They are important for undertaking regional population and industrial transfers. These zones are classified in carbon balance zoning as high carbon control zones or carbon emissions optimisation zones. High carbon emission zones should focus on constructing a regional innovation system; enhancing the capacity of industrial clustering; increasing the economic efficiency of carbon emissions of existing industries; strengthening the protection and restoration of the eco-environment; and enhancing the capacity for carbon sequestration in the region. Carbon emission optimisation zones need to explore the spatially synergistic carbon reduction paths of the two types of regions. First, for the urban centres of some cities, efforts should be made to give full play to the advantages of location; focus on the construction of strategic emerging industries and high-tech industrial bases; and actively expand green space at the same time. Second, some areas with better ecological bases and agricultural production conditions should give full play to the ecological advantages of the region in accordance with the characteristics of the lead; and optimise the development of green industries and modern agriculture and other low-pollution industries.

Agricultural production zones. These zones have good conditions for agricultural production, with the supply of agricultural products as their main function. They guarantee the security of production and the supply of agricultural products. These zones are classified in carbon balance zoning as high carbon control zones, carbon emission optimisation zones or carbon sink functional zones. Among them, the high carbon control zones should be careful to avoid the development of high-polluting and high-energy-consuming industries; focus on the development of green industries; and explore the development of low-carbon agriculture. Carbon emissions optimisation zones need to focus on improving agricultural infrastructure, guiding the development of advantageous agricultural products, and building nationally important production bases for cereals and advantageous agricultural products. Carbon sink functional zones need to safeguard the ecological advantages of their regions; enhance the comprehensive production capacity of cereals; and avoid excessive industrialisation and urbanisation.

Key Ecological Functional Zones. These zones possess high ecological sensitivity and a low resource and environmental carrying capacity, which makes them unsuitable for large-scale development and construction; as a result, they bear the important responsibility of maintaining the ecological security of the YRD region. These zones are classified as carbon emission optimisation areas or carbon sink functional areas in carbon balance zoning. Carbon emission optimisation zones should promote ecological restoration, strengthen ecological protection efforts, and explore the potential for realising the value of ecological products. Carbon sink functional zones should focus on enhancing their carbon sequestration capacity, develop characteristic industries that can be supported by resources and the environment according to local conditions, and guide the orderly transfer of overloaded populations.

5. Conclusions

In this paper, a carbon balance zoning study was conducted based on the estimation of carbon emissions and carbon absorption in the YRD region during 2005–2020. The driving factors of the regional carbon balance were analysed via GeoDetector, the impacts of various drivers on the regional carbon compensation rate were revealed, and spatially synergistic carbon reduction pathways for carbon balance zoning were proposed. The main conclusions are as follows:

1. In 2005–2020, the YRD region's carbon emissions and carbon absorption continued to increase, but the rise in carbon absorption did not offset the rise in carbon emissions. Consequently, the regional carbon imbalance will become increasingly prominent, and a significant spatial imbalance in the spatial development trend will occur. The high-value carbon emission zones are clustered in the optimised development zones and key development zones along the "Shanghai–Nanjing–Hangzhou–Ningbo" line, while the high-value carbon absorption zones are distributed mainly in the key ecological functional zones in the southeastern mountainous areas;
2. Based on the carbon accounting results and the NRCA index of each county unit attribute, the YRD region is divided into 87 high carbon control zones, 167 carbon emissions optimisation zones, and 51 carbon sink functional zones. Carbon emissions from the high carbon control zones account for 54.15% of the total carbon emissions in the YRD region, and the economic benefits of carbon emissions and carbon ecological support capacity are at a low level. The county units in the carbon emissions optimisation zones either have high levels of economic benefits related to carbon emissions or have compatible levels of economic benefits related to carbon emissions and ecological functions related to carbon sinks; the carbon sink functional zones contribute to 36.45% of the carbon sequestration, and they have a stronger carbon ecological support capacity;
3. From 2005 to 2020, there were spatial and temporal differences in the drivers of carbon balance changes in the YRD region. Furthermore, the main influences on the carbon compensation rate varied in the different functional zones due to regional differences in the direction of economic development and resource utilisation patterns. Overall, it is one of the most economically active regions in China, changes in territorial exploitation and utilisation patterns in the YRD region are the major drivers of changes in the regional carbon balance.

Considering the carbon balance zoning of the YRD region, differentiated spatially synergistic carbon reduction paths are proposed with respect to the carbon balance characteristics of different zones and the locations of the main functions to serve as a reference for realising sustainable low-carbon development in the YRD region.

Furthermore, because of limited access to county-level data, the county-level carbon emission estimations in this study are determined via an indirect method; the accuracy of carbon accounting still has room for improvement. Although the study of carbon balance zoning in the YRD region can offer ideas for regional low-carbon development, the realisation of China's dual-carbon goal needs to be based on the national scale, and appropriate spatially synergistic carbon reduction pathways need to be explored; therefore, the study of carbon balance zoning and drivers of carbon balance at the national scale should be strengthened in the future.

Author Contributions: Conceptualization, H.G.; Methodology, H.G.; Software, H.G.; Validation, H.G.; Formal analysis, H.G.; Resources, H.G.; Writing—original draft preparation, H.G.; Writing—review and editing, H.G.; Visualization, H.G.; Supervision, W.S.; Project administration, W.S.; Funding acquisition, W.S. All authors have read and agreed to the published version of the manuscript.

Funding: This research was funded by Marine Science and Technology Innovation Project of Jiangsu Province, grant number JSZRHYKJ202306; Science and Technology Planning Project of NIGLAS, grant number NIGLAS2022GS06; and Jiangsu R&D Special Fund for Carbon Peaking and Carbon Neutrality, grant number BK20220014.

Data Availability Statement: The original contributions presented in the study are included in the article, further inquiries can be directed to the corresponding author.

Conflicts of Interest: The authors declare no conflicts of interest.

References

1. IPCC. *Climate Change 2013: The Physical Science Basis. Contribution of Working Group I to the Fifth Assessment Report of the Intergovernmental Panel on Climate Change*; Stocker, T.F., Qin, D., Plattner, G.-K., Tignor, M., Allen, S.K., Boschung, J., Nauels, A., Xia, Y., Bex, V., Midgley, P.M., Eds.; Cambridge University Press: Cambridge, UK; New York, NY, USA, 2013; 1535p.
2. Van Soest, H.L.; den Elzen, M.G.J.; van Vuuren, D.P. Net-zero emission targets for major emitting countries consistent with the Paris Agreement. *Nat. Commun.* **2021**, *12*, 2140. [CrossRef] [PubMed]
3. Yang, J.; Sun, J.; Ge, Q.S.; Li, X.M. Assessing the impacts of urbanization-associated green space on urban land surface temperature: A case study of Dalian, China. *Urban For. Urban Green.* **2017**, *22*, 1–10. [CrossRef]
4. Fang, J.Y.; Tang, Y.H.; Son, Y. Why are East Asian ecosystems important for carbon cycle research? *Sci. China-Life Sci.* **2010**, *53*, 753–756. [CrossRef] [PubMed]
5. Wu, L.P.; Chen, Y.; Feylizadeh, M.R.; Liu, W.J. Estimation of China's macro-carbon rebound effect: Method of integrating Data Envelopment Analysis production model and sequential Malmquist-Luenberger index. *J. Clean. Prod.* **2018**, *198*, 1431–1442. [CrossRef]
6. Chen, T.; Chen, G.; Wang, Q. Spatiotemporal change patterns of carbon absorption/emission and decoupling effect with economy in Guizhou Province. *Acta Ecol. Sin.* **2024**, *44*, 915–929.
7. Friedlingstein, P.; O'Sullivan, M.; Jones, M.W.; Andrew, R.M.; Gregor, L.; Hauck, J.; Le Quéré, C.; Luijkx, I.T.; Olsen, A.; Peters, G.P.; et al. Global Carbon Budget 2022. *Earth Syst. Sci. Data* **2022**, *14*, 4811–4900. [CrossRef]
8. Xia, S.; Yang, Y. Spatio-temporal differentiation of carbon budget and carbon compensation zoning in Beijing-Tianjin-Hebei Urban Agglomeration based on the Plan for Major Function-oriented Zones. *Acta Geogr. Sin.* **2022**, *77*, 679–696.
9. Gu, Q. Concept Adjustment and Dilemma Relief of China's Territorial Space Governance under the "Dual Carbon" Goals. *China Land Sci.* **2023**, *37*, 12–19.
10. Doll, C.N.H.; Muller, J.P.; Elvidge, C.D. Night-time imagery as a tool for global mapping of socioeconomic parameters and greenhouse gas emissions. *Ambio* **2000**, *29*, 157–162. [CrossRef]
11. Xu, L.; He, N.P.; Li, M.X.; Cai, W.X.; Yu, G.R. Spatiotemporal dynamics of carbon sinks in China's terrestrial ecosystems from 2010 to 2060. *Resour. Conserv. Recycl.* **2024**, *203*, 107457. [CrossRef]
12. Huang, H.Z.; Jia, J.S.; Chen, D.L.; Liu, S.T. Evolution of spatial network structure for land-use carbon emissions and carbon balance zoning in Jiangxi Province: A social network analysis perspective. *Ecol. Indic.* **2024**, *158*, 111508. [CrossRef]
13. Liu, C.G.; Sun, W.; Li, P.X.; Zhang, L.C.; Li, M. Differential characteristics of carbon emission efficiency and coordinated emission reduction pathways under different stages of economic development: Evidence from the Yangtze River Delta, China. *J. Environ. Manag.* **2023**, *330*, 117018. [CrossRef] [PubMed]

14. Xu, Q.; Dong, Y.X.; Yang, R.; Zhang, H.O.; Wang, C.J.; Du, Z.W. Temporal and spatial differences in carbon emissions in the Pearl River Delta based on multi-resolution emission inventory modeling. *J. Clean. Prod.* **2019**, *214*, 615–622. [CrossRef]
15. Qin, Q.D.; Yan, H.M.; Li, B.X.; Lv, W.; Zafar, M.W. A novel temporal-spatial decomposition on drivers of China's carbon emissions. *Gondwana Res.* **2022**, *109*, 274–284. [CrossRef]
16. Change, I. 2006 IPCC guidelines for national greenhouse gas inventories. *Inst. Glob. Environ. Strateg. Hayama Kanagawa Jpn.* **2006**, *2*, 6.1–6.14.
17. Chuai, X.W.; Qi, X.X.; Zhang, X.Y.; Li, J.S.; Yuan, Y.; Guo, X.M.; Huang, X.J.; Park, S.; Zhao, R.Q.; Xie, X.L.; et al. Land degradation monitoring using terrestrial ecosystem carbon sinks/sources and their response to climate change in China. *Land Degrad. Dev.* **2018**, *29*, 3489–3502. [CrossRef]
18. Simmonds, M.B.; Di Vittorio, A.V.; Jahns, C.; Johnston, E.; Jones, A.; Nico, P.S. Impacts of California's climate-relevant land use policy scenarios on terrestrial carbon emissions (CO_2 and CH_4) and wildfire risk. *Environ. Res. Lett.* **2021**, *16*, 014044. [CrossRef]
19. Fang, J.Y.; Yang, Y.H.; Ma, W.H.; Mohammat, A.; Shen, H.H. Ecosystem carbon stocks and their changes in China's grasslands. *Sci. China-Life Sci.* **2010**, *53*, 757–765. [CrossRef] [PubMed]
20. Song, S.X.; Kong, M.L.; Su, M.J.; Ma, Y.X. Study on carbon sink of cropland and influencing factors: A multiscale analysis based on geographical weighted regression model. *J. Clean. Prod.* **2024**, *447*, 141455. [CrossRef]
21. Shan, Y.L.; Liu, J.H.; Liu, Z.; Shao, S.; Guan, D.B. An emissions-socioeconomic inventory of Chinese cities. *Sci. Data* **2019**, *6*, 1–10. [CrossRef]
22. Chen, J.D.; Gao, M.; Cheng, S.L.; Hou, W.X.; Song, M.L.; Liu, X.; Liu, Y.; Shan, Y.L. County-level CO_2 emissions and sequestration in China during 1997–2017. *Sci. Data* **2020**, *7*, 1–12. [CrossRef] [PubMed]
23. Liu, Q.W.; Wu, S.M.; Lei, Y.L.; Li, S.T.; Li, L. Exploring spatial characteristics of city-level CO_2 emissions in China and their influencing factors from global and local perspectives. *Sci. Total Environ.* **2021**, *754*, 142206. [CrossRef] [PubMed]
24. Wang, S.; Xie, Z.; Wang, Z. The spatiotemporal pattern evolution and influencing factors of CO_2 emissions at the county level of China. *Acta Geogr. Sin.* **2021**, *76*, 3103–3118.
25. Moore, F.C.; Lacasse, K.; Mach, K.J.; Shin, Y.A.; Gross, L.J.; Beckage, B. Determinants of emissions pathways in the coupled climate-social system. *Nature* **2022**, *603*, 103–111. [CrossRef] [PubMed]
26. Cao, Y.; Zhao, Y.H.; Wang, H.X.; Li, H.; Wang, S.; Liu, Y.; Shi, Q.L.; Zhang, Y.F. Driving forces of national and regional carbon intensity changes in China: Temporal and spatial multiplicative structural decomposition analysis. *J. Clean. Prod.* **2019**, *213*, 1380–1410. [CrossRef]
27. Hashmi, R.; Alam, K. Dynamic relationship among environmental regulation, innovation, CO_2 emissions, population, and economic growth in OECD countries: A panel investigation. *J. Clean. Prod.* **2019**, *231*, 1100–1109. [CrossRef]
28. Duro, J.A.; Padilla, E. International inequalities in per capita CO_2 emissions: A decomposition methodology by Kaya factors. *Energy Econ.* **2006**, *28*, 170–187. [CrossRef]
29. Shen, T.; Hu, R.P.; Hu, P.L.; Tao, Z. Decoupling between Economic Growth and Carbon Emissions: Based on Four Major Regions in China. *Int. J. Environ. Res. Public Health* **2023**, *20*, 1496. [CrossRef] [PubMed]
30. Zhao, R.Q.; Liu, Y.; Li, Y.X.; Ding, M.L.; Zhang, Z.P.; Chuai, X.W.; Jiao, S.X. An overview of regional carbon compensation: Mechanism, pattern and policy suggestions. *Areal Res* **2015**, *34*, 116–120.
31. Wang, W.X.; Wang, W.J.; Xie, P.C.; Zhao, D.Q. Spatial and temporal disparities of carbon emissions and interregional carbon compensation in major function-oriented zones: A case study of Guangdong province. *J. Clean. Prod.* **2020**, *245*, 118873. [CrossRef]
32. Xue, H.; Shi, Z.Q.; Huo, J.G.; Zhu, W.B.; Wang, Z.Y. Spatial difference of carbon budget and carbon balance zoning based on land use change: A case study of Henan Province, China. *Environ. Sci. Pollut. Res.* **2023**, *30*, 109145–109161. [CrossRef]
33. Fan, J. Draft of major function oriented zoning of China. *Acta Geogr. Sin.* **2015**, *70*, 186–201.
34. Fu, H.P.; Liu, J.; Dong, X.T.; Chen, Z.L.; He, M. Evaluating the Sustainable Development Goals within Spatial Planning for Decision-Making: A Major Function-Oriented Zone Planning Strategy in China. *Land* **2024**, *13*, 390. [CrossRef]
35. Kuriakose, J.; Jones, C.; Anderson, K.; McLachlan, C.; Broderick, J. What does the Paris climate change agreement mean for local policy? Downscaling the remaining global carbon budget to sub-national areas. *Renew. Sustain. Energy Transit.* **2022**, *2*, 100030.
36. Piao, S.L.; He, Y.; Wang, X.H.; Chen, F.H. Estimation of China's terrestrial ecosystem carbon sink: Methods, progress and prospects. *Sci. China-Earth Sci.* **2022**, *65*, 641–651. [CrossRef]
37. Zhou, X.; Li, L.; Liang, Y.; Yang, L. Spatiotemporal evolution of territorial spatial patterns and carbon emissions in the Yangtze River Delta from the perspective of main functional zones. *Trans. Chin. Soc. Agric. Eng.* **2023**, *39*, 236–244.
38. Yi, D.; Ou, M.; Guo, J.; Han, Y.; Yi, J.; Ding, G.; Wu, W. Progress and prospect of research on land use carbon emissions and low-carbon optimization. *Resour. Sci.* **2022**, *44*, 1545–1559. [CrossRef]
39. Wu, B.W.; Zhang, Y.Y.; Yang, Y.; Wu, S.D.; Wu, Y. Spatio-temporal variations of the land-use-related carbon budget in Southeast China: The evidence of Fujian province. *Environ. Res. Commun.* **2023**, *5*, 115015. [CrossRef]
40. Kong, M.; Hu, H.; Zhang, H.; Du, S. Spatio-temporal evolution of urban low-carbon competitiveness in the Yangtze River Delta from 2000 to 2020. *Geogr. Res.* **2023**, *42*, 2713–2737.
41. Wang, H.; Wu, X.Y.; Wu, D.; Nie, X. Will land development time restriction reduce land price? The perspective of American call options. *Land Use Policy* **2019**, *83*, 75–83. [CrossRef]

42. Nie, X.; Lu, B.; Chen, Z.P.; Yang, Y.W.; Chen, S.; Chen, Z.H.; Wang, H. Increase or decrease? Integrating the CLUMondo and InVEST models to assess the impact of the implementation of the Major Function Oriented Zone planning on carbon storage. *Ecol. Indic.* **2020**, *118*, 106708. [CrossRef]
43. Tan, X.; Lai, H.; Gu, B.; Tu, T.; Li, H. Carbon emission accounting from the perspective of main functional areas: A case study of Guangdong Province. *Acta Ecol. Sin.* **2018**, *38*, 6292–6301.
44. Xu, X.L.; Liu, J.Y.; Zhang, S.W.; Li, R.D.; Yan, C.Z.; Wu, S.X. *China Multi-Period Land Use Remote Sensing Monitoring Dataset (CNLUCC)*; Resource and Environmental Science Data Registration and Publication System: Beijing, China, 2018; Available online: http://www.resdc.cn/DOI/ (accessed on 29 January 2023). [CrossRef]
45. Nachtergaele, F.; Van Velthuizen, H.; Verelst, L.; Wiberg, D.; Henry, M.; Chiozza, F.; Yigini, Y.; Aksoy, E.; Batjes, N.; Boateng, E. *Harmonized World Soil Database Version 2.0*; Food and Agriculture Organization of the United Nations: Rome, Italy, 2023. [CrossRef]
46. Peng, S.Z. *1-km Monthly Precipitation Dataset for China (1901–2022)*; National Tibetan Plateau/Third Pole Environment Data Center: Beijing, China, 2020. [CrossRef]
47. Peng, S.Z. *1-km Monthly Mean Temperature Dataset for China (1901–2022)*; National Tibetan Plateau/Third Pole Environment Data Center: Beijing, China, 2019. [CrossRef]
48. Wang, Z.; Zhou, K.; Fan, J. County-level carbon emission accounting and Major Function Oriented Zones in western regions: Taking Sichuan Province as an example. *Acta Ecol. Sin.* **2022**, *42*, 8664–8674.
49. Shan, Y.L.; Guan, D.B.; Zheng, H.R.; Ou, J.M.; Li, Y.; Meng, J.; Mi, Z.F.; Liu, Z.; Zhang, Q. Data Descriptor: China CO_2 emission accounts 1997–2015. *Sci. Data* **2018**, *5*, 170201. [CrossRef] [PubMed]
50. Teng, F.; Wang, Y.; Wang, M.; Li, S.; Lin, Y.; Cai, H. Spatiotemporal coupling relationship between urban spatial morphology and carbon budget in Yangtze River Delta urban agglomeration. *Acta Ecol. Sin.* **2022**, *42*, 9636–9650.
51. Li, L.; Dong, J.; Xu, L.; Zhang, J. Spatial variation of land use carbon budget and carbon compensation zoning in functional areas:A case study ofWuhan Urban Agglomeration. *J. Nat. Resour.* **2019**, *34*, 1003–1015.
52. Yu, R.; Cai, J.C.; Leung, P.S. The normalized revealed comparative advantage index. *Ann. Reg. Sci.* **2009**, *43*, 267–282. [CrossRef]
53. Kohonen, T. Self-organized formation of topologically correct feature maps. *Biol. Cybern.* **1982**, *43*, 59–69. [CrossRef]
54. Wang, J.; Xu, C. Geodetector: Principle and prospective. *Acta Geogr. Sin.* **2017**, *72*, 116–134.
55. Song, Y.Z.; Wang, J.F.; Ge, Y.; Xu, C.D. An optimal parameters-based geographical detector model enhances geographic characteristics of explanatory variables for spatial heterogeneity analysis: Cases with different types of spatial data. *Giscience Remote Sens.* **2020**, *57*, 593–610. [CrossRef]
56. Zhao, R.; Liu, Y.; Ding, M.; Zhang, Z.; Huang, X.; Qin, Y. Theory, methods, and research progresses of regional carbon budget. *Prog. Geogr.* **2016**, *35*, 554–568.

Disclaimer/Publisher's Note: The statements, opinions and data contained in all publications are solely those of the individual author(s) and contributor(s) and not of MDPI and/or the editor(s). MDPI and/or the editor(s) disclaim responsibility for any injury to people or property resulting from any ideas, methods, instructions or products referred to in the content.

MDPI AG
Grosspeteranlage 5
4052 Basel
Switzerland
Tel.: +41 61 683 77 34

Land Editorial Office
E-mail: land@mdpi.com
www.mdpi.com/journal/land

Disclaimer/Publisher's Note: The title and front matter of this reprint are at the discretion of the Guest Editors. The publisher is not responsible for their content or any associated concerns. The statements, opinions and data contained in all individual articles are solely those of the individual Editors and contributors and not of MDPI. MDPI disclaims responsibility for any injury to people or property resulting from any ideas, methods, instructions or products referred to in the content.

www.ingramcontent.com/pod-product-compliance
Lightning Source LLC
LaVergne TN
LVHW072335090526
838202LV00019B/2427